# The Cybernetics Moment

T0146043

New Studies in American Intellectual and Cultural History
Jeffrey Sklansky, *Series Editor*

# The Cybernetics Moment

*Or Why We Call Our Age the Information Age*

RONALD R. KLINE

Johns Hopkins University Press

*Baltimore*

© 2015 Johns Hopkins University Press
All rights reserved. Published 2015
Printed in the United States of America on acid-free paper

Johns Hopkins Paperback edition, 2017
9  8  7  6  5  4  3  2  1

Johns Hopkins University Press
2715 North Charles Street
Baltimore, Maryland 21218-4363
www.press.jhu.edu

*The Library of Congress has cataloged the hardcover edition of this book as follows:*
Kline, Ronald R., author.
The cybernetics moment: or why we call our age the information age / Ronald R. Kline.
    pages cm. — (New studies in American intellectual and cultural history)
Includes bibliographical references and index.
ISBN 978-1-4214-1671-7 (hardcover : acid-free paper) — ISBN 1-4214-1671-9
(hardcover : acid-free paper) — ISBN 978-1-4214-1672-4 (electronic) — ISBN
1-4214-1672-7 (electronic)  1. Information theory.  2. Cybernetics—Social aspects.  I. Title.
II. Title: Cybernetics moment.  III. Title: Why we call our age the information age.
Q360.K56 2015
303.48'33—dc23       2014035091

A catalog record for this book is available from the British Library.

ISBN-13: 978-1-4214-2424-8
ISBN-10: 1-4214-2424-x

*Special discounts are available for bulk purchases of this book. For more information,
please contact Special Sales at 410-516-6936 or specialsales@press.jhu.edu.*

Johns Hopkins University Press uses environmentally friendly book materials, including
recycled text paper that is composed of at least 30 percent post-consumer waste,
whenever possible.

*To the memory of Alfred Motz, Margot Ruth Marcotte,*
*Maggie Marcotte Mattke, Raymond Orville Kline,*
*and Nellie Frank Motz*

Information is information, not matter or energy. No materialism which does not admit this can survive at the present day.

Norbert Wiener, *Cybernetics*, 1948

I think that the past is all that makes the present coherent.

James Baldwin, *Notes of a Native Son*, 1955

# CONTENTS

# ACKNOWLEDGMENTS

IT IS MY PLEASURE to acknowledge the many students, colleagues, and institutions who assisted me during the decade and a half I spent researching and writing this book.

I especially want to thank Terry Fine and Christina Dunbar-Hester for their advice, assistance, and support over the years. Terry, a professor emeritus in the School of Electrical and Computer Engineering (ECE) at Cornell University, a friend and colleague whose office was down the hall from mine for many years (in my joint appointment between ECE and the Science and Technology Studies Department in the Arts College), helped me understand the basic principles of information theory, discussed the often testy relationships between that field and cybernetics during his career, and critiqued my account of the field he loves. As a Ph.D. student in science and technology studies, Christina, now an assistant professor at Rutgers University, served as a sounding board and friendly critic of the ideas in this book when she was at Cornell, particularly when she took my seminar on cybernetics and helped me teach an undergraduate course on the history of information technology. In combing through the massive Warren McCulloch Papers at the American Philosophical Society as a research assistant, Christina deepened the research for the book at a critical time. Her comments on several chapters were insightful.

I would also like to thank other former students for their research assistance: Alec Shuldiner, for finding material on the development of information theory in the extensive AT&T archives; Albert Tu, for copying newspaper and magazine articles on cybernetics and information theory; Lav Varshney, for researching the acceptance of information theory in American electrical engineering journals and for alerting me to obscure published sources on Claude Shannon; and Daniel Kreiss at Stanford University, for gathering material on NASA's cyborg project at the Ames Research Center. Thanks also to Glen Bugos at NASA, for helping navigate the Ames Research Center archives, and to Rachel Prentice at Cornell and David Hounshell at Carnegie-

Mellon University, for providing copies of archival material from their own research. Rick Johnson at Cornell and Julian Reitman, a former officer of the IEEE Society on Systems, Man, and Cybernetics, shared their recollections of that society in the 1960s and 1970s. The late Dick Neisser, a founder of cognitive psychology, painted a vivid picture for me of information studies in the 1950s.

In addition to Terry and Christina, I wish to thank Bill Aspray, Michael Buckland, Peter Dear, Katie King, Daniel Kreiss, Kevin Lambert, Boyd Rayward, Eric Schatzberg, Phoebe Sengers, Suman Seth, Fred Turner, Ana Viseu, and Herb Voelcker for commenting on drafts of articles and presentations, and Arisa Ema, Bernard Geogehagen, Marge Kline, Lav Varshney, and Herb Voelcker for commenting on chapters of the book. I also thank audiences for commenting on presentations I gave at Bielefeld University in Germany, Cornell University, the University of Maryland, the University of Pennsylvania, Penn State University, Stanford University, the Chemical Heritage Foundation, and annual meetings of the Society for the History of Technology. Portions of chapters 6 and 8 appeared in an earlier form as articles in *IEEE Annals of the History of Computing*, *Social Studies of Science*, and *Technology and Culture*. I thank the publishers of those journals for permission to use this material and the editors and anonymous referees for their helpful comments at an early stage of my research.

Conversations with colleagues in the Science and Technology Studies Department at Cornell helped me strengthen my analytical voice in this book. I appreciated the lively discussions on the history of cybernetics with Cornell graduate students and thank my former Ph.D. student Honghong Tinn for preparing the index. Outside of Cornell, I benefited from correspondence and conversations with Eden Medina, Mara Mills, David Mindell, Jérôme Segal, and Jonathan Sterne.

The comments from an anonymous reader for Johns Hopkins University Press helped me to sharpen the argument in several places. Robert J. Brugger at the Press and Beth Gianfagna were exemplary editors.

The book could not have been completed without the resources of the Cornell University Library and access to manuscript collections at many institutions. These include the Niels Bohr Library at the American Institute of Physics, the American Philosophical Society, the AT&T Archives and History Center, the College Archives at Imperial College, London, the University of Illinois Archives in Urbana, the Library of Congress, the Institute Archives and Special Collections at the Massachusetts Institute of Technology, and the Rockefeller Archive Center. I thank the librarians at Cornell and the archivists in charge of those collections, especially Nora Murphy and Jeff Mifflin at MIT.

Generous financial support for research and writing came from the National Science Foundation, grant SES80689, and the Bovay Program in the History and Ethics of Engineering at Cornell University.

# The Cybernetics Moment

# Introduction

M ARGARET MEAD SAT IN THE FRONT ROW, waiting for the group photo to be taken. The most famous woman scientist of her time, an anthropologist whose books on adolescents in the Pacific Islands spoke to child-raising anxieties in modern America, Mead had attended all of the conferences on humans and machines. Sponsored by the Josiah Macy, Jr., Foundation and held from 1946 to 1953, the postwar meetings aimed to break down disciplinary barriers in the sciences. Mathematicians, engineers, biologists, social scientists, and humanists debated how the wartime theories of communications and control engineering applied to both humans and machines. They discussed, for example, if the new digital computers, which the media had dubbed "electronic brains," could explain how the human brain worked. After one of the group's founding members, Norbert Wiener, a mathematician at the Massachusetts Institute of Technology, published the surprisingly popular book *Cybernetics* in 1948, the group adopted the term *cybernetics* as the title of its conference series.

The proceedings of the Macy conferences, which Mead coedited, helped to establish the scientific fields of cybernetics and information theory in the United States. During contentious meetings filled with brilliant arguments, rambling digressions, and disciplinary posturing, the cybernetics group shaped a language of feedback, control, and information that transformed the idiom of the biological and social sciences, sparked the invention of information technologies, and set the intellectual foundation for what came to be called the information age. The premise of cybernetics was a powerful analogy: that the principles of information-feedback machines, which explained how a thermostat controlled a household furnace, for example,

FIGURE 1. Meeting of the last Macy Foundation Conference on Cybernetics, Princeton, New Jersey, 1953. The interdisciplinary Macy conferences, held from 1946 to 1953, developed and spread the new science in the United States. Leading lights included anthropologist Margaret Mead and neuroscientist Warren McCulloch (*sitting in the front row*) and anthropologist Gregory Bateson, mathematician Claude Shannon, and physicist Heinz von Foerster (*standing in the back row*). From *HVFCyb*, 10:6. Courtesy of the American Philosophical Society.

could also explain how all living things—from the level of the cell to that of society—behaved as they interacted with their environment.

The participants waiting with Mead for the group photo to be taken form a who's who in cybernetics and information theory at the time (fig. 1). Sitting next to Mead, on her left, is Warren McCulloch, the "chronic chairman" and founder of the cybernetics conferences. An eccentric physiologist, McCulloch had coauthored a foundational article of cybernetics on the brain's neural network. To his left sits W. Grey Walter, the British physiologist who built robotic "tortoises." Directly behind Mead, in the last row, stands the tall figure of anthropologist Gregory Bateson, Mead's former husband, who collaborated with her on fieldwork in New Guinea and Indonesia. Now divorced, they had organized the social science contingent of the cybernetic conferences from the beginning.

The group's physicists, engineers, and mathematicians stand together in the back row. At the end of the row, on the right, is Heinz von Foerster, an émigré physicist from Austria who founded the Biological Computer Laboratory at the University of Illinois and coedited with Mead the conference proceedings. To his right are mathematician and polymath Walter Pitts, who coauthored the paper on neural nets with McCulloch, worked with Wiener, and was now at McCulloch's lab at MIT; Claude Shannon, the mathematician at AT&T's Bell Laboratories who founded the American school of information theory; and Julian Bigelow, the chief engineer on mathematician John von Neumann's project to build a digital computer at Princeton University. He was also the engineer on Wiener's wartime project that started the cybernetics venture.

Some prominent figures were absent when the photo was taken at the last Macy conference on cybernetics in 1953. Three of the group's earliest members—Norbert Wiener, John von Neumann, and physiologist Arturo Rosenblueth—did not attend. Wiener and von Neumann had resigned the year before. Wiener was furious with McCulloch because he thought McCulloch had appropriated cybernetics for his own ends. Rosenblueth, who had coauthored a founding paper of cybernetics with Wiener and Bigelow, disliked the rambling discussions at the meetings; he stayed in his laboratory in Mexico City, the lab in which Wiener had written *Cybernetics*.

It might seem odd to today's readers that Margaret Mead sat in a prominent place at the now-famous Macy conferences and that she would be remembered a half-century later as one of the founders of cybernetics.[1] Why would a world-renowned anthropologist with no expertise or apparent interest in mathematics, engineering, and neuroscience attend all ten meetings, recruit social scientists for the meetings, and undertake the tedious job of editing the proceedings? When the group was first organized, Mead shared the enthusiasm of her husband Gregory Bateson that cybernetics would bring the rigor of the physical sciences to the social sciences. They thought cybernetic models could realistically explain the behavior of humans and society because they contained the information-feedback loops that existed in all organisms. This belief was reflected in the original title of the meetings: "Conference on Feedback Mechanisms and Circular Causal Systems in Biology and the Social Sciences." Everything that Bateson wrote after the Macy conferences—on a wide range of subjects, from psychiatry to animal learning—testified to his belief in the power of cybernetics to transform human ways of knowing. The conferences convinced Mead that the universal language of cybernetics might be able to bridge disciplinary boundaries in the social sciences. The presence of Mead and Bateson among the mathematicians, natural scientists, and engineers in the group photo symbolizes the interdisciplinary allure of cybernetics and information theory.[2]

In one respect, Mead and Bateson were right to be so enthused about the Macy conferences. There is little doubt that cybernetics and information theory were incredibly influential during the Cold War. In the 1950s, scientists were excited that Wiener and Shannon had defined the amount of information transmitted in communications systems with a formula mathematically equivalent to entropy (a measure of the degradation of energy). Defining information in terms of one of the pillars of physics convinced many researchers that information theory could bridge the physical, biological, and social sciences. The allure of cybernetics rested on its promise to model mathematically the purposeful behavior of all organisms, as well as inanimate systems. Because cybernetics included information theory in its purview, its proponents thought it was more universal than Shannon's theory, that it applied to all fields of knowledge.

This enthusiasm led scientists, engineers, journalists, and other writers in the United States to adopt these concepts and metaphors to an extent that is still evident today. Cognitive psychologists, molecular biologists, and economists analyze their subjects in terms of the flow, storage, and processing of information. Some quantum physicists have elevated information theory to a theory of matter.[3] In technology, the concepts of information theory are embodied in the invisible coding schemes for digital media (in cell phones, music files, and DVDs, for example), and in the equally invisible communications infrastructure of the Internet. Cybernetic ideas of circular causality—that information-feedback circuits allow a system to adapt to its environments—are used to model industrial, urban, transportation, economic, and social systems (such as family dynamics). Although the symbolic method of programming computers to exhibit artificial intelligence replaced the cybernetic, neural-net method in the 1960s, Grey Walter's way of building robots now informs the work done at the artificial intelligence laboratory at MIT. The noun *information* and the prefix *cyber-* mark the new vocabulary of our time. They inform how we talk, think, and act on our digital present and future, from the utopian visions invoked by the terms *information age* and *cyberspace* to the dystopian visions associated with enemy *cyborgs* and *cyber warfare*. The traces of cybernetics and information theory thus permeate the sciences, technology, and culture of our daily lives. Yet, those traces are often a vestige of the ideas that so excited Mead, Bateson, and their colleagues at the Macy conferences.

Institutionally, cybernetics and information theory did not attain the high status that their promoters envisioned for them in the 1950s; they did not become major scientific and engineering disciplines. Information theory, whose promoters drew sharp boundaries to separate their field from cybernetics, has enjoyed a better fate than cybernetics in the United States. It is a well-respected subfield in electrical engineering, with secure funding and a professional home in the Institute of Electrical and Electronics Engineers (IEEE) Society on Information Theory. Cybernetics has split into two subfields. First-order cybernetics—based on the work of Wiener and McCulloch—is a systems modeling discipline with a professional home in the IEEE Society on Systems, Man, and Cybernetics. Second-order cybernetics—based on the work of Bateson and von Foerster—leads a more precarious existence as a radical epistemology in the American Society for Cybernetics and in the journal *Cybernetics and Human Knowing*. Despite this marginality, the founders of cybernetics and information theory in the United States—Norbert Wiener and Claude Shannon—are now celebrated as progenitors of what is commonly called the information age.[4]

Yet we rarely examine why we call our era the information age. The origins of this discourse lie in the 1960s, when futurists, policymakers, journalists, social scientists, and humanists started writing about the coming of a new era based on computers and communications technology. They created a techno-revolutionary narrative that was popularized during the dotcom boom of the 1990s. Although social scientists and humanists criticized this discourse in the 1980s—Theodore Roszak, for example, called it the "Cult of Information"[5]—the phrases *information revolution, information economy, information society*, and *information age* have been woven into everyday speech. They describe the taken-for-granted belief that cell phones, personal computers, and the Internet are creating a new economic and social order. This discourse resembles what David Nye calls a utopian technological narrative, a variant of the nineteenth-century foundation stories in which the axe, watermill, and railroad turn frontier America into a second Eden. The utopian narrative typically emerges at the start of a foundation story and makes extravagant claims that a new technology will bring enormous social and economic benefits.[6]

*The Cybernetics Moment* examines the intellectual and cultural history of the information discourses of the Cold War, the latest of which celebrates how the merger of computers and communications is creating the next stage in American civilization: the information age. This way of talking has become so pervasive in the early twenty-first century that the venerable word *technology* no longer refers in popular media to all techniques and artifacts, from agriculture to nuclear power, but is shorthand for *information technology*.[7] I do not engage in debates about whether our time should be called an information age, whether previous information ages existed, or what makes our use of information technology different from pre-digital eras.[8]

Instead, I ask why we came to believe that we live in an information age, what work it took to make this a naturalized narrative, what scientific and engineering practices enabled the narrative, and what was at stake for groups in academia, government, industry, and the press to claim that an information revolution was at hand. Although I agree with critics who emphasize the narrative's ideology, I focus on its history. I ask what made this discourse possible and why alternative narratives were diminished during the Cold War.

I maintain that the invention of cybernetics and information theory is as important as the invention of information technologies in understanding the present age. At the heart of the book is what I call the "cybernetics moment"—the rise, fall, and reinvention of cybernetics that occurred alongside the rise of information theory in the United States. The cybernetics moment began when the two fields emerged shortly after World War II, reached its peak with their adoption and modification in biology, engineering, the social sciences, and popular culture in the 1950s and 1960s, and ended when cybernetics and information theory lost their status as universal sciences in the 1970s.[9] I relate that history to the invention of digital computers and communications systems, and the emergence of a utopian information narrative that thrived when the cybernetics moment ended. In adopting the language and concepts of cybernetics and information theory, scientists turned the metaphor of information into the matter-of-fact description of what is processed, stored, and retrieved in physical, biological, and social systems. Engineers used the theories to invent information technologies. The scientific concept of information and Wiener's notion of a second industrial revolution formed the intellectual foundations to talk about an *information age*. The merger of computers and communications, starting in the 1970s, reinforced that narrative and gave it staying power.

These transformations were contested at each step of the way. The definition of information, itself, was debated at length. Scientists, engineers, and humanists considered whether or not the new concept of information was semantic, whether or not it connoted meaning. Claude Shannon had discarded that idea and had proved mathematically that his theory of information measured uncertainty—for example, the uncertainty of what word would be transmitted next, not the meaning of the words. Yet many researchers tried to turn his theory into a theory of meaning. The profusion of interpretations of the word *information*—one social scientist counted thirty-nine meanings of the term in 1972[10]—was reduced in popular discourse to a transmission of commodified, equally probable bits in computer networks, or to such slogans as "information wants to be free."

Cybernetics, the "new science" with the mysterious name and universal aspirations, was interpreted even more broadly. In 1969, Georges Boulanger, the president of the International Association of Cybernetics, asked, "But after all what is cybernetics? Or rather what is it not, for paradoxically the more people talk about cybernetics the less they seem to agree on a definition." He identified several meanings of cybernetics: a mathematical theory of control; automation; computerization; a theory of communication; the study of analogies between humans and machines; and a philosophy explaining the mysteries of life. To the general public, Boulanger noted, cybernetics "conjures up visions of some fantastic world of the future peopled by robots and electronic brains!" His favorite definition was the "science of robots."[11] Cybernetics was a staple of science fiction and a fad among artists, musicians, and intellectuals in the 1950s and 1960s. Writer James Baldwin recalled that the "cybernetics craze" was emblematic of the period for him.[12]

The variety in the meanings noted by Boulanger points to a disunity of cybernetics, which was compounded by the different paths cybernetics took in different countries. Disunity was an ironic fate for a field that claimed to be an international, universal discipline that could unify the sciences.[13] Outside the United States, cybernetics encompassed several subfields: engineering cybernetics, management cybernetics, biocybernetics, medical cybernetics, and behavioral cybernetics. Europeans embraced the field in multiple ways. While researchers emphasized the performative character of cybernetic machines in Britain, a more philosophical style of cybernetics prevailed in France and Germany. Cybernetics became a state science in the Soviet Union after proponents overcame its reputation there as a reactionary pseudoscience in the 1950s. Chilean leaders looked to British cyberneticians to organize their socialist economy along cybernetic lines in the 1970s.[14] The collapse of cybernetics as a unifying science in the United States at this time—signaling the end of the cybernetics moment—occurred precisely when *information* was becoming the keyword of our time.

One result is that the rich discourse of cybernetics and information theory was flattened in the utopian information narrative. The basic analogy of cybernetics—that all organisms use information-feedback paths to adapt to their environment—is reduced to the adjective *cyber*. The scientific concept of information is reduced to digitized data. In today's discourse, information is no longer a measure of uncertainty in communications, nor is it related to biological, psychological, and physical processes.

The history of cybernetics, information theory, and the information age narrative can help us understand our era in several ways. *The Cybernetics*

*Moment* recovers the enthusiasm for the two "new sciences" in Cold War America, when scientists and humanists struggled to come to grips with the new relationships between humans and machines, and the meaning of "information." Those attempts persist today in academia and popular culture. Understanding their history can help us rethink the dominant technological narrative of our time, why we speak of an information age, rather than an age of knowledge or the computer age. What difference does it make that we chose to use the information metaphor? Why does it matter that we invoke the information-conduit model, rather than the ecological model from cybernetics, to talk about the digital era?[15] I take up that task by considering the relationship between technology and ideas, the interplay between the invention of computers and communications systems and the invention of cybernetics and information theory. In turn, the development of those technologies and ideas helped to construct each other.[16]

The dominance of today's discourse about an information age, which emerged at the end of the cybernetics moment, would have surprised Margaret Mead and her colleagues who attended the last Macy conference on cybernetics if they had lived long enough to see it. In the 1970s, Bateson became a cult figure of the counterculture, which thought that his philosophy of cybernetics—which embraced the unity of mind and nature—could solve the ecological crisis of the day. Like most cyberneticists, Bateson did not pay much attention to the utopian narrative of information. Norbert Wiener was an exception. From the very beginning, he warned about the social consequences of cybernetics, the vast unemployment that might accompany the second industrial revolution. He made the connection between cybernetics and what came to be known as the information age in the context of competing with his former student Claude Shannon to invent a theory of information at the end of the World War II, a story to which we now turn.

# CHAPTER ONE

# War and Information Theory

NORBERT WIENER WAS FURIOUS with Walter Pitts. The brilliant but erratic graduate student had lost an important manuscript of Wiener's that he was supposed to edit. Pitts had checked it at Grand Central Station, along with his laundry and other personal items, while attending a Macy Foundation conference on cybernetics in New York City in October 1946, then forgot to pick it up when he returned to the Massachusetts Institute of Technology. His friends and fellow graduate students, Oliver Selfridge and Jerry Lettvin, failed to retrieve it for him—Selfridge a week and a half later, Lettvin during a trip to the city in December. Meanwhile, the threesome managed to put Wiener off until they enlisted Giorgio de Santillana, a humanities professor at MIT and friend of Wiener's, to track down the parcel, now classified as unclaimed property by the contractor who ran the checkroom. Wiener finally got his manuscript back in the first week of April 1947.[1]

Wiener had overlooked past high jinks by Walter and his friends in this male, homosocial environment. They spent all of their money on an auto trip to Mexico to work with Wiener's collaborator, neurophysiologist Arturo Rosenblueth; stayed up late rather than working at MIT; and indulged in dubious moneymaking schemes with de Santillana. Wiener had to hound Pitts to fulfill his obligations under a Guggenheim fellowship he had vouched for. But the escapade of the missing manuscript, which dragged on for five months, had crossed a line. In one of the outbursts he was famous for, Wiener wrote to Pitts in Mexico, "Under these circumstances please consider me as completely disassociated from your future career." Wiener apologized after Lettvin took the blame for losing the manuscript, but a week later Wiener criticized Pitts for sending him an "impertinent and evasive" postcard about the matter.[2] Wiener wrote Rosenblueth that the affair showed the "total irresponsibleness of the boys." The long delay in revising the manuscript "meant that one of my competitors, [Claude] Shannon of the Bell Telephone Company, is coming out with a paper before mine." Wiener complained to Warren McCulloch, "The manuscript has turned up but in the meantime I lost priority on some important work."[3]

The work in question was soon called "information theory" in the United States and Britain, a scientific theory that measured the amount of information (not data) transmitted in communications systems and processed by computers.[4] Wiener knew Shannon was a competitor because, in March 1947, before Wiener had recovered his manuscript, he heard Shannon give a talk titled "The Transmission of Information" at the Harvard mathematics colloquium. Shannon, who earned a Ph.D. in mathematics at MIT, based his talk on an unclassified section of a classified report on cryptography that he had written at AT&T's Bell Telephone Laboratories in 1945. It clearly stated a definition of information that was similar to Wiener's. Wiener told Pitts that "Shannon's work on Amount of Information is following our lines precisely."[5] In April 1947, after Wiener had recovered his manuscript—and his composure—he and Shannon gave papers at the same session of a mathematics conference held at Columbia University. After the meeting, Wiener wrote to McCulloch with some satisfaction, "The Bell people are fully accepting my thesis concerning the relationship of statistics and communication engineering."[6]

Wiener's fear of losing priority to Shannon was lessened when they simultaneously published similar theories of information in 1948. Shannon's theory appeared in the summer and fall, in a two-part article in the *Bell System Technical Journal*, Wiener's appeared in the fall, in his surprisingly popular book *Cybernetics*. Both men measured information in regard to the patterns transmitted in communication processes, whether these occurred in humans or machines. Both men bridged communication theory and physics by defining information with an equation that was similar to the formula for the physical concept of entropy, the unavailability of a system's energy to do work.[7]

Faced with this simultaneity, contemporaries and historians have disagreed about the contributions made by Wiener and Shannon to information theory. One camp gives equal credit to each for mathematically equating information with entropy and, sometimes, for creating the theory of information based on this concept. The attributions "Shannon-Wiener" or "Wiener-Shannon" are common in these accounts.[8] John von Neumann, who knew both men, disputed this pedigree by noting that a physicist, Leo Szilard, had equated information with entropy in the 1920s.[9] Many commentators acknowledge that Shannon drew on Wiener's statistical theory of communication, as Shannon himself stated in the 1948 paper, but credit Shannon with founding the discipline of information theory because of how extensively he mapped out the subject in that paper.[10] Some American information theorists went further and wrote Wiener out of a narrow history of their field during the celebration of its twenty-fifth anniversary in 1973 and its fiftieth anniversary in 1998.[11]

My aim is not to settle these priority and boundary disputes. Instead, I discuss how Wiener and Shannon proposed similar definitions of information by treating the engineering problem of providing accurate communications in different ways, ways that depended on differences in their mathematical expertise and their research projects. In those contexts, Wiener and Shannon did much to lay the foundations for cybernetics and information theory, which contemporaries viewed as two "new sciences" coming out of World War II.[12]

## Cybernetics and Information Theory: Wiener versus Shannon

Because of the competition between Wiener and Shannon, it should not surprise us that the texts that established the two new fields—Wiener's book *Cybernetics* and Shannon's article "A Mathematical Theory of Communication"—were published in the same year, 1948. Although they contain similar theories of information, the texts map out two markedly different views of science and engineering, which stem from different intellectual agendas and research cultures.

In his book, Wiener boldly announced the birth of a "new science of cybernetics," a term he derived from the Greek word for "steersman." Wiener praised his colleagues in mathematics, engineering, and physiology for recognizing the "essential unity of the set of problems around communication, control, and statistical mechanics, whether in the machine or in the living tissue." The interdisciplinary research conducted by Wiener and his colleagues—in developing control and communications systems, mathematically modeling the nervous system, inventing prostheses, and creating electronic computers—convinced Wiener of the extraordinary promise of cybernetics for the postwar era.[13]

According to Wiener, it all began when he worked with Julian Bigelow on an antiaircraft project during World War II. Wiener consulted with Arturo Rosenblueth, to whom *Cybernetics* is dedicated, and realized that the engineering theories of control and communication could explain behavior in both humans and machines. That analogy, expressed in the article "Behavior, Purpose and Teleology," coauthored by Rosenblueth, Wiener, and Bigelow, became the main tenet of cybernetics.[14] From 1947 to 1952, Wiener collaborated with Rosenblueth in Mexico on a five-year grant on "Mathematical Biology" from the Rockefeller Foundation, during which time *Cybernetics* was written and published. In the book, Wiener defined cybernetics as the "entire field of control and communication theory, whether in the machine or in the animal." In an article in *Scientific American*, he explained that "cybernetics attempts to find the common elements in the functioning of automatic machines and of the human nervous system."[15]

A tour de force that defined the field for decades, *Cybernetics* described the wide scope of the new field, and its philosophical and mathematical underpinnings, to experts and lay readers alike. In the nonmathematical chapters, which constitute about one-half of the book, Wiener dealt with a wide range of topics, including the aims and history of cybernetics, the philosophy of time, and the role of information in society. Wiener explained the central analogy of cybernetics in terms of a generalized feedback control system. This system could model animals and automatic machines because they both have sensors, effectors, brains, and feedback-paths with which they communicate (exchange information) with the outside world and operate in and on that world. To illustrate this basic principle of cybernetics, Wiener described similarities between the operation of digital computers and the human brain such as pattern recognition and "psychiatric" problems. He noted, for example, the similarity between giving electric shock treatments to humans and erasing a computer's memory. In making these analogies, he drew on the physiological concept of homeostasis: the ability of a body's feedback-control systems to maintain constant its blood pressure, temperature, and other vital signs. He even mentioned the fusion of humans and machines, what would later be called cyborgs, by describing Warren McCulloch's research on prosthetics that would allow the blind to see by aural means. An appendix outlined his and Shannon's suggestions on how to program computers to play chess.[16]

The mathematical chapters of the book dealt with the core sciences of cybernetics: statistical mechanics, information theory, and the theory of control systems. Undecipherable by most readers, these chapters were directed at Ph.D. mathematicians. The material upon which Wiener staked his claim as a founder of information theory, the manuscript mislaid by Pitts, formed the basis of the chapter "Time Series, Information, and Communication."[17] Information was a key concept in *Cybernetics*. When used quantitatively, it measured what was communicated in the messages flowing through feedback control loops that enabled all organisms, living and nonliving, to adapt to their environments. When used metaphorically, information related the principles of cybernetics to wider social issues.

In developing his theory of information, Wiener employed the mathematics of predicting events in a time series (a sequence of data distributed in time, in which data is recorded continuously, as by a seismograph, or in discrete intervals, as by a stock ticker). Wiener reasoned that the signals handled by communications, control, and computer systems could be analyzed in a time series. "One and all, time series and the apparatus to deal with them whether in the computing laboratory or the telephone circuit," Wiener said, "have to deal with the recording, preservation, transmission and use of information." He first treated this information as the "recording of a choice between two equally probable simple alternatives, one or the other of which is bound to happen," such as the tossing of a coin. The alternatives could be represented by the binary values "1" and "0." He then derived a general formula to measure the amount of information in a time series. He noted that this quantity was the "negative of the quantity usually defined as entropy in similar situations."[18] Entropy measures the degradation of energy in a closed thermodynamic system, the disorder of the system. For example, if the energetic and less energetic molecules in a system containing a hot body and a cold body become mixed (disordered), its entropy will increase, indicating that there is less energy available to do work. Wiener defined information as negative entropy, which means, for communications systems, that if the selection of messages becomes more random (more disordered), less information will be transmitted. For Wiener, information thus measured the pattern and order of the messages sent, not how much data was sent.

Wiener related his theory to the physical concept of entropy by treating communication as the retrieval of a message in the presence of noise. He noted that "the information carried by a precise message in the absence of noise is infinite. In the presence of noise, however, this amount of information is finite and it approximates 0 very rapidly as the noise increases in intensity." He concluded that the "processes which lose information are, as we should expect, closely analogous to the processes which gain entropy," i.e., those that become disordered. In contrast, Shannon restricted his concept of information to being mathematically equivalent to the entropy equation; he did not speak in terms of physics. Yet when Wiener turned to communications problems, he derived the same equation Shannon had obtained for the rate of information of a noisy channel in the analog case. Wiener remarked, "This is precisely the result which the author and Shannon have already obtained," thus revealing his familiarity with Shannon's unpublished work.[19]

In *Cybernetics*, Wiener applied his theory of information beyond engineering problems to comment on the importance of information in society. In regard to national groups, Wiener argued that "whatever means of communication the race may have, it is possible to measure the amount of information available to the race, and to distinguish it from the amount of information available to the individual. Certainly, no information available to the individual is also available to the race, unless it modifies the behavior of one individual to another." In this instance, and in similar statements about the relationship between information and the size of communities, Wiener added the linguistic dimensions of semantics (meaning) and pragmatics (how communication affected its recipients) to his syntactic definition of information.[20]

Even more broadly, *Cybernetics* became well known for its philosophy of information. When comparing the human brain to the digital computer, he noted that vacuum-tube computers used much more energy than their biological counterparts, but the energy spent on each calculation was very small. "The mechanical brain does not secrete thought 'as the liver does bile,' as the earlier materialists claimed, nor does it put it out in the form of energy, as the muscle puts out its activity," Wiener explained. "Information is information, not matter or energy. No materialism which does not admit this can survive at the present day." That is, the amount of information was related to a choice among messages (a pattern), not to the material basis or the energy involved in its communication. In discussing the societal implications of cybernetics, Wiener noted that the centralized "control of the means of communication is the most effective" factor reducing stability (homeostasis) in society. "One of the lessons of the present book is that any organism is held together in this action by the possession of means for the acquisition, use, retention, and transmission of information." Finally, Wiener claimed that the technology of the present age—the digital computer and the automated control system—was creating an "age of communication and control," a second industrial revolution.[21] In 1949, a science writer interpreted Wiener to mean that the "19th-century [industrial] revolution was based on the transformation and transmission of energy. . . . The 20th-century revolution is based on the transformation and transmission of information."[22]

Shannon's article contained none of these mathematical, sociological, and philosophical musings about information. Published in two installments, "A Mathematical Theory of Communication" was Shannon's technical masterpiece, a lengthy treatise whose many proofs and coding theorems were aimed at mathematicians and communications engineers. The first installment, published in July 1948, dealt with the digital case of transmitting

individual symbols in a message (as in the telegraph). The second installment, appearing in October, the same month in which *Cybernetics* was published, dealt with the analog case of transmitting time-varying signals (as in
the telephone and radio).[23]

Although Shannon published a theory of information similar to Wiener's,
they approached this topic from different points of view. First, Shannon
defined amount of information to be mathematically equivalent to *positive*
entropy, indicating the amount of disorder (unpredictablity) in the communication of messages. In contrast, Wiener defined amount of information as
*negative* entropy, indicating the amount of order (predictability) in a set of
messages. Shannon commented on this difference in a letter he wrote to
Wiener in October 1948, after reading an advance copy of *Cybernetics*. "It
was interesting to note how closely your work has been paralleling mine in
a number of directions. I was somewhat puzzled by one or two points. You
consider information to be *negative* entropy while I use the regular entropy
formula . . . with no change in sign. I do not believe this difference has any
real significance but is due to our taking somewhat complementary views of
information. . . . We would obtain the same numerical answers in any particular question. Does this diagnosis seem correct to you?" Wiener agreed
that it did: "I think that the disagreements between us on the sign of entropy
are purely formal, and have no importance whatsoever."[24]

Second, while Wiener broadened this definition of information to include
semantics and pragmatics, Shannon held tightly to his nonsemantic position. In an often quoted passage from the 1948 paper, Shannon said, "Fre-
quently the messages [transmitted by a communications system] have *meaning*; that is, they refer to or are correlated according to some system with
certain physical or conceptual entities. These semantic aspects of communication are irrelevant to the engineering problem."[25] That problem was how
to transmit messages accurately in the presence of noise, regardless of
whether or not the messages made any sense. If a sender wanted to transmit
gibberish (or a message in a private code that seemed like gibberish), it was
the communications engineer's duty to transmit it faithfully. Wiener's definition of information was nonsemantic in this manner because it measured
the amount of choice in an ensemble of possible messages. But Shannon,
who developed his mathematical theory of communication while working
at AT&T, adhered to the engineer's injunction to ignore the meaning of
messages and value their accurate transmission.

Third, Shannon developed a much more extensive theory of information
than Wiener's by creating a general theory of communication. At the heart
of Shannon's project is the model shown in figure 2, which applies to any

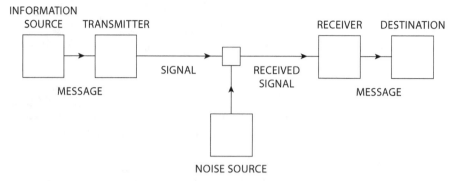

FIGURE 2. Claude Shannon's general communication model of 1948 became the blueprint used by natural scientists, engineers, and social scientists to apply Shannon's information theory to any communications system, from living organisms to intelligent machines. From Claude E. Shannon, "A Mathematical Theory of Communication," *Bell System Technical Journal* 27 (1948): 381. © 1948 *The Bell System Technical Journal.*

communication process, whether in machines, living organisms, or a combination of these. An information source selects a message from a set of possible messages. The transmitter encodes that message and converts it into a signal, which is sent along a channel affected by a noise source. The received signal (transmitter signal plus noise) enters the receiver, which decodes it and converts it into a message for the destination. Shannon defined information as the amount of uncertainty involved in the selection of messages by the information source, not as the amount of data selected to be transmitted. The greater the uncertainty of what the information source will select from an ensemble of messages, the more information it produces. The maximum amount of information is generated when the choice of messages is random (i.e., when there is an equal probability of choosing any message). No information is generated when the selection of messages is known. This made sense intuitively because the more *uncertain* we are of what message will be selected, the more information we receive. The more *certain* we are, the less information we receive. Using probability theory, Shannon derived a formula to measure "information, choice, and uncertainty" and noted its equivalence to the formula for entropy in statistical mechanics.[26] This definition of information is similar to Wiener's, except for the sign, and it explains why many commentators referred to it with the attributions "Shannon-Wiener" or "Wiener-Shannon."

Shannon cites Wiener's statistical theory of communication for the approach he used in considering the selection of a message from an ensemble of messages. But he says in the 1948 paper that he developed his theory of information from work done at AT&T in the 1920s. In 1928, physicist Ralph Hartley at Bell Labs published "Transmission of Information" in the *Bell Labs Technical Journal*, in which he defined the "amount of information" that can be transmitted in a noiseless communications system, such as an ideal telegraph, in terms of the number of symbols transmitted and the size of the alphabet (number of possible symbols). Hartley did not take into account the probability of which symbols would be selected and thus did not treat the problem statistically. In his 1948 paper, Shannon transformed Hartley's nonstatistical definition of information into a powerful statistical theory of how to code information to transmit it in the presence of noise.[27] Hartley's theory was better known to engineers than to mathematicians; Wiener, for example, did not cite Hartley's paper in *Cybernetics*. Yet the Institute of Radio Engineers awarded Hartley its medal of honor in 1946, partly for his research on the transmission of information.[28]

What sets apart the work of Shannon and Wiener—besides the reversal of sign in their equations—is Shannon's extensive analysis of coding. In that regard, Shannon introduced the important new concept of "channel capacity." In the digital case, channel capacity gives the maximum rate of transmission of information in bits per second. The paper seems to mark the first appearance in print of the term *bit* (a contraction of "binary digit"), which Shannon attributed to Princeton University mathematician John Tukey, who consulted for Bell Labs.[29]

Shannon argued that the concept of channel capacity allowed the communications engineer to devise codes that would protect the transmission of information from the harmful effects of noise. Shannon used the concept to posit and prove several coding theorems in what he first called "information theory" in the 1948 paper.[30] The Fundamental Theorem for a Discrete Channel with Noise, for example, states the remarkable result that if the information rate of a source is less than the channel capacity, a code can be devised to transmit information over such a channel with an arbitrarily small error. The trade-offs are complex codes and long delays in the transmitter and receiver (in order to do error-checking, for example).[31] These and other theorems established fundamental limits on data compression (encoding) and the transmission of information in communications systems.

Shannon's scheme uses the same tactic that inventor Samuel F. B. Morse employed to develop his famous telegraph code in the mid-nineteenth century. Morse devised shorter codes for the most commonly used letters in the English language and longer ones for the rarest letters (e.g., a dot for an *e*, and a dot, dash, dash, dash for a *j*). The Morse code enabled more messages to be sent over crowded telegraph lines than would a code that used equal numbers of dots and dashes for each letter.[32] Shannon's method put the encoding of messages on a scientific basis. But calculating the information parameters for human languages and devising coding schemes to send messages written in those languages, in the presence of noise, proved to be exceedingly difficult.

## Wiener and the Antiaircraft Project

To understand why Wiener and Shannon presented similar definitions of information in the contexts of two very different publications in 1948, we need to consider the different research contexts in which they developed their theories. The types of mathematics Wiener and Shannon preferred, their experiences working with engineers on wartime research projects, and their interactions with each other illuminate the social construction of cybernetics and information theory as two new postwar sciences.

A generation older than Shannon, Wiener made his mark in mathematics between the world wars. He joined MIT in 1919 as a mathematical prodigy—he received his Ph.D. in the philosophy of mathematics from Harvard at the age of eighteen—and became the star of MIT's mathematics department. He earned an international reputation in "pure" mathematics for his work in harmonic analysis (the study of complex, continuous functions), the statistical analysis of time series, and other areas.[33] Because he taught at MIT, Wiener also worked on applied mathematics by collaborating with faculty in its electrical engineering department. In the late 1920s, he helped Vannevar Bush put an engineering form of calculus on a firmer mathematical foundation. The experience prompted Bush to invent a series of analog computers, primarily to analyze electrical transmission networks, which culminated in the general-purpose differential analyzer in 1931.[34] In 1930, Wiener supervised a Ph.D. thesis in electrical engineering on the synthesis of electrical networks by Yuk-Wing Lee, a Chinese student. After Lee returned to China, Wiener worked with him there on an electrical network (filter), whose patents they sold to Bell Labs for a small sum in the late 1930s. The main reason AT&T, the parent company of Bell Labs, bought their patents was to protect the company's patent position, not to use the Lee-Wiener network.[35] The experience soured Wiener on the U.S. patent system and AT&T for years to come.[36]

During World War II, Wiener supported the war effort by working on research projects funded through the National Defense Research Committee, headed by Bush, then a prominent figure in science policy in Washington. Karl Compton, the president of MIT, headed the NDRC's Division D (Detection, Controls, and Instrumentation). In the fall of 1940, impressed by how Britain's low-frequency radar had helped it win the Battle of Britain by detecting German bombers flying across the channel from occupied France, the NDRC established the Radiation Laboratory (Rad Lab) at MIT to develop high-frequency, microwave radar, and it set up a Fire Control section to design antiaircraft systems. Both units were under Compton's division. Warren Weaver, an applied mathematician who directed the natural sciences division of the Rockefeller Foundation, headed the Fire Control section.[37]

In late 1940, Wiener received a small contract from Weaver's unit, for about thirty thousand dollars, to bring his mathematical expertise to bear on the antiaircraft problem. While the Rad Lab, Bell Labs, and industrial companies employed large staffs of scientists, engineers, and technicians to research the antiaircraft fire-control problem, Wiener worked with a staff of three—engineer Julian Bigelow; technician Paul Mooney; and a "Miss Bernstein," who did the calculations as the group's "computer"—to design a device that could predict the evasive actions of an enemy plane and direct a gun to shoot it down.[38] By all accounts, studying human operators as links in the control systems of the plane and the antiaircraft director—that is, the pilot and the tracker—led Wiener and Bigelow, in collaboration with Arturo Rosenblueth, to realize that humans and machines could be analyzed using the same principles of communications and control engineering, the basic idea of cybernetics.[39]

We know a good deal about Wiener's wartime research because of the considerable interest in the history of cybernetics and of control systems during World War II. But we know much less about how Wiener created a theory of information by designing an electrical network to shoot down airplanes.[40] Wiener gives the following version of events in *Cybernetics*.

> To cover this aspect of communication engineering, we had to develop a statistical theory of the *amount of information*, in which the unit amount of information was that transmitted as a single decision between equally probable alternatives. This idea occurred at about the same time to several writers, among them the statistician R. A. Fisher, Dr. Shannon of the Bell Telephone Laboratories, and the author. Fisher's motive in studying this subject is to be found in classical statistical theory; that of Shannon in the problem of coding information; and that of the author in the problem of noise and message in electrical filters.[41]

Although Wiener says that his theory of information grew out of the anti-aircraft project, he does not say how this occurred, nor whether the work of Fisher and Shannon played a role.

Wiener and Bigelow attacked the mathematical part of the antiaircraft problem, predicting the path of a piloted plane taking evasive actions, at MIT in early 1941 with funding from a two-year grant from the NDRC. Samuel Caldwell, head of MIT's Center for Analysis and a member of the NDRC, relayed the grant proposal to the NDRC after he and Wiener modeled a curved-flight predictor on MIT's differential analyzer. When Wiener and Bigelow turned that model into an electrical network, it ran into problems of stability. They then adopted a statistical approach that utilized Wiener's considerable expertise in this branch of mathematics. During the period of the grant, Wiener's group presented its statistical design to Bell Labs, which was working on a new director; built a prototype; demonstrated it to the NDRC; evaluated it against actual flight data obtained by visiting military installations; built a tracking system in the lab; and created a statistical theory of prediction and filtering.[42]

Their prediction device was an electrical network made up of capacitors, resistors, and vacuum tubes. Upon receiving an input from an optical or radar tracker, the predictor statistically correlated past positions of the airplane to the present and estimated future positions in order to calculate a target position at which to aim a gun. The Air Force specified that these actions had to occur within thirty seconds—ten seconds for data collection and analysis, plus twenty seconds for aiming and firing the gun. Wiener and Bigelow treated the data as a continuous time series that could be analyzed statistically.[43] They used the language of communications engineering to design the predictor. The tracking "signal," a time series representing observations of the plane, was assumed to consist of two other time series added together. The "message" represented the true path of the plane; "noise" represented tracking errors, electrical transients, and deviations the plane made from a geometrical path. Wiener and Bigelow used correlations between past, present, and predicted positions to filter out the noise. They used the remaining signal, the message, to obtain an optimum target position by minimizing errors between predicted and actual positions.[44]

Although they recognized problems with the director, Wiener wanted to continue the project, whose grant was due to expire in December 1942. The NDRC was more skeptical about the project and about curved-flight predictors in general. In early 1943, Warren Weaver, the head of the Fire Control section, reported that Bell Lab's electrical director—a straight-line predictor that was modified to predict curved paths—worked nearly as well as the

Wiener-Bigelow network on actual curved flights. It was much simpler and, thereby, more practical to produce for the war effort. Weaver terminated Wiener's project in February 1943, and Bigelow was transferred to another fire-control project at Columbia University.[45]

As noted by David Mindell, Wiener was an outsider to the fire-control traditions in place before the war—the mechanical systems approach at the Sperry Gyroscope Company, the communications systems approach at Bell Labs, and a control theory approach at MIT. When Wiener made his proposal to the NDRC, he independently sought to merge communications and control, a prewar trend evident in the work of Henrik Bode at Bell Labs and Harold Hazen at MIT, which the NDRC accelerated with wartime research contracts.[46] But Wiener was not a complete outsider. He became familiar with the prewar culture of communications engineering when he worked with Bush and Lee at MIT. He discussed the Lee-Wiener network with Bode and Bode's boss, Thornton Fry, at Bell Labs during the patent negotiations with AT&T in the 1930s.[47] During the war, Fry, a member of the NDRC's Fire Control section, and Gordon Brown, head of the newly formed Servomechanism Laboratory at MIT, thought Wiener's mathematical expertise fit the area nicely.[48] In fact, Wiener bragged to Weaver that researchers at Brown's Servo Lab "are coming to us all the time with problems that are right down our alley."[49]

Although Wiener and Bigelow's antiaircraft project was rejected for the war effort, it produced a mathematical theory of prediction and filtering that is used today and also forms a basis for cybernetics. The theory of prediction and filtering was distributed as a classified NDRC document in February 1942 and gained the racist World War II–era nickname, the "Yellow Peril," among engineers because of its yellow cover and abstruse mathematics.[50] Bigelow was not a mere technician. He was educated at MIT in electrical engineering, had worked for the Sperry Gyroscope Company and for IBM, and had done graduate work in mathematics at MIT during the antiaircraft project. Bigelow recalled that he wrote the first draft of the NDRC manuscript based on notes he took of equations Wiener wrote on the blackboard. Wiener revised and published the manuscript as sole author, a common practice at the time.[51] The NDRC report, published essentially verbatim as a book in 1949, merged the statistical theory of time series with communications engineering. With the prior work of Andrei Kolmogorov in Russia and Herman Wold in Sweden, it helped establish the foundations for the mathematical field of prediction theory.[52]

The word *information* appears several times in the NDRC report, but it is not related to entropy. Instead, the term refers to information gained by correlating time series, whether they represent a signal, message, or noise, to each other (cross-correlation) or to themselves (autocorrelation), rather than to the transmission of possible alternatives—the concept behind the theory he and Shannon independently published in 1948. Wiener comes close to stating that concept by saying the "transmission of a single fixed item of information is of no communicative value. We must have a repertory of possible messages, and over this repertory a measure determining the probability of these messages."[53] Wiener did not make explicit the connection between a time-series analysis and information as a measure of possible alternatives until after the war. In fact, there are only a few instances where he wrote about information as a scientific concept before then. Even the protocybernetic article he published with Bigelow and Rosenblueth in 1943—which discussed the cybernetic ideas of feedback, behavior, and communication—refers only once to information in a specialized manner, as the (nonquantified) "spatial information" a bloodhound picks up from a scent.[54]

Yet Wiener did expand his study of information at this time. In June 1944, he wrote Rosenblueth that he had "a number of papers in various stages of completion—statistical mechanics, prediction, multiple prediction, non-linear prediction, amount of information, etc., together with some philosophical work on the nature of time—with [Giorgio] De Santillana" at MIT.[55] This is the earliest instance I have found of Wiener using the technical term *amount of information*. The term was known in electrical engineering theory through the work of Ralph Hartley at Bell Labs in the 1920s,[56] but mathematicians commonly associated it with statistician Ronald Fisher who made his mark analyzing agricultural experiments. Fisher introduced the phrase into statistics in the 1920s and popularized it and its equivalent, *quantity of information*, among scientists in the *Design of Experiments* (1935).[57] That was probably the source for Wiener's usage of the term. A report on the work of the MIT mathematics department in 1944 juxtaposed Wiener's research on analyzing time series with an allusion to Fisher's work on determining the "amount of information" to be obtained from experimental observations.[58]

Whatever the source, all of the topics listed in Wiener's 1944 letter to Rosenblueth found their way into *Cybernetics*. The collaboration with his close friend de Santillana resulted in the chapter "Newtonian and Bergsonian Time," which contrasted reversible with irreversible time. Statistical mechanics and amount of information were behind Wiener's claim that information was negative entropy. Prediction and filtering were the methods he used to combine communications and control engineering, the basis of cybernetics. Tellingly, Wiener prepared a research memo on this subject for a meeting of the short-lived Teleological Society, a protocybernetics group he helped organize in late 1944, whose members included John von Neumann, Warren McCulloch, and Walter Pitts (see chapter 2).[59]

Pitts, who had coauthored with McCulloch an influential paper on neural nets in 1943, assisted Wiener on his research memo.[60] On McCulloch's recommendation, Pitts—an eccentric mathematical prodigy—joined Wiener—another eccentric mathematical prodigy—as a graduate student at MIT in the fall of 1943. Finished with the antiaircraft project, Wiener wanted to build up a group with Rosenblueth, who was then at Harvard's medical school, on what came to be called cybernetics. They wanted Pitts to represent the field of mathematical biophysics.[61] McCulloch told Rosenblueth that Pitts's "brain is so much like Wiener's that it's funny."[62] After the war, Pitts helped with a book (which was never published) that Wiener was writing on prediction and time series, an extension of the NDRC report that would combine most of the research topics listed in Wiener's 1944 letter to Rosenblueth. A hundred-page paper Wiener wrote on this project in early 1946 was the subject of the lost manuscript episode. It formed the basis for the chapter titled "Time Series, Information, and Communication" in *Cybernetics*.[63]

Pitts was a valued member of the Macy conferences on cybernetics, where Wiener first presented his theory of information and linked that theory to his research on time series. Wiener made these comments at several meetings of the Macy group. McCulloch reported that at the first meeting, held in March 1946, "Wiener pointed out that energy was the wrong concept to use with respect to communication systems which worked in terms of information." McCulloch does not say if Wiener related information to entropy at that time.[64] A transcript of an interim Macy meeting, devoted to the social sciences and held in September 1946, shows that Wiener said the yes-no character of nerve impulses could be analyzed as a discrete time series. Information was the logarithm of the number of impulses per second.[65] Difficult-to-read notes taken by Margaret Mead show that Wiener discussed "amount of information," its logarithmic measure, coding, and entropy at the second regular conference, held in October 1946. At the third conference held in March 1947, Wiener said, "Information is a neg[ative] am[oun]t of entropy."[66]

The subject was certainly on Wiener's mind in early 1947, the key period for working out this concept. That February he agreed to give a luncheon talk to a group of high school mathematics teachers on the subject of "Notions of Message and Entropy." In May, he told Warren Weaver, now back at the Rockefeller Foundation, which funded his postwar research with Rosenblueth on mathematics and neurophysiology, "Information is equivalent to negative entropy." Electrical engineer Robert Fano at MIT, recalled that sometime in 1947, Wiener "walks into my office, energetically sucking and puffing on his cigar. His belly sticking out and slightly bent backward as he walks, [and] announced 'information is entropy' and then walks out again. No explanation, no elaboration."[67]

Wiener presented these ideas in a nonmathematical form in "Time, Communication, and the Nervous System," a manuscript he wrote in 1947 for a chapter in a proposed book on the philosophy of culture, edited by Filmer S. C. Northrop, a regular member of the Macy group. A version of the paper was available at the Macy conference in March 1947 and was sent to conferees after the meeting. McCulloch told Northrop that "more than any other man he [Wiener] has helped to separate the notions of energy and communication so that information begins to take on, quite properly, the semblance of formal cause." Pitts, who was then working with McCulloch at the University of Illinois in Chicago, also commented on the paper. Having just dealt with the episode of the lost manuscript, Wiener gave strict instructions to McCulloch that Pitts was not to take the Macy manuscript home with him and only to read it inside McCulloch's lab! Wiener revised the paper as his contribution to the published proceedings of a symposium on teleological mechanisms sponsored by the New York Academy of Sciences in conjunction with the Macy Foundation. When "Time, Communication, and the Nervous System" was published in October 1948, it was promptly overshadowed by the publication of *Cybernetics* that same month.[68]

In the Macy manuscript, Wiener uses the term *amount of information* in three related senses: to measure statistical data gained from correlating messages and noise (as in the NDRC report); to refer to meaningful statements; and to measure choice among possible messages, the definition independently derived by Shannon.[69] Wiener gave the following example to illustrate the latter meaning of information. "If I send one of these elaborate Christmas or birthday messages favored by our telegraph companies containing a large amount of sentimental verbiage coded in terms of a number from one to one hundred," Wiener said, "then the amount of information which I am sending is to be measured by the choice among the hundred alternatives, and has nothing to do with the length of the transcribed 'message.' "[70]

Unlike Shannon, however, Wiener interpreted entropy in a physical manner. He declared that information in a message could only be degraded, noted that an organism's exchange of information with the environment enabled it to organize itself around greater energy, and used such phrases as "amount of information socially available" and the "amount of communal information" when discussing society as a communications system.[71]

In many ways, *Cybernetics* is the Macy manuscript writ large. Although the manuscript does not contain any equations—unlike the highly mathematical book—it integrates the scientific concept of information into Wiener's evolving cybernetic ideas that communication and control processes guide behavior in both humans and machines. Wiener related the three meanings of information in the Macy manuscript to a set of ideas that would become central to the book: the arrow of time, message, entropy, feedback, and the analogy between digital computers and the human nervous system.

Yet Wiener modified the equation defining "amount of information" before *Cybernetics* was published. In April 1948, he explained to the publisher why he was making significant mathematical changes at the galley proof stage. "The reason for this extensive modification is that on talking matters over with [John] von Neumann it has become clear to me that my earlier definition of entropy, though qualitatively alright, was not precisely the thing usually called by that name, and was not as easy to handle."[72] To lend further scientific support to the claim that his definition of information applied in the animal or the machine, Wiener cited physicist Erwin Schrödinger's *What Is Life?* (1944), which he also added at the galley stage of *Cybernetics*. For Schrödinger, an organism "can only keep aloof from it [death, the state of maximum entropy], i.e., alive, by continually drawing from its environment negative entropy," for example, food and water, to maintain itself, to maintain its order.[73] For Wiener, information helped all organisms, living and nonliving, adapt to their environment. Thus an appropriate definition of information for him was negative entropy, as opposed to Shannon's positive entropy.

While Wiener's antiaircraft project led directly to the feedback and control ideas in cybernetics, it led indirectly to his definition of information. Of the three meanings of "amount of information" that appear in the Macy manuscript and in *Cybernetics*, only one—a statistical correlation—came directly from the wartime NDRC report. McCulloch recalls Wiener saying that Bigelow impressed upon him the importance of the information conveyed, rather than a physical voltage or current in the antiaircraft predictor. That usage comports with the meaning of "information" that Bigelow employed a dozen times in an unpublished report on the predictor: as data fed into or retrieved from a machine, a meaning that would have been familiar to Bigelow as a former IBM engineer.[74] Yet computer engineers did not think of information as a measure of choice. For Wiener, that idea seems to have grown out of the notion in the NDRC report that deterministic time series carry no information. The idea of equating a choice of possible messages with entropy probably came later—sometime between 1944 and 1947—when Wiener started to link information with the ideas that would form the basis for *Cybernetics*.

Wiener was probably stimulated to think in those terms while Walter Pitts worked for him as a graduate student. A section of the 1944 report of the MIT mathematics department noted that Wiener's wartime work "has led him to further study of control engineering as a field analogous to communication engineering, and also of control in the living organism and the nervous system." The mathematical technique for this research was time series. "In this work, he is joined by Mr. Walter H. Pitts, who independently has made an analysis of the nervous system as a set of switching or relay controls, using the theory of Boolean algebra, as Mr. Shannon had done before him [in his master's thesis], but going beyond Mr. Shannon in treating the problem in time as a dynamical and not a statistical problem."[75] In the crucial period from 1944 to 1947, Pitts worked closely with Wiener on the analogy between the digital computer and the brain, neural nets, digital nerve impulses, and multiple prediction.[76]

The consummate collaborator, Wiener thus had access to his own Claude Shannon in the person of Walter Pitts. The result was a theory of information that, like Shannon's, was indebted to the statistical theory of communication arising from Wiener and Bigelow's antiaircraft project. But Wiener related his theory to time series and neural nets, rather than to cryptography, the stimulus for Shannon's theory.

### Shannon and Cryptography

Even though Claude Shannon had studied briefly under Norbert Wiener at MIT, he developed his theory of information in the context of a different mathematical and engineering culture. Shannon was steeped in the discrete mathematics of symbolic logic, in the culture of analyzing such digital devices as telephone switches and logic circuits. Wiener was expert in the continuous mathematics of harmonic analysis, in the culture of analyzing such analog devices as prediction and filter circuits. In his autobiography, Wiener said, "Shannon loves the discrete and eschews the continuum." In all of his work, "Shannon has been true to his first intellectual love for problems of a sharp yes-and-no nature, such as those of the wall switch, in preference to problems which seem to suggest the continuous or roughly continuous flow of electricity."[77] Although Wiener rather snobbishly discounted Shannon's love for the discrete (digital) case and overlooked his analysis of the continuous (analog) case in the 1948 paper, Shannon's preference for the digital is evident in his graduate education and the steps leading up to his theory of information.

After receiving dual bachelor degrees in mathematics and electrical engineering at the University of Michigan in 1936, Shannon studied for his master's degree in electrical engineering at MIT as a research assistant on the Rockefeller Differential Analyzer.[78] An updated version of Vannevar Bush's differential analyzer, the new machine was funded by the Rockefeller Foundation through the ubiquitous Warren Weaver. Two thousand vacuum tubes and a larger number of relays (electromagnetic switches) were added to digitally control the analog, electromechanical machine. When completed in 1942, it computed ballistic trajectories faster than the better known digital computers developed during World War II: the electronic ENIAC at the University of Pennsylvania, and the relay-driven Harvard Mark I. Larry Owens calls the Rockefeller analyzer "the most important computer in existence in the United States at the end of the war."[79]

In studying the machine's digital control unit—digital because its relays were either in the on or off position—Shannon realized that he could apply the Boolean algebra he had learned at Michigan to analyze digital logic circuits in general.[80] He developed the idea further during a summer job with the mathematics department at Bell Laboratories in 1937, then located in New York City. It was a propitious place to work, because the telephone system used countless relays to switch calls. Shannon finished his master's thesis on relay logic, supervised by Frank Hitchcock in MIT's mathematics department in the summer of 1937. "A Symbolic Analysis of Relay and Switching Circuits" won a major award from the American Institute of Electrical Engineers, which published the paper in 1938.[81] Engineers widely adopted Shannon's method because it could analyze any logic circuit, whether built from relays or vacuum tubes, or later from transistors and integrated circuits. As with other successful mathematical theories of electrical circuits, Shannon's master's thesis transcended the specific technology for which it was developed.[82]

Shannon owed a good deal of his success to the fact that Vannevar Bush, a powerful figure in American science and engineering, took him under his wing. As dean of engineering at MIT, Bush advised Shannon on the relay paper and greatly admired him. Bush wrote a recommendation letter to get the paper published and hired Shannon to work on his Rapid Selector, an early information-retrieval machine, in the summer of 1938.[83] When Shannon transferred from electrical engineering to MIT's Ph.D. program in mathematics, Bush suggested that he apply his proven ability in symbolic algebra to Mendelian genetics for his dissertation.[84] Bush did not drop his interest in Shannon when he left MIT to head the Carnegie Institution in Washington in 1939. In that year he wrote to Barbara Burks, a mathematically minded psychologist at the institution's Eugenics Record Office at Cold Spring Harbor on Long Island, asking her for a frank opinion of an exploratory paper Shannon had written on his thesis topic. Bush prejudged the matter, saying, "Apparently Shannon is a genius. I am not yet quite sure, but he has done some very unusual things with unusual types of algebra."[85] Burks agreed with her boss, thanking Bush "for letting me have a part in 'discovering' Shannon."[86]

Shannon's nominal Ph.D. adviser was Frank Hitchcock. A longtime member of MIT's mathematics department, Hitchcock did not have the international reputation of a Norbert Wiener. Shannon presumably chose to work with the "almost self-effacing" Hitchcock because he was an algebraist and had supervised Shannon's master's thesis.[87] Shannon did take a course from Wiener, recalling fifty years later that Wiener "was an idol of mine when I was a young student at M.I.T."[88] Wiener's recollection was more subdued. "While Shannon was an M.I.T. man, and while Bush was among the first of our staff to understand him and to value him, Shannon and I had relatively little contact during his stay here as a student."[89] Yet Wiener knew enough about Shannon to recommend him, with some reservations, for an applied mathematics job at the University of Pennsylvania in 1941, after he had finished his Ph.D. Wiener thought Shannon was "a man of extraordinary brilliancy and intelligence but I think can profitably spend some time on improving his general mathematical background instead of following his present tendency to limit himself in engineering problems to discrete as opposed to continuous methods. He has already done work of great originality and is with no doubt a coming man."[90]

Shannon did the research for his dissertation on genetic algebra with Burks at Cold Spring Harbor in the summer of 1939. Bush again made the arrangements.[91] Although Burks seems to have been genuinely impressed with Shannon, their correspondence leaves the impression that Vannevar Bush, a conservative New England patrician, thought it was the duty of an experienced female scientist to aid an up-and-coming male scientist, especially a shy genius.[92]

Shannon received his master's degree in electrical engineering and his Ph.D. in mathematics at the same commencement ceremony in the spring of 1940. Neither dealt with communication theory.[93] Yet Shannon had begun to consider that topic while working on the dissertation. In early 1939, he wrote Bush, "Off and on I have been working on an analysis of some of the fundamental properties of general systems for the transmission of intelligence, including telephony, radio, television, telegraphy, etc." He considered a communications system in terms of the mathematical transformation of one function "representing the intelligence to be transmitted," into a second function, which was affected by noise during transmission, then transformed into a third function by a receiver. Shannon was trying to prove a theorem that, "roughly speaking, it is impossible to reduce bandwidth times transmission time for a given distortion. . . . The idea is quite old; both [Ralph] Hartley and [John] Carson [at Bell Labs] have made statements of this type without rigid mathematical specifications or proof."[94]

Here are the seeds of the theory of information that Shannon would express in 1945 and develop more fully in 1948. A general communications system is represented by mathematical transformations at the transmitter and the receiver, and the effect of noise on bandwidth for reliable transmission is considered. What is missing from this scheme, of course, is the main element: the "amount of information" transmitted. Although Shannon referred to Hartley's ideas on bandwidth, he did not mention, in the letter to Bush, Hartley's related idea of using a logarithmic formula to measure the amount of information transmitted in a communications system.[95]

Shannon put aside these ideas, finished his Ph.D., and worked at Bell Labs in the summer of 1940. Shannon wrote Bush that he feared he would not have enough freedom of inquiry if he took a permanent job in industrial research. Bush recommended that Shannon stay in academia for a while and used his considerable influence to get him a fellowship under Hermann Weyl, a world-renowned mathematician at the newly established Institute for Advanced Study at Princeton.[96] Weyl wrote Shannon in April, "I shall be glad indeed to have you work on your biomathematical problems at our Institute" and to supervise Shannon as much as he could. "You will have the chance to discuss the matter also with [John] von Neumann and other algebraists who happen to be with us next year."[97]

Shannon gained a better appreciation for industrial research while working at Bell Labs that summer. He wrote to Bush, "I got quite a kick when I found out that the Labs are actually using the relay algebra in design work and attribute a couple of new circuit designs to it." In June, Bush encouraged Shannon to move toward applied mathematics and advised him to finish the work at Bell Labs before deciding what project to undertake at Princeton. "The only point I have in mind is I feel that you are primarily an applied mathematician, and that hence your [research] problem ought to lie in this exceedingly broad field rather than in some field of pure mathematics."[98]

After Shannon arrived at Princeton in the fall of 1940, the mobilization of science and engineering for World War II pushed him toward applied mathematics, and from there into a theory of information. In November, Warren Weaver, as head of the Fire Control section of the National Defense Research Committee, convinced Shannon to postpone his fellowship and work on one of Weaver's projects at Princeton.[99] In the typical style of the "old-boy" networks endemic to research contracting during the war, Thornton Fry—Shannon's former boss at Bell Labs, a member of the Fire Control section, and Weaver's former colleague in the mathematics department at the University of Wisconsin—had recommended Shannon to Weaver, who awarded him a ten-month contract at a little over three thousand dollars. Titled "Mathematical Studies Relating to Fire Control," Shannon's contract began on December 1, 1940, the same date as Wiener's much larger contract on the antiaircraft predictor at MIT. Weaver supervised both contracts.[100] After the war, Weaver—an applied mathematician—recalled that he persuaded Shannon "to give up a fellowship which he then had at Princeton (and under which he was in the process of being a pure topologist!) to undertake some studies in fire control design and prediction theory for [the NDRC]. He did some really stunning work for us."[101]

Shannon was less successful in his personal life. He wrote Bush in the spring of 1940, "As you may have heard, I was married in January and my wife and I have started housekeeping in Cambridge. I did not, as you may have anticipated, marry a lady scientist, but rather a writer. She was helping me with my French (?) and it apparently ripened into something more than French. She's a very intelligent girl, and I hope you can meet her some time."[102] Shannon moved with his wife, nineteen-year-old Radcliffe junior Norma Levor, from Cambridge to Greenwich Village to work at his summer job at Bell Labs. According to Norma, they were happy there, frequenting the jazz clubs. But the marriage fell apart when they moved to the IAS at Princeton, where Shannon became reclusive. Norma left Claude in 1941 when he refused to seek psychiatric help. Weaver recalled that Shannon "had a tragic break-up in his personal life . . . and for a time it looked as though he might completely crack up nervously and emotionally. It is Thornton Fry who deserves the primary credit for getting him out of that state, and for offering him work in the Bell Laboratories. The rest of it is history."[103]

When Shannon joined Bell Labs as a permanent employee in Fry's mathematics department in the fall of 1941, he was immediately set to work on war contracts. Although the full extent of Shannon's activities during the war is unknown because some of it is still classified, we know that he worked in three areas: antiaircraft directors, a secret communications system, and the theory of cryptography.

Participating in the fire-control project is probably how Shannon learned about Wiener's statistical theory of communication. In October 1941, Weaver suggested that Shannon sit in on a meeting at Bell Labs with John Atanasoff, a visiting computer inventor, about the antiaircraft director being designed by Henrik Bode's group at the labs. The meeting included George Stibitz, an inventor of fire-control relay computers at Bell Labs and Weaver's technical aide at the NDRC. Weaver made an appointment for Atanasoff to go to MIT after the meeting to see how Wiener and Bigelow were coming along with their competing antiaircraft project. Later, in July 1942, Stibitz and Fry later saw a demonstration of the Wiener-Bigelow predictor as representatives of the NDRC.[104]

Apparently, Shannon made similar trips to MIT. Bigelow recalled in an interview, "In the time I was associated with Wiener [at MIT], Shannon would come up and talk to Wiener every couple of weeks and spend an hour or two talking with him. . . . Wiener would exchange ideas with him in a most generous fashion, because Wiener had all the insights of what information theory would be like and he spewed out all these ideas and his comments and suggestions to Shannon."[105] If Bigelow's memory is correct, these visits add a new dimension to Shannon's acknowledgment of Wiener in his 1948 paper on information theory: "Credit should also be given to Professor N. Wiener, whose elegant solution of the problems of filtering and prediction of stationary ensembles has considerably influenced the writer's thinking in this field."[106]

Shannon's second wartime project at Bell Labs was a top-secret communications system that allowed U.S. President Franklin Roosevelt and British Prime Minister Winston Churchill to communicate securely by radio-telephone. Declassified in 1975, Project X—also known as Sigsaly or Ciphony I by the U.S. Army Signal Corps—began in the fall of 1940. Two years later, a group at Bell Labs had turned a 1930s invention of the labs, the vocoder, a multichannel frequency-compression device, into a secret communications system. The researchers digitized each of the vocoder's signals and added to them a random cryptographic key, consisting of noise recorded on a phonograph disc. An experimental model of a twelve-channel Project X system was tested over the Atlantic in the fall of 1942. AT&T's manufacturing subsidiary, Western Electric, made the sets for the Signal Corps.[107]

Two prominent communication theorists at Bell Labs, whom Shannon later drew on for his theory of information, were involved with Project X. Harry Nyquist worked on it in early 1941; Ralph Hartley wrote a technical report relating the system to his theory of information in March 1941. No longer full-time staff members, Nyquist and Hartley consulted for Bell Labs during the war. Shannon does not appear to have worked directly with either of them on the project.[108] His main task was to mathematically prove the security of the Project X system. In May 1943, Shannon wrote a confidential report on the Vernam system, a mechanized method of cryptography invented during World War I for sending secret messages over telegraph lines. Shannon's report employed probability theory to prove the "perfect secrecy of the Vernam system," on which Project X was based.[109]

Working on Project X, which used a digitizing scheme known as pulse-code modulation (PCM), probably stimulated Shannon to think about channel capacity, a novel aspect of his theory of information. While the vocoder reduced the bandwidth required to send a telephone signal, PCM required more bandwidth and less power per signal, but lent itself to time-division multiplexing (an efficient way to send multiple signals on a line). Jonathan Sterne stresses the importance of working on PCM to Shannon's invention of information theory, noting that Bell Labs was keenly interested in developing this and other techniques to increase the efficiency of telephonic communications and thus the profits of AT&T.[110]

It is not coincidental that Shannon wrote his report on the Vernam system shortly after talking with mathematician Alan Turing, Britain's top cryptographer, during a secret visit Turing made to the United States during the war. As the Anglo-American liaison on cryptography, Turing arrived in Washington in November 1942 to discuss the techniques he and others at Bletchley Park, England, had invented to break the Enigma code used by the Germans to transmit messages to and from U-boats in the Atlantic.[111] After completing his work in Washington, Turing went to Bell Labs in January 1943 for two months, where he analyzed Project X. Turing wanted to ascertain for the British government that telephoning by Project X was completely secure, especially because one of its main callers would be Churchill. He was not completely satisfied and suggested some alterations. At Bell Labs, Turing spent a good deal of time talking privately with Shannon "every day at teatime in the cafeteria," according to Turing's biographer Andrew Hodges. Among other topics, they discussed machines that could imitate the human brain, a project Shannon had mentioned to Bush in 1939. Hodges draws a connection between Turing's and Shannon's complementary ideas about information, but says "they were not free to discuss them" at the time, a point Shannon corroborated in a later interview.[112]

In 1945, Shannon drew together the intellectual strands of his wartime research on fire-control systems and Project X to produce "A Mathematical Theory of Cryptography," the classified report that formed the basis for the theory of information he published in 1948. Shannon recalled, "The work on both the mathematical theory of communications and the cryptology went forward concurrently from about 1941. I worked on both of them together and I had some of the ideas while working on the other. I wouldn't say one came before the other—they were so close together you couldn't separate them."[113] In another interview, Shannon said, "In one case [you are] trying to conceal information, and in the other case [you are] trying to transmit it." He worked at home on information theory and "used cryptography as a way of legitimatizing the work." He "was not yet ready to write up information theory. For cryptography you could write up anything in any shape, which I did." Furthermore, "Working on cryptography led back to the good aspects of information theory." Part of the 1948 paper "was taken verbatim from the [declassified 1945] cryptography report, which had not been published at that time." It was published in 1949.[114]

The close connection between Shannon's research on cryptography and that on information theory is clear from the 1945 report, "A Mathematical Theory of Cryptography." Figure 3 shows the diagram of a general cryptographic system from the report. Although Shannon restricted his analysis to systems using discrete symbols (e.g., letters of the alphabet), he thought it could "be generalized to study continuous cases, and to take into account the special characteristics of speech secrecy systems," an allusion to Project X.[115] Initially, a *key* (a code) is selected from a set of possible keys in the key source and given to the encipherer and the decipherer. Then the encipherer applies the key to a *message* (encodes it), which is selected from a set of possible messages in the message source, to produce a *cryptogram*. The cryptogram is then transmitted to the decipherer who uses the key to decode the cryptogram and read the message. The enemy cryptanalyst, who does not have the key, intercepts the cryptogram and tries to decode it.[116]

A major goal of Shannon's report was to create methods to judge the relative merits of secrecy systems. By treating cryptology from the viewpoint of probability theory, Shannon assumed that the selection of a message and a key were statistical processes, whose a priori probabilities were known to the enemy cryptanalyst. Upon intercepting the cryptogram, the enemy could calculate new probabilities. "This set of *a posteriori* probabilities constitutes his knowledge of the key and message after interception." Perfect secrecy was defined as the situation where the enemy's a posteriori probabilities equaled his a priori probabilities; he would be no better off after intercepting the cryptogram than beforehand. Shannon created two technical criteria with which to compare cryptographic systems: equivocation, a "theoretical secrecy index" that indicated "how uncertain the enemy is of the original message after intercepting a cryptogram of N letters"; and the unicity distance, "how much intercepted material is required to obtain the solution to

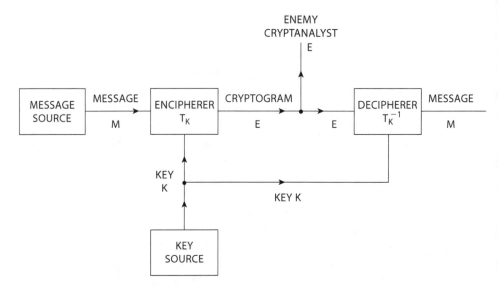

FIGURE 3. Claude Shannon's 1945 model of a general secrecy system, declassified by 1949, reveals the debt his theory of information owed to his work on cryptography during World War II (compare with fig. 2). From Claude E. Shannon, "Communication Theory of Secrecy Systems," *Bell System Technical Journal* 28 (1949): 661. © 1949 *The Bell System Technical Journal.*

a secrecy system." The more redundancy in a language, the shorter the unicity distance. According to David Kahn, Shannon's theory "made possible, for the first time, a fundamental understanding of the process of cryptogram solution."[117]

Shannon introduces his definition of information early in the 1945 report. He proposes a mathematical quantity, the equation for the statistical form of entropy for discrete events, as a "numerical measure of this rather vague notion" of the uncertainty, or choice, in making a selection from possible events whose probabilities of occurring are known. He then says, "Quantities of this kind appear continually in the present paper and in the study of the transmission of information," an allusion to his parallel work on information theory. The quantity, he continues, "measures in a certain sense how much 'information is generated' when the choice is made." He concludes by noting "that quantities of this type . . . have appeared previously as measures of randomness, particularly in statistical mechanics. . . . Most of the entropy formulas contain terms of this type."[118] Shannon copied large chunks of this section of the report verbatim into the 1948 paper.[119]

The nonsemantic approach, a hallmark of the 1948 paper, was also common in cryptography. "The only properties of a language of interest in cryptography are statistical properties," Shannon said in the 1945 report. "What are the frequencies of the various letters, of different digrams (pairs of letters), trigrams, words, phrases, etc.? What is the probability that a given word occurs in a certain message? The 'meaning' of a message has significance only in its influence on these properties." The quantification of information as uncertainty played a key role in Shannon's theory of cryptography. He used these quantities to calculate the information rate (in digits per letter or second) of a source, redundancy of a source's language, the unicity distance, and the equivocation (uncertainty in decoding the original message). From these parameters, he judged the relative secrecy of specific cryptographic systems. Equivocation was "almost identical with information, the difference being one of point of view."[120]

By comparing the system diagrams in the two papers (in figs. 2 and 3), we can see how Shannon drew on cryptography to develop his theory of information. If the goal of the cryptographer is to outwit the enemy cryptanalyst, the goal of the communications engineer is to outwit noise. Codes conceal information from the enemy in cryptology and allow information to be reliably transmitted in the presence of noise in communications. While the 1945 paper defined criteria, based on the information measure, to compare coding schemes, the 1948 paper proved several coding theorems in what Shannon called "information theory."[121]

## Credit for Information Theory

Wiener was more interested than Shannon in who should get credit for equating information with entropy in communications systems. In late 1948, Warren Weaver thanked Wiener for sending him a copy of *Cybernetics*, which he thought was "fascinating and important," then turned to the subject of Shannon. "I am sure that you have been following the work which Claude Shannon has recently published in the Bell technical journal. It seems to me that this is also first-rate. He is so loyal an admirer of yours that I find it difficult to decide how much of this was really inspired by you, and how much he deserves individual credit for. That is probably a very bad question to put to you, but I would nevertheless be interested in your comments." Wiener apparently did not answer the letter.[122] In 1953, Wiener told a science writer that he had worked out a statistical definition of information, transmitted as binary digits, a few months to a year before hearing Shannon give his paper at Harvard in March 1947.[123]

But Wiener usually spoke in terms of joint credit. In his review of a book published in 1949 by Shannon and Weaver on information theory, Wiener said the idea of equating information with entropy "was developed about the same time by the author of the present review." (He did not mention Fisher in this account.) Wiener further noted that Shannon drew on his statistical theory of communication.[124] In his autobiography, Wiener emphasized the influence he and Shannon had jointly exerted on communications engineering. "In introducing the Shannon-Wiener definition of quantity of information (for it belongs to the two of us equally), we made a radical departure from the existing state of the subject" by considering the problem of noise. "Thus, the statistical point of view in communication theory, which I had anticipated so many years before with my generalized harmonic analysis, and which Shannon and I had jointly made fundamental at the beginning [sic] of World War II, became inevitable and basic shortly after the war had begun."[125]

Scientists and engineers had an opportunity to compare the work of Shannon and Wiener before "A Mathematical Theory of Communication" and *Cybernetics* were published. At the annual meeting of the Institute of Radio Engineers, held in March 1948, a plenary symposium on "Advances Significant to Electronics" featured three of the most prominent scientists of the day. John von Neumann spoke on "Computer Theory"; nuclear physicist I. I. Rabi, who had worked at the Rad Lab and Los Alamos, spoke on "Electronics and the Atom"; and Norbert Wiener gave a talk on "Cybernetics." Two lesser figures filled out the session. E. M. Deloraine, an engineer with the International Telephone and Telegraph Corporation, spoke on "Pulse Modulation"; and Claude Shannon talked on "Information Theory." Both Wiener and Shannon dealt with the new concept of information.[126] The talks of Wiener and Shannon seem to have been well received. Henry Wallman, a professor in MIT's mathematics department, wrote to Shannon that the "set of three talks by Wiener, [von] Neumann, and you was very good, and I heard many favorable comments." Wallman reported that Wiener's book, *Cybernetics*, then at the galley proof stage, "is mainly philosophical and never gets down to a precise and detailed an analysis as the one you presented, but it's certainly interesting."[127]

The IRE session and Wallman's remarks foreshadow a good deal of the subsequent history of cybernetics and information theory: their popularization as two "new sciences" in the 1950s; the contested boundaries drawn around the two disciplines then and now; and persistent questions about how and why they reshaped the physical and social sciences in Cold War America.

# CHAPTER TWO

# Circular Causality

T HE FAULT LINES between the natural sciences and the social sciences
ran deep at the Macy conferences on cybernetics. At a special meeting
on the social sciences, held between the first two conferences in 1946, Norbert Wiener flatly stated that the social sciences did not have long enough
runs of consistent data to which to apply his mathematical theory of prediction.[1] Matters got worse in 1947. The cybernetics group, chaired by Warren
McCulloch, refused to elect psychoanalyst Erik Erikson as a permanent
member because Walter Pitts thought he lacked rigor and logical reasoning.
Furthermore, the group's mathematicians and physiologists roundly criticized a presentation by Wolfgang Köhler, a founder of Gestalt psychology,
for not being grounded in neurological data. Gregory Bateson complained
to McCulloch, "I doubt whether any of us social scientists will come up to
his [Pitts's] standards for another thirty years."[2]

Bateson and Margaret Mead were determined that the social scientists
who attended the next meeting, as guests, would make a better showing in
front of the group's leaders, whom Bateson called the "physico-math boys."[3]
Although Gregory and Margaret were separated as husband and wife, they
coached the guests, all of whom were linguists, at an evening party held
in early 1948 at the apartment they formerly shared in Greenwich Village
in New York City. To appease McCulloch, who held a low opinion of linguistics, Bateson told him that "none of them [are] professional linguists."
Bateson also held a preconference meeting in Boston, where he was teaching, which included social psychologist Alex Bavelas at MIT. Bateson and
Mead had wanted to invite five or six linguists to the 1948 conference, but
Bateson told McCulloch, "we are willing to sacrifice two linguists to get me
Bavelas: making four linguists and one Bavelas."[4] In a letter inviting one of

the linguists to the affair at Greenwich Village, Mead wrote: "I think you have heard something of the Conference of people interested in teleological mechanisms in which Gregory has been so much interested and of which he and I are members. . . . This is the thing that Gregory cares more about than any other intellectual endeavor and it is a most exciting focus of thought."[5]

The 1948 Macy meeting seems to have gone well. Bavelas was elected as a permanent member, and the natural scientists did not complain about the guest linguists: Dorothy Lee from Vassar, Roman Jakobson and John Loetz from Columbia, and Charles Morris from the University of Chicago.[6] One reason was that Mead and Bateson had worked hard to prepare them for the meeting. Mead later told the cybernetics group, "We select the guests and it is our job to brief them. Now, let me tell you, when Gregory and I planned that linguistics meeting two years ago, we spent about twenty hours before that meeting with each person; that is, first Gregory spent about three hours with each person, then a whole day with a group of them, with Jakobson and Dorothy Lee and Loetz, working it over beforehand, explaining what it was all about, giving them a view of what happened in the past." Mead thought the group's custom of not briefing guests and then complaining about what they said at the conference was "abominably unfair."[7]

In working through these difficulties of speaking across disciplines, by focusing on the communication rather than the research collaboration aspect of interdisciplinarity,[8] the Macy conferences became a crucial site for the development and spread of cybernetics and information theory into the social sciences and biology. They provided an interdisciplinary space for prominent scholars from a range of disciplines to meet regularly over an extended period of time—in ten conferences held from 1946 to 1953—to discuss how to apply concepts from the two fields outside of mathematics and engineering.[9] Other venues had more limited aims. Information theory was promoted as a discipline by the London Symposium on Information Theory and by the U.S.-based Institute of Radio Engineer's Professional Group on Information Theory, both of which were established in the early 1950s (see chapter 4). Some members of the Macy group strove to combine cybernetics and information theory into a universal discipline. The contentious interdisciplinary communication at the Macy conferences—among mathematicians, engineers, physiologists, and social scientists—reveal how difficult that task was and the many issues at stake when the founders and the skeptics of cybernetics and information theory debated the promise and limitations of the two new fields.

## Organizing the Macy Conferences on Cybernetics

Established in 1930 to fund medical research, the Josiah Macy, Jr., Foundation was part of a movement to promote interdisciplinary research in the United States. That path had been charted by the Social Science Research Council in the 1920s, which sought to integrate the social science disciplines. In the 1930s, Warren Weaver, as head of the Natural Sciences Division of the Rockefeller Foundation, funded the application of physics and chemistry to biology to create the research area of molecular biology.[10] The success of physics in World War II created new opportunities and problems in promoting interdisciplinarity. In 1949, Frank Fremont-Smith, medical director of the Macy Foundation, told the cybernetics group that the foundation's conference program addressed the "need to break down the walls between the disciplines and get interdisciplinary communication." This was "particularly difficult between the physical and biological sciences on the one hand, and the psychological and social sciences on the other." The problem had some urgency, he said, because the "physical sciences have developed to such a point and have gotten so far ahead of the social sciences that there is grave possibility that social misuse of the physical sciences may block or greatly delay any further progress in civilization."[11] The organizers of the cybernetics group thought their field would help in this regard by providing a common language with which to bridge the communications gulf between the natural sciences and the social sciences.[12]

The order in which presenters spoke at the early meetings indicates the hierarchical assumptions behind this belief. During the first four meetings, held twice a year in 1946 and 1947, the mathematicians and engineers—led by Norbert Wiener and John von Neumann—spoke first about the war-inspired digital computers and antiaircraft fire-control systems. Wiener explained how control systems behaved in a purposeful manner by feeding back the negative value of a system's output to its input. The circular flow of information allowed the system to compare its current state with a preset goal and take action to achieve that goal. The principles applied to living organisms, as well, such as the self-regulation of body temperature. The group's leaders referred to this process as *circular causality*. Next to speak were the neurophysiologists, led by McCulloch and Arturo Rosenblueth. Their research sanctioned the cybernetic analogy between computing and feedback systems, on the one hand, and human brains and nervous systems on the other. The social scientists—the psychiatrists, psychologists, sociologists, and anthropologists, led by Bateson and Mead—spoke last, about how to apply these analogies to their fields.[13]

Although McCulloch abandoned this order of speaking midway through the series of meetings,[14] the epistemological hierarchy that subjected the biological and social sciences to the physical sciences (in this case, mathematics and engineering) was evident at the Macy conferences. The hierarchy is reflected in the original title of the meetings: "Feedback Mechanisms and Circular Causal Systems in Biology and the Social Sciences." After Wiener popularized the word *cybernetics* in 1948, the group called its proceedings, *Cybernetics: Circular Causal and Feedback Mechanisms in Biological and Social Systems*, which published transcripts of the last five meetings. The wording of both titles reflects the hierarchy of the sciences assumed by the organizers, the prestige of the physical sciences in Cold War America, and the desire of the social sciences to emulate them.[15]

The founders of the cybernetics conferences derived their structure and goals from two meetings held during the war. The first meeting, a Macy conference on Cerebral Inhibition (conditioned reflexes and hypnosis) held in 1942, resembled the postwar cybernetics conferences in many respects. It had a similar interdisicplinary profile, and several attendees became charter members of the cybernetics conferences, including McCulloch, Rosenblueth, Bateson, and Mead. Remarkably, early versions of the two founding articles of cybernetics were presented at the meeting. Rosenblueth outlined how the wartime work by Wiener and Bigelow on antiaircraft systems could provide a general explanation of purposeful (teleological) behavior in both animals and machines. McCulloch presented the research he and Pitts had done to prove mathematically that neural nets in the human brain could perform the logical functions of a generalized digital computer (i.e., of a Turing machine).[16] Both articles were published in 1943. In using mathematics, engineering, and neurophysiology to blur the boundaries between humans and machines, the presentations created quite a stir at the 1942 meeting. Mead recalled, "That first small conference was so exciting that I did not notice I had broken one of my teeth until the conference was over."[17]

Mead and Bateson drew on their connections with the Macy Foundation to play leading roles in the 1942 meeting and in the cybernetics conferences. They were colleagues and friends of psychologist Lawrence (Larry) Frank, a former executive secretary of the Macy Foundation who attended the 1942 meeting and all of the cybernetics conferences. A founder of the interdisciplinary conference movement in the United States, Frank had introduced Mead to the value of these small assemblies in the 1930s.[18] Their lives were intertwined socially as well. As a married couple, Mead and Bateson rented part of Frank's townhouse in Greenwich Village and spent summers with him in New Hampshire. During World War II, the Frank household raised their young daughter Cathy (Mary Catherine Bateson) while they were busy doing war work, Margaret in Washington, DC, and Gregory in the Pacific.[19] As adherents of the interdisciplinary culture and personality school in the social sciences, Mead and Frank thought the cybernetic model of mind could aid their broader work, assisted by Fremont-Smith at the Macy Foundation, to improve mental hygiene as a way to ease the tensions of the Cold War. Fremont-Smith had also attended the 1942 meeting.[20]

The second protocybernetics meeting held during the war was dominated by the natural sciences. In January 1945, a dozen mathematicians, physicists, engineers, and neurophysiologists met at the Institute for Advanced Study in Princeton at the invitation of Wiener, von Neumann, and Howard Aiken, head of the Navy's Mark I digital computer project at Harvard, to discuss postwar plans. The invitation letter reflects the feedback-systems research conducted by Wiener, Bigelow, and Rosenblueth during the war. "A group of people interested in communication engineering, the engineering of computing machines, the engineering of control devices, the mathematics of time series in statistics, and the communication and control aspects of the nervous system, has come to a tentative conclusion that the relations between these fields of research have developed to a degree of intimacy that makes a get-together meeting between people interested in them highly desirable." Although the organizers did not invite any social scientists, they said that the targeted research areas had "even economic and social interest," which probably referred to the recent book by von Neumann and economist Oskar Morgenstern on game theory.[21] Wiener and Aiken suggested "that the group be known as the Teleological Society," a name suggested by the title of the 1943 article published by Rosenblueth, Wiener, and Bigelow, "Behavior, Purpose and Teleology."[22]

The first day of the two-day meeting was devoted to engineering, with Wiener speaking on communications systems and von Neumann on computers. The second day was devoted to neurophysiology, with McCulloch and neurophysiologist Lorente de Nó of the Rockefeller Institute speaking about the human brain. Wiener wrote Rosenblueth, who could not attend the meeting, "In the end we were all convinced that the subject embracing both the engineering and neurology aspects is essentially one, and we should go ahead with plans to embody these ideas in a permanent program of research." The group decided not to organize a "formal permanent society" because their research was classified, but Wiener noted that the group planned to establish a professional society, journal, and research center after the war and said the Rockefeller and Guggenheim Foundations had indicated their support.[23] The group assigned research projects to its members in the areas of filtering and prediction; computerization of statistical problems; using the computer to solve differential equations in such fields as ballistics and hydrodynamics (probably for the atomic bomb project); and the relationship between engineering and neurophysiology.[24]

War pressures intervened, however, and the group disbanded. It did not hold its second meeting, planned for the spring, probably because von Neumann was busy in "the West," code for working on the classified atomic bomb project at Los Alamos.[25] Complicating matters further, a plan to entice von Neumann to MIT, to work with Wiener in a new interdisciplinary research center on computers, control systems, and neurophysiology, failed to materialize.[26] Wiener lamented in the fall of 1945, after the war had ended and the MIT mathematics department was starting to reorganize, that von Neumann "remains an unknown quantity." It seems that von Neumann used the MIT offer and another one from the University of Chicago to convince Princeton to allow him to build a computer, the Institute for Advanced Study machine, which was funded by the military and the Atomic Energy Commission.[27]

In the face of these setbacks, Wiener sought other means to bridge mathematics, engineering, and physiology. As noted earlier, he and Rosenblueth, who had moved to the Cardiology Institute in Mexico City, secured a multiple-year grant from the Rockefeller Foundation to enable them to collaborate on "mathematical biology," part of its ongoing program to promote interdisciplinary fields. Wiener also helped Pitts obtain a grant from the Guggenheim Foundation to work on the "mathematics of the nervous system." The Macy conferences on cybernetics, established in early 1946, were in many respects, an intellectual successor to the short-lived Teleological Society. In fact, the informal name of the meetings was the "Conference on Teleological Mechanisms." The Macy conferences also complemented the institutional support that Wiener was able to secure after failing to establish a protocybernetics center at MIT.[28]

All accounts agree that organizing the conferences was a joint effort between McCulloch, Fremont-Smith, Bateson, Mead, and Larry Frank. But the central figure was McCulloch, who was well-positioned to lead the Macy group.[29] Trained in psychiatry, philosophy, and neurophysiology at Yale, McCulloch had set for himself early in his career the ambitious and rather poetic research question, "What is a number that a man may know it, and a man that he may know a number?" He came to think of the lifetime project as experimental epistemology.[30] Holding an appointment at the Illinois Neuropsychiatric Institute at the University of Illinois Medical School in Chicago, McCulloch had close ties with Wiener and fellow neurologists Rosenblueth and Fremont-Smith. The only person to attend both protocybernetics meetings during the war, McCulloch was impressed with Rosenblueth's presentation on feedback mechanisms at the Macy conference on cerebral inhibition in 1942. He had convinced his coauthor Pitts to become a graduate student under Wiener at MIT during the war, and by 1946, McCulloch had received three small research grants from the Macy Foundation to study neurophysiology.[31]

After McCulloch urged Fremont-Smith to establish the feedback conferences, he worked with him and Bateson to select members from several disciplines to attend the first meeting. McCulloch wrote to Pitts that the agreed-upon list was "Wiener, von Neumann, and you on Mathematical Engineering, and Rosenblueth, Lorente de Nó and Ralph Gerard for Neurophysiology. [Lawrence] Kubie (first to suggest importance of feedback in psychiatric problems) and Hank Brosin (a dandy) for Psychiatry. Meade [sic], Bateson and someone else for Sociology—Bateson, you will remember, is the man who insists on the importance and lack of theory in Sociology. [Donald] Marquis, [Heinrich] Kluver or [Kurt] Lewin and [Molly] Harrower for Psychology and half a dozen others from various fields not well known to me."[32]

McCulloch ordered the list of speakers to reflect the consensus reached at the protocybernetics meetings: that the computer-and-feedback analogy between humans and machines should set the agenda for biology and the social sciences. In suggesting the talks to be given at the first conference, held in spring 1946, McCulloch told Fremont-Smith that the "final, or purposive, aspect can scarcely be formulated until the formal, logical, or go-no-go [digital] aspect has been clearly stated and the men most competent to handle these statements are undoubtedly the mathematicians and engineering brains in the group." They should be followed by a neurophysiologist to indicate applications of these ideas to the nervous system. "If we can get this much clear, I believe we will be ready to attempt formulation in wider fields." Bateson agreed with this hierarchy. In fact, he published an editorial in *Science* that summer on the necessity for the natural sciences to come to the aid of the social sciences.[33]

Wiener did not play a big role in organizing the conferences. After McCulloch gave him a list of two dozen invitees in early 1946, Wiener suggested adding Julian Bigelow, then chief engineer on von Neumann's computer project at the Institute for Advanced Study in Princeton, and humanist Giorgio de Santillana, Wiener's colleague at MIT. But Fremont-Smith and McCulloch selected only Bigelow, whom McCulloch knew from the wartime Teleological Society.[34] Even though von Neumann had turned down MIT's offer, Wiener wanted to push their old scheme in this new venue: "I think this is our great opportunity to present our point of view and that we ought to be in a position to correlate our talks before we begin," which they did when Wiener attended the public celebration of the ENIAC computer in early 1946. Wiener told McCulloch, "This meeting is going to be a big thing for us and our cause."[35]

The cybernetics conferences followed the pattern of the other small conferences on medical topics run by the Macy Foundation at the time. McCulloch, Fremont-Smith, and Bateson chose two dozen charter members from a range of disciplines. There were five members from mathematics and engineering, six from physiology and medicine, one from ecology, eight from the social sciences, and one from philosophy. They invited guests and replaced permanent members who had dropped out. In all, seventy members and guests attended the conference series. McCulloch chaired the two-day meetings, all but one of which was held at a conference room in the Hotel Beekman in midtown Manhattan; the last meeting was held at the Nassau Tavern in Princeton. Macy staff recorded the wide-ranging discussions (critics called them "rambling") on the taping machinery of the day, a SoundScriber wire recorder. Edited transcriptions of the last five conferences were published from 1950 to 1955. The group selected physicist-engineer Heinz von Foerster from the University of Illinois as chief editor of the proceedings, and Mead and psychologist Hans Lukas Teuber as coeditors.[36] McCulloch recalled that he "could count on Margaret Mead's keeping a flowchart of the discussion in her head and on Walter Pitts' understanding everybody." Mead was more critical, recalling that McCulloch "had a grand design in mind [for how the conversation should proceed from meeting to meeting]. He got people into that conference, who he then kept from talking. . . . He wouldn't let Ralph Gerard talk. He said 'You can talk next year.' He was very autocratic."[37]

Although the cybernetics group followed the pattern of the other Macy conferences, its interdisciplinary reach was extraordinary. Rather than analyze the full range of those discussions, I focus on three issues that were debated at length at the meetings: the role of analogies and models in cybernetics; the relationship between cybernetics and information theory; and the attempt to turn cybernetics into a universal discipline with a universal language.[38] The issues proved to be of lasting concern to cyberneticians and information theorists for decades to come, as well as to the historians and sociologists who have studied these fields.

## Analogies and Models

The intellectual foundation of the conferences was the cybernetic analogy relating humans to machines, and machines to humans. In this analogy, the nervous system was deemed to work like a feedback-control mechanism, the brain like a digital computer, and society like a communications system—metaphors that Wiener had popularized in *Cybernetics* (1948) and the *Human Use of Human Beings* (1950). The analogy had become so closely associated with the cybernetics conferences that one regular member, Leonard Savage, a statistician from the University of Chicago, felt compelled to remind the group of its roots when it strayed from them in 1951: "We used to be a seminar or meeting on the subject of computing machines and we naturally do revert to the theme of the analogy between computing machines and human and social behavior, because at one time, at any rate, it was one of our most important theses, that we might find something fruitful in that analogy."[39]

By trying to turn the discussion back to the original center of the cybernetics enterprise, Savage raised a basic question: what made analogies and models fruitful in cybernetics? Were they created to represent mechanisms common to humans, machines, and society, to stimulate research into those mechanisms, or to mediate between science and nature?[40] Rosenblueth and Wiener came down on the representational side of the question in an article they had published on the role of models in science in the journal *Philosophy of Science*. Distinguishing between theoretical models and material models, they thought the value of the former was their logical structure and ability to solve black-box problems, where the box's inputs and outputs are known, but not its inner workings. The material model made experimentation easier to pursue by simplifying the "original complex system" to consider phenomena of interest. In contrast, psychologist Hans Lukas Teuber questioned the literalness of machine analogies. In 1947, Teuber complained to McCulloch that at the last conference, "psychology was getting into a squeeze between neurophysiology and robotology—until there was little room left for matters psychological." Robots might "become capable of doing innumerable tricks the nervous system is able to do; [but] it is still unlikely that the nervous system uses the same methods as the robot in arriving at what may look like identical results. Your models remain models." The psychologist should be the "mediator between neurophysiology and robotology" at the meetings.[41]

Margaret Mead agreed that cybernetic models were not strictly representational, but they stimulated her thinking in fruitful ways. When briefing psychologist Erik Erikson for his guest appearance at the 1948 meeting, Mead told him that the conferences were "an experiment in using conceptual models—drawn from the field of 'feedback' mechanics—to think about human behavior. So far most of the emphasis has been upon such concepts as positive and negative feedback. . . . Now, we want to do more with the conceptual model provided by these giant calculating machines [the digital computers], in which there are a lot of problems analogous to those of human perception, memory, problem solving, etc." That is where Erikson's work would come in. "The value of the conceptual model is of course only methodological, there is no trap of saying the human body is a machine, but merely that the methods, especially the mathematics used in these machine problems, may be available tools for thinking more precisely about human behavior."[42]

McCulloch arrived at a similar result by starting from theoretical models. An experimental physiologist who had helped create a mathematical theory of neural nets, McCulloch wrote to Teuber, "I look to mathematics, including symbolic logic, for a statement of a theory in terms so general that the creations of God and man must exemplify the processes prescribed by that theory. Just because the theory is so general as to fit robot and man, it lacks the specificity required to indicate mechanism in man to be the same as mechanism in the robot. But every robot suggests a mechanistic hypothesis concerning man, and it is the glory of a hypothesis that it leads to experiments that do reveal the mechanisms actually at work in man."[43]

The debate about the role of models in cybernetics heated up during a lengthy discussion about the analogy between the human brain and the digital computer at the 1950 conference. Ralph Gerard, a charter member of the group and a neurophysiologist at the University of Chicago, questioned the digital representation of the brain as primarily a collection of on-and-off neurons and of the nervous system as primarily controlled by on-and-off electrical impulses. Gerard criticized the group for switching from the "as if" idiom to the "is" idiom, for treating analogies as reality. "To take what is learned from working with calculating machines and communication systems, and to explore the use of these insights in interpreting the action of the brain, is admirable; but to say, as the public press says, that therefore these machines are brains, and that our brains are nothing but calculating machines, is presumptuous." Gerard reminded the group that it had considered the digital *and* analog aspects of the brain, and stated the scientific consensus that "chemical factors (metabolic, hormonal, and related) which influence the functioning of the brain are analogical, not digital." He argued that "digital functioning is not overwhelmingly the more important of the two," a minority position in the cybernetics group.[44]

The discussion that followed Gerard's presentation addressed three issues: whether to model the brain as primarily a digital or an analog organ; the meaning and usefulness of the terms *analog* and *digital*; and the value of analogies in science.[45] Several members had a lot riding on the digital side of the debate. McCulloch and Pitts had made their reputation creating a logical model of the brain as a computer composed of digital neurons. Von Neumann and Bigelow had used this model to design and build the electronic computer at Princeton.[46] During the discussion, von Neumann agreed with Gerard that "it is very plausible, indeed, that the *underlying* mechanism of the nervous system may be best, although somewhat loosely, described as an analogical mechanism." The digital logic function of the brain then rested on that chemical, analog foundation. But it was unclear if messages sent in the nervous system were coded in a digital or analog manner. Von Neumann thought the proposed coding scheme was a "very imperfect digital system." Although von Neumann agreed that the simple model of the neuron as an on-off device did not capture its biological complexity, he did not want to give up its digital representation.[47] Bigelow explained that mathematicians and physicists preferred to ignore the biological structure of neurons and interpret them as having the "property of carrying out processes like computation, that is, the property of carrying out operations which are in fact digital." Gerard replied, "I think the physiologists would be likely to say that that is just like a physicist."[48]

McCulloch tried to mediate the debate as a neurophysiologist who had represented neurons as digital devices. He suggested that the analog (continuous) alpha rhythm of the brain could be the envelope of digital (discrete) firings of impulses. He tried to settle the question of when to use an analog or a digital representation by saying that researchers should ask how information is coded. A continuous method of coding, as in chemical processes, would be classified as analog; a discrete method of coding, as in nerve impulses, would be classified as digital.[49]

McCulloch's mediation did not settle the dispute. Instead, the group delved into the question of how to define the terms *analog* and *digital*. In practice, scientists and engineers labeled a system as "digital" if it dealt with a signal or a numerical quantity in a discrete fashion, and "analog" if it dealt with them in a continuous fashion. For example, the telegraph was digital because it transmitted discrete signals representing discrete dots and dashes, while the telephone of that period was analog because it transmitted continuous signals representing continuous changes in sound levels and frequency. By a similar token, Von Neumann's Princeton computer was digital because it operated on discrete numerical quantities, while Vannevar Bush's differential analyzer was analog because it operated on continuous functions. This distinction seemed clear-cut in those realms, but it was contested by the interdisciplinary Macy group. When asked by Bateson to explain the difference between analog and digital, Gerard made the usual distinction between a continuous representation and a discrete representation, which he had learned from the group's mathematicians and engineers. He gave the example that a dimmer switch for electric lighting was an analog device, while an on-off switch was a digital device.[50]

The ensuing discussion did not satisfy Bateson, who broke in to say, "It would be a good thing to tidy up our vocabulary." He questioned the strict duality between the terms *analog* and *digital, continuous* and *discontinuous*, and also the concept of analogy itself. Bateson's probing elicited a surprising admission from von Neumann that the terms *analog* and *digital* were ambiguous:

> BATESON: . . . First of all, as I understand the sense in which "analogical" was introduced to this group by Von Neumann, a model plane in a wind tunnel would be an "analogical" device for making calculations about a real plane in the wind. Is that correct?
>
> WIENER: Correct.
>
> VON NEUMANN: It is correct.
>
> BATESON: It seems to me that the analogical model might be continuous or discontinuous in its function.
>
> VON NEUMANN: It is very difficult to give precise definitions of this, although it has been tried repeatedly. Present use of the terms "analogical" and "digital" in science is not completely uniform.
>
> MCCULLOCH: That is the trouble. Would you redefine it for him? I want to make it as crystal clear as we can.[51]

Von Neumann gave examples of different types of analog devices. Wind tunnels simulated aeronautical conditions (by analogy); the differential analyzer calculated continuous (analog) functions. In "almost all parts of physics the underlying reality is analogical," von Neumann said. "The digital procedure is usually a human artifact for the sake of description."[52] That is, humans often digitized nature's analog functions to make them easier to work with.

Further elaboration by von Neumann, Wiener, McCulloch, and others did not satisfy the social scientists. J. C. R. Licklider, a Harvard psychologist who would later head the government office that developed the predecessor of the Internet, understood how the terms, *analog, digital, continuous*, and *discontinuous* were used in mathematics and electronics, "but we would like explained to us here, to several of us and to many on the outside, in what sense these words are used in reference to the nervous system." McCulloch asked Pitts, a mathematician, rather than the neurophysiologists, to answer the question. Pitts gave a lengthy explanation of how to decide where a continuous mathematical series ended and a discrete one began, which did not settle matters, nor did interventions by Wiener, Stroud, Bigelow, and others.[53] Licklider asked, "Is it then true that the word 'analogues' applied to the context of the computer's brains, is not a very-well chosen word; that we can do well if we stick to the terms 'discrete' and 'continuous,' and that when we talk about analogy we should use the ordinary word 'analogy' to mean that we are trying to get substitution?" Savage replied, "We have had this dichotomy with us for four or five years, Mr. Licklider. I think the word [analogue] has worked fairly well on the whole. There would be some friction for most of us in changing it now." Licklider became more frustrated as the discussion wore on. "These names confuse people. They are bad names, and if other names communicate ideas they are good names." When the group could not clarify the meaning nor the origins of the terms analog and digital, Licklider thought the debate had gone on long enough. "We really ought to get back to Gerard's original problem. We will use the words as best we can."[54]

The contentious debate raised, once again, the question of the role of models in cybernetics. Once again, the debate split along the lines of representation versus the stimulation of research. Gerard said in his opening remarks that it was admirable to use the computer model to gain insights into the functioning of the brain. Teuber supported Gerard by saying, "To assume digital action is permissible as long as we remember that we are dealing with a model. The only justification for using the model is its heuristic value."[55] McCulloch, von Neumann, and Wiener thought models were more representational than that. Yet this time Wiener tried to mediate between the two positions using concepts he and Rosenblueth had developed in their article on models in science. Whether to use a digital or analog representation was really a matter of convenience. "I say that the whole habit of our thinking is to use the continuous where that is easiest and to use the discrete where the discrete is the easiest. Both of them represent abstractions that do not completely fit the situation as we see it. One thing that we cannot do is to take the full complexity of the world without simplification of methods. It is simply too complicated for us to grasp."[56] Wiener's agnosticism failed to settle matters.

The group made more progress on the role of models in cybernetics by considering material models—the feedback mechanisms that became the hallmark of cybernetics. The group discussed three mechanisms at the conferences: Wiener's hearing glove; Shannon's electromechanical mouse Theseus; and W. Ross Ashby's homeostat. Only Shannon displayed a physical model to the group; Wiener and Ashby read papers about their devices.

FIGURE 4. Claude Shannon demonstrated his maze-running "mouse," Theseus, an early form of artificial intelligence, to the Macy conference on cybernetics in 1951. It showed that the logic elements used in the telephone system and in early computers (electromechanical relays) could simulate animal learning. Courtesy of AT&T Archives and History Center.

Here, I discuss the two robotlike mechanisms, Theseus and the homeostat, and leave the cyborglike hearing glove for the next chapter.

Shannon demonstrated the electromechanical maze-solving Theseus at the 1951 meeting. He built the robotlike device from about seventy-five electromechanical relays of the type his employer, the Bell System, used to switch telephone systems (fig. 4). Theseus, a three-inch-long, mouse-shaped piece of wood hollowed out to contain a bar magnet, moved on three wheels by being attracted to a motor-driven electromagnet moving on the underside of the metal flooring of the maze. The maze consisted of twenty-five squares with rearrangeable partitions and a movable metallic goal (the "cheese"). Programmed by the logical circuitry of the relays, Theseus would explore the maze, bumping into walls, which it sensed with copper "whiskers," and clearing passages, until it reached the cheese at the center of one of the cubicles, thereby ringing a bell. The digital relays allowed Theseus to "remember" a clear path for the next run, thus showing how a feedback process could enable machines to learn, a primary interest of the Macy group. A trial-and-error exploration algorithm enabled Theseus to find the cheese, after which a goal algorithm enabled it to repeat the performance without bumping into the walls. Both strategies were fixed rather than random. Its many feedback loops allowed Theseus to keep searching if no cheese was present. Shannon designed an "antineurotic circuit" that would, after a few tries, break the mouse out of a "vicious circle" induced by the experimenter's moving the partitions after the mouse had learned the previous configuration. During his presentation, Shannon used the terms *sensing finger* and *goal*, rather than *mouse*, Theseus (his nickname for the mouse), and *cheese*, which were favored by the popular press.[57]

The published discussion of Shannon's presentation was short and dealt mostly with technical questions (from Pitts and Bigelow), rather than the implications of his model for the biological and social sciences. After the meeting, McCulloch acknowledged that Shannon had "refused to talk about it [social implications] but he agreed to bring his gadget to show us, and I think we are getting him into the conversation." A few discussants had brought up these topics briefly at the meeting. Upon hearing that the mouse would keep fruitlessly searching the maze if the goal was removed, Larry Frank said, "It is all too human." Gerard, Savage, and Mead asked a few questions about the possible "neurotic" behavior of the mouse, but did not push the subject.[58]

Theseus illustrates the blurring of boundaries between animals and machines that has fascinated commentators on cybernetics since the 1950s.[59] But the editors of the conference proceedings—von Foerster, Mead, and Teuber—noted a major problem with Shannon's model. Goal-seeking devices such as guided missiles had "intrigued the theorists [of cybernetics] and prompted the construction of such likeable robots as Shannon's electronic rat." Yet the "fascination of watching Shannon's innocent rat negotiate its maze does not derive from any obvious similarity between the machine and a real rat; they are, in fact, rather dissimilar. The mechanism, however, is strikingly similar to the *notions* held by certain learning theorists about rats and about organisms in general." Theseus thus modeled a theory of learning, rather than how real mice learned to run mazes. The editors concluded that the "computing robot provides us with analogues that are helpful as far as they seem to hold, and no less helpful whenever they break down." Empirical studies on nervous systems and social groups were necessary to test the relationships suggested by the models. "Still, the reader will admit that, in some respects, these models are rather convincing facsimiles of organismic or social processes—not of the organism or social group as a whole, but of significant parts [of it]."[60]

The cybernetics group took up this issue in-depth at the next meeting, in 1952, when W. Ross Ashby, dean of the English cyberneticians, discussed the homeostat, a model of the adaptive capacities of the human brain. A clinical psychiatrist and pathologist, Ashby constructed the homeostat out of war-surplus parts to illustrate the theory of adaptation through feedback mechanisms that he had developed before Wiener published *Cybernetics*. McCulloch had met Ashby at the Ratio Club (which one member called a "Cybernetics Dining Club") on a visit to England in 1951 and was keen to invite him to the 1952 Macy conference. The criteria for membership in the Ratio Club was that one had "had Wiener's ideas before Wiener's book [*Cybernetics*] appeared," and was not a full professor.[61] McCulloch wrote to Mead that "Ashby is a must for this meeting." He told the group that Ashby would help them "keep to the main topic of cybernetics, namely inverse feedback and homeostasis," from which the group had strayed.[62]

In his book on the homeostat, *Design for a Brain* (1952), which became a classic in cybernetics, Ashby assumed the guise of an experimenter who tried to determine the hidden, inner workings of the human nervous system by observing only its behavior. He thus treated the brain as an unknown "black box" having known inputs and outputs, a practice and terminology well-known in electrical engineering, then and now. Using this method, Ashby deduced that an adaptive system must contain variables, step functions, and the ability to randomly reorganize itself to adjust to its environment.[63] From those principles, he built the homeostat.

The homeostat consisted of four interconnected boxes filled with electronic gear and switches. On top of each box sat an electromagnet, whose pivoted needle made contact with a metallic liquid in a trough. The electrical output of each unit was fed to the input of another unit in a reconfigurable manner. If one of the needles moved out of range—deflected purposely by the experimenter for example—all four needles would move back and forth until they all reached a steady state, each within a middle range. By designating one or more of the units the "environment" and the other units the "organism," Ashby argued that the homeostat modeled how an organism adapted to a new environment or changes in the environment, for example, how a body's blood pressure and temperature could be kept stable via homeostasis. The homeostat used a random search process to reach what Ashby called "ultrastability." This random feedback process, similar to that in natural evolution, was Ashby's answer to the question, how can a deterministic mechanism exhibit nondeterministic adaptive behavior? Ashby thought his solution applied to both nonorganic and organic mechanisms.[64]

Despite the vast differences between the electromechanical hardware of the homeostat and the biology of the brain, Ashby maintained that his device accurately represented the adaptive processes of the nervous system. In fact, Ashby said he designed the homeostat to copy precisely the behavior of the brain, not to improve it. "In particular, if the living brain fails [to adapt to its environment] in certain characteristic ways, then I want my artificial brain to fail too; for such failure would be valid evidence that the model was a true copy."[65] Ashby opened his talk on the homeostat at the 1952 conference by saying, "We can consider the living mouse as being essentially similar to the clockwork mouse and we can use the same physical principles and the same objective method in the study of both."[66] Wiener could not have stated the basic principle of cybernetics more clearly.

The discussion following Ashby's brief remarks was contentious. The group did not deal with the general question of model building, but the more specific questions of whether or not the homeostat operated as Ashby claimed it did, and if it was of value to psychology. A major point of dispute was Ashby's claim that the homeostat adapted to its environment (by reorganization) through a random process. Pitts, who had worked with McCulloch on the theory of random neural nets, and Bigelow criticized the homeostat for not being truly random in its operation and for not having a random network, as did the brain. Bigelow satirically concluded, "It may be a beautiful replica of something, but heaven only knows what."[67]

The debate over whether or not the homeostat learned was just as heated. Ashby argued that the homeostat behaved differently after being presented with a new environment, which showed learning. Ashby agreed with a fellow guest at the meeting, Henry Quastler, a biologist at the University of Illinois, that the homeostat had a "serious fault" because its memory was reset after the environment was disconnected, and thus it could not remember how it had obtained stability when facing the same environment again. Bigelow insisted that memory was part of the learning process, as did another guest, Jerry Wiesner, director of MIT's Research Laboratory of Electronics and McCulloch's new boss. Pitts and McCulloch sided with Ashby, because if Shannon's Theseus learned, so did the homeostat.[68]

The psychologists were not impressed with the homeostat as a model of learning. Teuber said that mice, cats, and humans adopt "highly stereotypical forms of behavior" to solve a problem. "Shannon's rat does not do that, and neither does Ashby's homeostat. In that sense, these mechanisms are much more appropriate to certain simple learning theories than they are to real rats in mazes," a view that he and his coeditors had expressed in the Macy proceedings. Heinrich Klüver, a psychologist at the University of Chicago and a charter member of the cybernetics group, said that learning was "one of the least interesting" aspects of the interactions between the animal and the environment for psychologists. But if "Dr. Ashby's machine is of some help in understanding pathology of behavior, it may be of greater value than merely another model for a learning process."[69]

Bateson thought the homeostat modeled learning in an ecological system. He asked his ecologist friend George Evelyn Hutchinson, another charter member of the group, "if an environment consists largely of organisms, or importantly of organisms, is not the learning characteristic of Ashby's machine approximately the same sort of learning as that which is shown by the ecological system?" Hutchinson replied that it was, and also agreed with Bateson's further point that the homeostat showed how such environments are "likely to settle down to various degrees of greater stability."[70] These were the seeds of Bateson's idea of the "ecology of mind," which became popular among the counterculture in the 1970s.

A guest anatomist at the meeting, John Z. Young from University College, London, was not convinced that Ashby's "statistical machinery" could explain his own research on how an octopus learned to discriminate visually after surgery removed an optical lobe. Young preferred to think of the homeostat's "memory as largely a function of a reverberating circuit," a common computer analogy employed at the Macy conferences. Ashby insisted that the "analogy with the homeostat could be worked out" and made some suggestions to that effect. He admitted the superiority of the octopus to the homeostat, but, remarkably, he thought that both were products of evolution—one natural, the other human-made. "Obviously, the octopus has evolved a complete stage ahead of the homeostat, and has some arrangements by which it can use the homeostat principle" of random searching by step functions. Young was skeptical, but moved closer to Ashby's point of view by the end of the discussion.[71]

These divergent interpretations of the homeostat—by mathematicians, engineers, neurophysiologists, anatomists, ecologists, psychologists, and anthropologists—and the unresolved debate about the value of the homeostat as a model question the characterization of cybernetics, then and now, as a unified discourse. The homeostat became an icon of early cybernetics, as argued by Katherine Hayles,[72] but ironically it was not a stable artifact supporting a stable discourse in the Macy group. After the meeting, McCulloch reported the raucous discussion of Ashby's paper to an English colleague. "Ross Ashby did a superb job of presenting complicated affairs simply and clearly and has made for himself many friends and admirers, even among those who heckled him most at the meeting. He took it beautifully and made his point in the face of their opposition."[73]

Ashby also gained the admiration of Wiener, who was not at the 1952 meeting. Although Wiener privately questioned Ashby's mathematical ability, he touted the homeostat in the second edition of the *Human Use of Human Beings* (1954). "I believe that Ashby's brilliant idea of the unpurposeful random mechanism which seeks for its own purpose through a process of learning is not only one of the great philosophical contributions of the present day, but will lead to highly useful technical developments in the task of automatization."[74] Statements like these did more than the Macy conferences to make the homeostat an icon of cybernetics in the 1950s.

The cybernetics group took up the question of the value of scientific models for the last time at the 1952 meeting. McCulloch repeated the position of the editors of the conference proceedings: "The best model of the behavior of the brain is the behavior of the brain." Ashby defended his style of model making, saying "We learn things only by building the model within a wide conceptual field, so that we can see what is the relation of the brain as it is to what the brain would be if it were built differently." Pitts replied that models served two purposes: proving that a function can be performed by a mechanism; and gaining insight into how an organism works.[75]

These reflections on scientific modeling, a key tenet of cybernetics, fell short of what philosophers of science wrote on the subject in the 1950s. In 1958, for example, Max Black at Cornell University analyzed scientific models under the categories of scale, analogue, mathematical, theoretical, and implied models. In his terminology, Theseus and the homeostat would be analogue models, and cybernetics, itself, would be a theoretical model.[76] Although several members of the cybernetics group published occasional articles in *Philosophy of Science*, the field's major journal in the United States, they were much better at creating cybernetic models—theoretical and material—than in studying their role in cybernetics.[77]

### The Turn to Information Theory

If the analogy between humans and machines was the founding metaphor of the Macy conferences, the idea of circular causality was the mechanism behind the metaphor. The concept of negative feedback as a causal mechanism bridged such disparate disciplines as mathematics, engineering, biology, and the social sciences at the conferences and in the wider field Wiener had christened as "cybernetics." Despite Wiener's efforts from the very first meeting to educate the Macy group on the concept of "amount of information," its equivalence to negative entropy, and the central role information played in all communications systems, living and nonliving, information took a back seat to the concept of feedback during the first five Macy conferences. Most regular members had become comfortable with the idea of feedback, even though it came from control engineering, because it resonated well with previous ideas about circular causal systems in biology and the social sciences. Before the war, charter members Lorente de Nó and Lawrence Kubie had advanced the concept of circular causality in neurophysiology and Gregory Bateson had developed a similar idea in anthropology: schismogenesis, the generation of schisms between groups, caused, for example, by escalating their reactions to each other's actions.[78] Information, however, was a new and counterintuitive scientific concept.

The Macy group began to shift its focus from feedback to information at the 1949 meeting, the first one held after Wiener and Shannon published their theories of information. Pitts referred to that work by proposing that "Wiener's information theory" could help settle a dispute about whether three hundred "thinking neurons" could account for an ant's behavior.[79] Psychologist John Stroud continued that trend by using the terms *amount of information* and *unit of information* in regard to his experiments on human operators acting as servo systems. But neither Stroud, nor Bateson in the discussion of the presentation, understood that Wiener and Shannon defined information in regard to the number of decisions made by humans (or machines). Talking about information theory continued to the point where McCulloch thought it had gone on too long. Although McCulloch himself had referred to Shannon's theory during the discussion and was a fan of it, he thought the meeting should "get back to the central idea of this conference, namely, feedback." At that point McCulloch introduced the next speaker by reminding the group that Lawrence Kubie "was the first ever to propose that there were such things as closed paths in the central nervous system."[80]

Yet the Macy group, including McCulloch, kept bringing up information theory during the discussion of Kubie's paper on human neurosis and adaptation.[81] In regard to metaphors in psychiatry, McCulloch thought the "question of psychic energy has always seemed to me better quantified if one thought not in terms of energy, which is certainly wrong for the nervous system, but in terms of the amount of information that can be handled, and is being handled by those circuit elements which are still free to work," i.e., neurons not "hung up" by a neurosis. The debate over which metaphor to use, information or energy, both of which were drawn from physics, prompted Wiener to say at the 1949 meeting that information "is a concept which can go over directly from the study of the nervous system to the study of the machine; it is a perfectly good physical notion, and it is a perfectly good biological notion."[82] It was, therefore, central to cybernetics.

The discursive shift from feedback to information intensified when Mc-Culloch invited Claude Shannon to attend the next two Macy meetings as a guest. McCulloch had met him when Shannon gave a talk on communication theory in Chicago in 1949. In exchanging reprints of their papers, Shannon wrote to McCulloch, "There is considerable activity here [at Bell Labs] in information theory, computing machines and other 'Cybernetic' questions, and a fresh viewpoint [from McCulloch's papers] is stimulating."[83] When McCulloch invited Shannon to attend the 1950 conference, he said Shannon would have a chance to edit the transcription of the remarks he made at the meeting "because we desire to have discussion as free as possible, and we often say in our shirtsleeves things we do not like to see in print. You may find even your chairman driven to profanity."[84]

McCulloch invited Shannon to the 1950 meeting, which was devoted to language, because his theory of information dealt with optimal coding and the statistical structure of language. Shannon presented new research on quantifying the redundancy of the English language. He had done a "prediction experiment," asking readers (his wife and a colleague) to predict the subsequent letters that would occur in sentences drawn randomly from randomly chosen books. The readers guessed until they predicted the correct letter, then the subsequent ones. As in his 1948 paper on information theory, Shannon studied the statistical relationship between symbols, that is, the probabilities of their occurrence (what linguists called syntactics, the relation between signs), and ignored their meaning (semantics, the relation between signs and what they signified), and their effects (pragmatics, the relation between signs and their interpreters).[85] Shannon assumed that the prediction experiment would retrieve the knowledge that experienced readers held about the statistical structure of a natural language. Using the probabilities produced in the experiment, he calculated the redundancy of English to be about 75 percent; it could be compressed by that amount while being communicated and still be understood.[86]

The cybernetics group treated Shannon's presentation with respect. Pitts peppered him with technical questions, while Wiener compared Shannon's results to his own method of predicting time series. Shannon replied that he was working with an "uncorrelated time series," which satisfied Wiener. Savage questioned whether picking out a book from a book shelf was a random selection process, as Shannon had claimed, but that was the extent of the criticism of Shannon's research method.[87]

Instead, the group turned to the issue of semantics, which Shannon had insisted was outside the scope of his theory of information. When the group veered from syntactics to semantics, Savage reminded them "that Shannon has talked about redundancy at the presemantical level. . . . I think it would require special and difficult experimentation to measure the semantical redundancy of the language, though the experiment last reported by Shannon does bear on the subject to some extent in that the guesser knows English and is utilizing that knowledge in his guesses." Savage's intervention didn't help. The group persisted in this vein by relating Shannon's theory to what Bavelas called "second-order information" (information about the relationship between the speaker and the hearer) and further issues in semantics, pragmatics, and how to distinguish messages from noise.[88]

Savage noticed that Shannon had been quiet throughout the discussion, and suggested he speak up because his "ideas have been discussed for half an hour." Shannon replied from the point of view of a theorist of communications systems, not the sender or receiver. "I never have any trouble distinguishing signals from noise because I say, as a mathematician, that this is a signal and that is noise. But there are, it seems to me, ambiguities that come at the psychological level," such as identifying the useful part of a telephone conversation as signal, and the nonuseful part as noise. But "that is hardly a mathematical problem. It involves too many psychological elements."[89]

Licklider, a psychologist who had applied Shannon's theory in his presentation to the group on speech distortion, agreed with Shannon and pointed out difficulties in applying Shannon's theory outside of engineering. "It is probably dangerous to use this theory of information in fields for which it was not designed, but I think the danger will not keep people from using it." There "may have to be modifications, of course," to use it in psychology. (Licklider was prescient on both counts.) When Licklider asked Shannon to define the "concept of information—not just amount of information, but information itself," as he had done recently in a speech Licklider heard him give, Shannon gave a technical definition that probably did not satisfy Licklider nor other nonmathematicians at the meeting.[90]

The debate over Shannon's theory of information continued when McCulloch invited Shannon and Donald MacKay to attend the 1951 conference as guests. A lecturer in physics at King's College in the University of London, MacKay had received a travel fellowship from the Rockefeller Foundation, awarded by the ubiquitous Warren Weaver, to study American research at the nexus between communication theory, neurophysiology, and psychology.[91] McCulloch had met MacKay in England in 1949 and arranged for him to visit many members of the cybernetics group in the United States. These included Pitts, Wiesner, and Bavelas at MIT; Licklider at Harvard; Shannon at Bell Labs; and von Neumann and Bigelow at Princeton. The industrious MacKay made his headquarters at McCulloch's laboratory in Chicago and saw more than his share of researchers, visiting 120 laboratories in all! He and McCulloch became great friends; they corresponded often and visited each other after MacKay returned to Britain at the end of 1951.[92]

At the 1951 Macy meeting, MacKay presented a comprehensive theory of information that combined approaches from physics and engineering.[93] MacKay defined scientific information as the vector sum of "metrical information" and "structural information." He derived the former concept from Ronald Fisher's research in statistics, the latter from Denis Gabor's research in communications. MacKay relabeled Shannon's and Wiener's definition of information as "selective information." While scientific information indicated the precision and degrees of freedom of an experiment, selective information gauged the uncertainty of transmitting messages. MacKay gave the example that "two people, A and B, are listening for a signal which each knows will be either a dot or a dash. A dash arrives." A, a physicist, measures the signal and reports receiving a "good deal of information" about it. B, an engineer, reports receiving "little information," because he or she knew it would be either a dot or a dash. "For lack of a vocabulary," MacKay explained, "they are using the phrase 'amount of information' to refer to different measurable parameters of the different representational activities in which they are engaged."[94] MacKay created his general theory to provide that vocabulary.

In regard to semantics, MacKay drew a vector diagram indicating the amount of structural, metrical, and selective information in any proposition, body of data, or representation. The projection of this "information vector" on an axis indicated meaning. In regard to what later became known as artificial intelligence, MacKay suggested that "trains of thought (if you like the term) correspond to successive transformations of information vectors."[95]

MacKay's comprehensive theory—ranging from scientific experimentation to thought itself—went far beyond other theories of information. Even Wiener, who was not at this Macy meeting, had not gone that far in his wide-ranging book, the *Human Use of Human Beings* (1950). Because Shannon did attend the meeting and because he had insisted that his theory excluded meaning, his remarks on MacKay's presentation generated much interest then and have continued to do so.[96] Although commentators have emphasized the inflexibility of Shannon's interpretation of his own theory, he took a rather ecumenical stance before the Macy group. During the long discussion of MacKay's paper, Shannon replied, "it seems to me that we can all define 'information' as we choose; and, depending on what field we are working in, we will choose different definitions. My own model of information theory, based mainly in entropy, was formed precisely to work with the problem of communication. Several people have suggested using the concept in other fields where in many cases a completely different formulation would be more appropriate."[97]

Shannon's admission that different concepts of information had value did not satisfy the cybernetics group, who continued to debate the relative merits of and the relationships between the theories of information that MacKay had combined. Bigelow asked MacKay and Shannon to comment on his observation that information theory as applied to communications defined an engineering parameter such as channel capacity rather than the general "concept of information in any concrete or abstract sense." Pitts interjected that the term *amount of information*, used by Shannon, was complete, should not be divided, and "possibly there should be some Greek word invented to convey it by itself." Shannon agreed with Pitts, then admitted, "I think perhaps the word 'information' is causing more trouble in this connection than it is worth, except that it is difficult to find another word that is anywhere near right." He then repeated his own definition of information, saying "it is only a measure of the difficulty in transmitting the sequences that are produced by some information source."[98]

As chair, McCulloch let the speaker, MacKay, have the last word. "I was not," said MacKay, "presenting in any way an alternative theory to Shannon's, but a wider framework into which I think his ideas fit perfectly. In fact, as I tried to make clear, what I was doing was to show how Shannon's ideas were validly applicable, I think, universally, though such an analysis is not necessarily *appropriate* to all problems. I think he is too modest in his disclaimer, because as long as you define your ensemble [of possible messages], then his definition of the entropy or selective-information content of the ensemble, it seems to me, is always applicable without any contradictions."[99] The characterization of Shannon as being too modest became a common theme in later attempts to apply his theory in the physical and social sciences.

After the meeting, the cybernetics group recognized that information had supplanted feedback as its guiding metaphor. The editors of the proceedings of the 1951 meeting remarked that all of the regular members "have an interest in certain conceptual models which they consider potentially applicable to problems in many sciences. . . . Chief among these conceptual models are those supplied by the theory of information." They relegated feedback to second place: a "second concept, now closely allied to information theory, is the notion of circular causal processes."[100] Several participants at the next meeting, in 1952, had referred to the scientific concept of information despite the fact that neither Shannon nor MacKay attended the meeting. After the meeting, McCulloch told the inner group, "when we started out, it was with a clearcut basis on inverse feedback, on negative feedback. The growth of information theory, in lots of places and in lots of ways, has more and more tended to crowd out the feedback component."[101]

McCulloch was ambivalent about the shift to information. At the start of the last Macy meeting in 1953, he tried to explain why the shift had occurred. By the time of the 1949 meeting, the group "had become so weary of far-flung uses of the notion of feedback that we agreed to try to drop the subject for the rest of the conference." The group "had already discovered that what was crucial in all problems of negative feedback in any servo system was not the energy returned but the information about the outcome of the action to date. Our theme shifted slowly and inevitably to a field where Norbert Wiener and his friends still were the presiding genii." The group then turned its attention to computers, coding, language, and information theory, even considering the "amount of information conferred upon us by our genes." McCulloch ended his summary with a warning. "It is my hope that by the time this session is over, we shall have agreed to use very sparingly the terms 'quantity of information' and 'negentropy' [negative entropy]."[102]

The group did not heed McCulloch's warning. At the 1953 meeting, philosopher Yehoshua Bar-Hillel, then at MIT, talked about the theory of semantic information that he had developed with logical positivist Rudolf Carnap at the University of Chicago, which extensively used terms like *quantity of information*. Bar-Hillel noted that "the concepts of entropy and negentropy have popped up in the discussion of this group more than once." For another guest at the meeting, English cybernetician W. Grey Walter, cybernetics embraced both feedback and information. "To me, as a British neurophysiologist, it means, on the one hand, the mechanical apotheosis of reflexive action [as in his robot tortoises], on the other, the incarnation of information."[103]

## Interdisciplinary Blues

The discursive shift from feedback to information at the Macy conferences challenges the claim made by several historians that cybernetics had become a universal discipline with a universal language in the 1950s. According to Geoff Bowker, cyberneticians used several rhetorical strategies to achieve this result. They claimed that the new age demanded a new universal language, based on communications and control engineering, to span the divide between humans and machines. The universal discipline (or metascience) of cybernetics would subsume other sciences through its wide applicability, support them through computerization, and reorder the physics-dominated hierarchy of science. Slava Gerovitch has argued that the language of cybernetics, in the United States, Europe, and the Soviet Union, was marked by such linked terms as *feedback, control, information, entropy,* and *homeostasis.*[104]

Although many cyberneticians aimed to create a universal language, the Macy group had a more limited goal of maintaining an interdisciplinary discourse during the conferences. The group thus treated cybernetics as an "interdiscipline"—a hybrid field of knowledge existing between and within disciplines, similar to such new postwar fields as operations research and materials science—rather than as a metadiscipline.[105] In 1951, the editors of the Macy proceedings noted, "One of the most surprising features of the group is the almost complete absence of an idiosyncratic vocabulary. In spite of their six years of association, these twenty-five people have not developed any rigid, in-group language of their own. Our idioms are limited to a handful of terms borrowed from each other: analogical and digital devices, feedback and servomechanisms, and circular causal processes. Even these terms are used only with diffidence by most of the members, and a philologist given to word-frequency counts might discover that the originators of 'cybernetics' use less of its lingo than do their more recent followers. The scarcity of jargon may perhaps be a sign of genuine effort to learn the language of other disciplines, or it may be that the common point of view provided sufficient basis for group coherence."[106]

The published proceedings bear out the editors' claims. Although Wiener, Bigelow, Rosenblueth, and McCulloch often applied the language of *feedback, information,* and *entropy* to humans and machines, they did so less extensively than did some guests invited to the meetings. Psychologist John Stroud is a case in point. During the discussion of his research on human perception, presented at the 1949 conference, Stroud referred to the human operator as a servo mechanism, as the "machine . . . we placed in the middle" in an antiaircraft fire-control system. Although Wiener and other founders of cybernetics had popularized this way of speaking, Stroud expanded it by considering God to be an electrical engineer. In regard to regulating the brightness of human vision, Stroud "assumed that the fellow who made the human being put in A.V.C. [automatic volume control] circuits [like those in radio sets] in addition to the ordinary ones of chemical adjustments and adjustments of pigment materials, pupil adjustments, and so forth."[107] When Teuber complained to McCulloch about the "robotology" way of talking at a meeting in 1947, he noted that a guest psychologist, Wolfgang Köhler, was "trying to outrobot every one of your robotologists."[108] The group saw common concepts, models, and idioms as a means to break down the barriers between disciplines and improve communication between them, not as elements of a universal discipline.[109]

Despite these tensions, the group tried to communicate across disciplines at the research level. The problem was not easy. McCulloch noted at the end of the first year that the group's members had "come together from such unlike disciplines of science that they have had to learn each other's vocabularies before they could hope to understand one another. This has made it difficult to follow arguments which in themselves were often relatively simple."[110] At the end of the conferences, McCulloch did not think much progress had been made on this score. Research results had "been gathered by extremely dissimilar methods, by observers biased by disparate endowment and training, and related to one another only through a babel of laboratory slangs and technical jargons. Our most notable agreement is that we have learned to know one another a bit better, and to fight fair in our shirt sleeves."[111]

The editors of the 1951 proceedings thought the group had found one way to be interdisciplinary. "This ability to remain in touch with each other, to sustain the dialogue across departmental boundaries and, in particular, across the gulf between natural and social sciences is due to the unifying effect of certain key problems with which all members are concerned: the problems of communication and of self-integrating [i.e., feedback] mechanisms."[112] As noted earlier, the editors said this unification occurred by creating a limited idiom, not a jargon or a universal language.

Communicating across disciplines was achieved through hard work. Mead and Bateson spent several hours informing guest social scientists about the group's beliefs and ways of talking. McCulloch did his share as well. In 1948, he consoled psychologist Molly Harrower, a charter member of the group, "It is painfully apparent in all [inter]disciplinary meetings that not one of us has an adequate background to understand fully what all members of the group produce. . . . In a sense, a mathematician may sail gracefully along in his high level of abstraction into almost any field, but to us [experimentalists] that always seems empty and inadequate." McCulloch asked Harrower, rather paternalistically, to "stick with the gang" and help him moderate the next meeting by paying attention to the "poetical side" of the discourse, matching the sounds of words to emotions.[113]

Bateson worked alongside McCulloch to promote interdisciplinary communication. In addition to selecting members and coaching guests, he intervened in group discussions. At the 1952 meeting, Bateson felt it was a "challenge to the whole group to discuss" Ralph Gerard's presentation on neurophysiological mechanisms. Bateson asked Gerard to explain how his talk could help those who were not neurophysiologists "think about problems in personal communication, in evolution, in all fields to which the cybernetic approach is applicable. We are not here as a neurophysiological technical group, trying to solve neurophysiological problems, but we are profoundly interested in how those who solve neurophysiological problems do it. We are interested in the nature of such problems, because similar problems occur in a vast number of other fields."[114] Although the group's neurophysiologists ignored this plea, Bateson extensively applied the cybernetic approach to the social sciences.

Cybernetics, of course, had also become a way of thinking for other leaders of the Macy conferences, including Wiener, McCulloch, and von Foerster, who did collaborative interdisciplinary research. But cybernetics did not attain the status of a universal language at the meetings. Instead, the group contested its vocabulary. The term *information* overshadowed *feedback* as the meetings progressed, and the group hotly debated the meanings of *analog* and *digital*. McCulloch could not convince Shannon to discuss the social aspects of *information*. Wiener later blamed the lack of progress toward interdisciplinarity at the meetings on the lack of a common language such as mathematics, especially among the social sciences. "These semantic difficulties resided in the fact that on the whole there is no other language which can give a substitute for the precision of mathematics, and a large part of the vocabulary of the social sciences is and must be devoted to the saying of things that we do not yet know how to express in mathematical terms."[115]

Frank Fremont-Smith had hoped that psychiatry would provide an common language for the cybernetics group. During the discussion of Kubie's paper on neurosis in 1950, he was overjoyed that the group had finally taken the psychiatrists' notion of the *unconscious* seriously: "I do feel that this time we have come closer to a discussion in which there was a common denominator for every discipline here . . . and I think it is the first time we have had it. I feel very pleased." The patron was happy on that day, but the group did not adopt the language of psychoanalysis for its common idiom. McCulloch, for one, barely tolerated the discipline.[116]

The goal of interdisciplinary communication remained elusive. After the conference series ended, McCulloch and the editors of the proceedings debated whether or not to publish the group discussions at the last meeting.

The editors did not want to publish them because they were more diffuse than usual. McCulloch favored publication, even though book reviews had satirized the rambling way of talking at previous meetings. McCulloch told Fremont-Smith that such reviews had "only persuaded me that it was important to let other scientists in on the difficulties of communication in your wildest interdisciplinary venture. The record preserves the flow-sheet of our confusions and of such communications as actually occurred." The proceedings showed "to what extent science suffers from the babel of its many tongues and hence how necessary it is for us to learn to speak again to one another." Margaret Mead disagreed, saying that "we have already published quite enough volumes" to show the successes and failures of communicating across disciplines.[117] The editors got their way and published a thin volume containing three formal papers by guest members Grey Walter, Bar-Hillel, and linguist Yuen Ren Chao, which were bookended by McCulloch's introductory and concluding remarks. The thin volume symbolizes the difficulty of achieving interdisciplinarity primarily by communication, since most members of the Macy group did not engage with each other in the second technique of interdisciplinarity, research collaboration.

## Wiener versus McCulloch

The legacy of the Macy conferences on cybernetics was marred by the absence of several charter members at the tenth and final meeting in 1953, despite attempts by the Macy Foundation to bring them back for a reunion. John von Neumann, Norbert Wiener, and sociologist Paul Lazarsfeld had resigned and chose not to attend the last meeting. Arturo Rosenblueth and Lorente de Nó had not attended for some time and missed the last meeting as well. Because Wiener had done so much to establish the Macy group and the field of cybernetics, his absence left a gaping hole. It was not smoothed over by a letter that McCulloch and Fremont-Smith sent to him, signed by the conferees under the heading: "Affectionate greetings to our Father in Science from his Cybernetics children."[118]

Undoubtedly, the letter was heartfelt, but it masked the fact that Wiener had resigned from the group in early 1952 after cutting off relations with McCulloch, Pitts, and Jerome Wiesner at MIT. Wiener was troubled that MIT had hired McCulloch to establish a cybernetics research unit at its military-funded Research Laboratory of Electronics, headed by Wiesner. In December 1951, Wiener complained to the president of MIT that the university had allowed McCulloch to set up a lab with lavish government funding, a level of support it had not given him, and about the irresponsibility of McCulloch's researchers Pitts and Jerry Lettvin, who had sent Wiener an "impertinent" and "gloating" letter about the lab's facilities. Wiener also disliked McCulloch's expanding role in cybernetics. He described McCulloch, who wore a full beard and affected a Bohemian lifestyle, as a "picturesque and swashbuckling figure in science whose attractiveness is considerably in advance of his reliability." Wiener concluded, "It is probably impossible for me to prevent my field of work being made into a rat race by every eager beaver that wishes to appropriate it, but I can definitely say that the present atmosphere is not one in which I can continue my work in cybernetics." The president assured Wiener that MIT valued him highly. Bell Labs provided the majority of the support for McCulloch's lab, and Wiener should take it as a compliment that McCulloch was working on the "practical implications of your fundamental concepts" in cybernetics.[119]

Wiener was not moved. He wrote a colleague at this time that "Professor McCulloch is a somewhat expansive personality and has taken such measures to aggrandize his role in cybernetics at the Macy meetings and elsewhere, that I feel that I am gradually being elbowed out of public participation in them."[120] Wiener formally resigned from the group in February 1952, telling the foundation that the "Macy meetings on Cybernetics have served their purpose, which was that of establishing this discipline as a matter of general interest." He did not mention his feud with McCulloch, but he did bring it up shortly before the last meeting in 1953. Wiener threatened to sue Fremont-Smith and McCulloch over remarks McCulloch had made, in his precirculated address to that meeting, "concerning the direction of my work and my intentions which are in fact injurious to me. . . . The fact that these statements are made in a tone of pretended friendship does not remove the damage they are capable of doing, nor exonerate either of you in any way."[121] Fremont-Smith and McCulloch probably orchestrated the letter from the "Cybernetics children" to mollify Wiener.

In previous disputes with colleagues, Wiener usually made up with them after letting off steam and receiving some sort of apology. This time he kept up the feud with McCulloch, Pitts, and Lettvin until his death a decade later, in 1964.[122] Wiener's biographers, Flo Conway and Jim Siegelman, argue that MIT's support of McCulloch and his expanding role in cybernetics only partially explains why Wiener broke up with McCulloch and his boys. Wiener's daughter Peggy claimed in an interview that the breakup was due to a family matter that was kept secret for more than fifty years. She said that Wiener's wife, Margaret, jealous of McCulloch's growing influence at MIT, told Norbert a made-up story about sexual improprieties that Pitts and Lettvin had taken with their other daughter, Barbara, when she worked as a lab assistant for McCulloch in Chicago. Wiener apparently never learned that the story was false.[123]

Although some scholars have decried the break between Wiener and Mc-Culloch for sabotaging the future of cybernetics,[124] the rupture raised an immediate question in the minds of cyberneticians: who spoke for cybernetics? At the time of the last Macy conference, Wiener complained to Grey Walter that McCulloch "has manipulated himself into a position at M.I.T. in order to give some color to his tendency to speak in my name about Cybernetics."[125]

The question of who spoke for cybernetics, who proscribed its principles and how they should be applied, loomed large after the Macy conferences ended. The issues debated with such passion at the meetings—the role of models and analogies in cybernetics, the relationship between cybernetics and information theory, and whether or not cybernetics represented a universal discipline with a universal language—were just as enthusiastically debated by the disciplines represented at the Macy conferences, which adopted, modified, and extended cybernetics in their fields. The popular press shared that enthusiasm and thus helped create the cybernetics craze of the 1950s.

# CHAPTER THREE

# The Cybernetics Craze

I N THE SPRING OF 1949, the American poet Muriel Rukeyser wrote to Norbert Wiener, whom she had met during the war, that his recently published book *Cybernetics* "has been meaning a great deal to me." The biographer of American physicist J. Willard Gibbs—whom Wiener celebrated as a founder of cybernetics—Rukeyser enclosed a poem she had written that "might interest you in its suggestion of information and its answer to the symbols of entropy." Wiener was "much complimented." He thought her book on Gibbs had expressed a deep relationship between information and entropy. "As for me, I am utterly confused by the success of the Cybernetics book, and feel that within a very short time I must get back from the false position of being a newspaper figure to new work on mathematical physics."[1]

It was disingenuous of Wiener to depict himself as the victim of press sensationalism. *Time* magazine reported in 1950 that some critics thought *his* writings, which equated the electronic computer with the human brain, were sensational, a view shared by some colleagues.[2] Yet there is little doubt that the mathematically dense *Cybernetics* surprised everyone when it became a best seller. Published in the fall of 1948, the book sold seven thousand copies in its first four months and fifteen thousand by the end of 1949, when sales finally slowed down.[3] Booksellers were confounded. The *New York Times Book Review* noted, "Off the beaten track, two books have amazed dealers by their sales—Dr. Norbert Wiener's 'Cybernetics' and Al Capp's 'The Life and Times of the Shmoo' [an armless, pear-shaped cartoon character]. 'Cybernetics' is a serious scientific treatise on the workings of the human and electronic brains, while the 'Shmoo'—well, everyone knows what Shmoos are." *Business Week* compared the surprising success of *Cybernetics* to that of another scientific book published in 1948, Alfred Kinsey's *Sexual Behav-*

*ior in the Human Male.* "The public response to it [*Cybernetics*]," noted *Business Week*, "is at least as significant as the content of the book itself."[4]

Numerous magazines and newspapers reviewed Wiener's book, ranging from the middle-brow *Time* magazine to the high-brow *Saturday Review of Literature*, and from the national *New York Times* to the local Norfolk, Virginia, *Ledger-Dispatch*.[5] The popularity of *Cybernetics* on college campuses surprised two of Wiener's colleagues in mathematics. Will Feller wrote to Wiener a month after the book was published, "Wherever you go across the Cornell campus you find someone reading Cybernetics." Joseph Doob told Wiener, "I am amazed to discover how many copies of Cybernetics are wandering around here [the University of Illinois]. I do not know whether people understand it, but surprisingly many think that ownership is necessary."[6] In early 1949, *American Speech* took notice of the new word *cybernetics*, and a radio station in Siloam Springs, Arkansas, asked Wiener how to pronounce it. The editors of *Webster's Dictionary* made a similar request that summer. Wiener preferred "SIGH-ber-nee-tics," which did not catch on.[7] In 1952, a firm in Long Island, New York, called itself the National Cybernetics Company without asking Wiener's permission to do so. An incensed Wiener told the Better Business Bureau that he had defined cybernetics as a science and thus did not trademark the term, but he still thought his permission should be sought by those trying to make money from his ideas.[8] Scientists and philosophers debated at length what promise, if any, the new science held for their fields, while writer James Baldwin recalled that the "cybernetics craze" was emblematic of the 1950s for him.[9]

Why was cybernetics taken up so enthusiastically by a wide variety of social groups in the 1950s? The Macy conferences were crucial to spreading the gospel of cybernetics in the biological sciences, the social sciences, and the humanities. Many researchers and promoters followed their lead and wanted to emulate the prestige that the physical sciences had gained by developing new weapons in World War II. Others thought cybernetics would unify the sciences, which had become too specialized. More broadly, the cybernetics craze fed off a lively public discourse about the changing relationship between humans and machines, a discourse stimulated by the invention of electronic computers (electronic brains) in the Cold War and the fear that automation would cause mass unemployment. The many reasons to embrace cybernetics encouraged multiple interpretations of a field that claimed to be universal, as the public and experts alike tried to fathom the new science with the mysterious name.

Far from being a neutral spectator, a detached scientist who sat by while his work was popularized and adopted, Wiener extensively promoted cybernetics in the media and to scientists. He gave interviews to reporters, spoke to scientific conferences, appeared on radio and television programs, wrote encyclopedia articles, and published a popular book, *The Human Use of Human Beings* (1950), which discussed the social implications of cybernetics. Through these herculean efforts, Wiener blurred the sharp boundaries that supposedly exist between genuine science and simplified popularization, in an attempt to control the fate of field he had named.[10]

## Robot Brains and Automatic Factories

We can begin to understand the surprising popularity of *Cybernetics* in the United States despite its forbidding mathematics, by noting that it was the first book that described the new digital computers. Newspapers had touted them as giant "electronic brains" since the unveiling of the army's secret wartime project, the ENIAC, at the University of Pennsylvania in 1946. They touted the ENIAC, whose vertical racks of electronic gear lined a U-shaped room, surrounding its (often invisible) female programmers with eighteen thousand vacuum tubes cooled by its own air-conditioning system, as a machine that could think faster than Einstein.[11]

In the nonmathematical sections of *Cybernetics*, Wiener not only described how the new machines worked as high-speed calculators, the focus of most newspaper coverage, but how the science of cybernetics explained the inner workings of human and electronic brains. He described computers as the central nervous system for robotlike devices that received information from sensors, such as photo cells, and acted on the world through effectors, such as electric motors. He took the position, unusual for a scientist at the time, of warning the public about the dangers of his own research, that an unchecked cybernetics might throw people out of work by creating computerized "automatic factories." The *Saturday Review of Literature* remarked, "before this book was published it was not known to the ordinary man the lengths to which the human imagination could carry these massive combinations of tubes, meters, gauges, photo-sensitive cells, electronic devices, scanners, and other magical devices known or still in the womb of time."[12] Science writer John Pfeiffer, a friend of Warren McCulloch's whom Wiener did not want to publicize his work, said it was the "first book to describe some of the technological and philosophical implications of such machines, and to indicate what they mean to future generations." Pfeiffer thought that Edmund Berkeley's *Giant Brains: Or Machines That Think* (1949), a more detailed account of how computers worked, "should make an ideal companion volume to Norbert Wiener's much-discussed 'Cybernetics.' "[13]

Another factor in the book's success is what we might call the "Hawking effect," after the huge popularity of physicist Stephen Hawking's *A Brief History of Time*, published forty years later in 1988. Although its mathematics was not nearly as formidable as that in *Cybernetics*, in both cases many people seem to have purchased the volumes because they were touted as *the* book to read—or display on one's bookshelf—in order to keep up on a newsworthy science. In 1949, David Dietz, the science editor for the Scripps-Howard news service, praised *Cybernetics* as one of three "brilliant" science books of the year. It was "Mathematical in spots. Not easy reading. But a 'must' if you would be abreast in science."[14] The students carrying a brand-new copy of *Cybernetics* across colleges campuses would have probably agreed. The odd title, "Cybernetics," from the Greek word *Kybernetes*, meaning "steersman" or "governor," which Wiener mistakenly thought he was the first person to apply to a science, no doubt helped sales as well. Reviewers alluded to this reason for the book's success when they referred to its "formidable name" and "mystifying title," the "strangely titled best-seller" that was *Cybernetics*.[15]

Newspapers and magazines, which usually covered the book in their science sections, focused on three interrelated themes. First, they explained that cybernetics was a "new science" founded on the analogy between human brains and the "electronic brains" of digital computers. Second, they discussed Wiener's warning that computer-controlled "automatic factories" would throw people out of work in a coming "second industrial revolution." And third, they noted that the scientific concepts of information and feedback were central to understanding the robotlike machines. All three themes are evident in a news release distributed by Science Service in the fall of 1948. Wiener approved the release himself, which was probably written by his daughter, Barbara, who had worked for the service that summer.[16]

The three themes are evident to varying degrees in my survey of forty newspaper and magazine articles regarding cybernetics that were published in the first year and a half after the book appeared. Nearly all of them engaged with the first theme, which the press had already popularized by calling computers "electronic brains." The titles of book reviews in the news magazines follow this trend: "The Brain Is a Machine (*Newsweek*); "In Man's Image" (*Time*), and "Machines That Think" (*Business Week*).[17] The second theme, the automatic factory of the second industrial revolution, received less coverage, occurring in about one-half of the articles. It was reflected in such newspaper headlines as "Devaluing Brains in Industry" and "Mechanical Slaves Forecast"; both phrases came directly from *Cybernetics*.[18] The third theme, principles of information and feedback systems, was not well covered. About one-fourth of the articles mentioned the scientific theories of information or feedback control. No headline carried the words "information" or "feedback." Although Wiener did not refer to "robots" in *Cybernetics*, about one-fourth of the articles used this language. The "sensational" term was not confined to the popular press: *Science Digest* titled its extract from *Cybernetics*, "World of Robot Brains."[19] Wiener, himself, contributed to this discourse by stating in a 1949 interview, that the "second industrial revolution of electronic robots is here." He also wrote a prologue in 1950 for MIT's revival of Karel Capek's play *R.U.R.* [Rossum's Universal Robots], which had coined the term "robot" in the 1920s.[20]

The computer-brain analogy was so prevalent in the early press coverage that it set the terms for defining the popular meaning of *cybernetics* in the United States. William Laurence, a science writer for the *New York Times*, captured this meaning in the headline, "Cybernetics, a New Science, Seeks the Common Elements in Human and Mechanical Brains." The *Los Angeles [Daily] News* described cybernetics as a "kind of synthesis of physics and physiology." The "crux of its theory is based on the close analogy between robot thinking machines and the human nervous system." But journalists often dispensed with the physiological element and reduced the meaning of cybernetics to the science of computers and robots. *The Saturday Review of Literature* said "Cybernetics is about the development, theory, and philosophy of what science fantasy writers have long known as the mechanical brain." The Wilmington, Delaware, *Star* put it succinctly: cybernetics was the "new science of robot brains."[21]

The specter of cybernetic unemployment caused a great deal of angst in the newspapers. The *Philadelphia Inquirer* expressed this fear by shifting the meaning of cybernetics from a science to a technology. Cybernetics "would substitute machines for brains, much as the industrial revolution of the last century substituted machines for muscles."[22] The prediction came straight from an often-quoted passage in *Cybernetics*. The "first industrial revolution, the revolution of the 'dark satanic mills,' was the devaluation of the human arm by the competition of machinery," Wiener wrote. "The modern industrial revolution is similarly bound to devalue the human brain at least in its simpler and more routine decisions." Highly skilled white-collar jobs such as the scientist and administrator might survive. "However, taking the second revolution as accomplished, the average human being of mediocre attainments or less has nothing to sell that is worth anyone's money to buy."[23]

These dire statements gained currency during a national debate about automation. Historian Amy Bix has shown that the fear of technological unemployment, widespread during the Great Depression, lingered beneath the affluence of the early postwar era.[24] On the optimistic side, advocates promoted automation as a "new gospel of postwar economics." In 1949, *Fortune* magazine extolled the advantages of the "automatic factory," predicting that automatic "eyes" (photo cells), "noses" (gas detectors), and "electronic brains" would perform better and be more reliable than their human counterparts. On the pessimistic side of the debate, Wiener discussed his predictions with U.S. congressmen and met with labor leader Walter Reuther, head of the United Automobile Workers (UAW), to advise him about how to deal with the onset of computer-controlled factories. Yet Wiener's advice was also sought by management. *Time* magazine described Wiener in 1950 as a long-haired "prophet who is listened to by short-haired, hardheaded businessmen."[25]

Information was a minor theme in the early media coverage of *Cybernetics* for a variety of reasons. The subject was apparently too technical for newspapers and magazines, despite the fact that the news release for the book gave a nonmathematical exposition of information theory and that Wiener described the term "quantity of information" in an article in *Scientific American*.[26] Nearly all of the dozen articles I surveyed that discussed "information" used the term in a general sense to refer to what was processed by computers or nervous systems. The sole exception was *Scientific Monthly*, which gave a mathematical definition of "amount of information."[27]

I have found only one article in the popular or scholarly press that combined the rhetoric of the "second industrial revolution" with that on "information" before the 1960s. Science writer Harry Davis began his 1949 piece on "mathematical machines" (computers) in *Scientific American* by claiming, "A new revolution is taking place in technology today. It both parallels and completes the Industrial Revolution that started a century ago. . . . The 19th-century revolution was based on the transformation and transmission of energy. . . . The 20th-century revolution is based on the transformation and transmission of information." Although Davis did not cite Wiener, the statement was clearly inspired by *Cybernetics*, which loosely linked the idea of a second industrial revolution with computers that transformed information, not energy. Davis was familiar with the book, having interviewed Wiener for the *New York Review of Books*. But it was Davis, not Wiener, who made the explicit connection between information and the second industrial revolution, thus creating the rhetorical basis for what would later be called the "information age."[28]

The press promoted cybernetics as a science that blurred the boundaries between humans and machines by publicizing cybernetic mechanisms. These ranged from models of behavior, such as Claude Shannon's mouse, Theseus, which he presented at a Macy conference on cybernetics, to prosthetic devices. All of the mechanisms used information and feedback to pursue goals in either deterministic or random ways.[29]

An early prosthetic mechanism was the "hearing glove." Wiener helped initiate the project in 1948 at MIT's Research Laboratory of Electronics, where he advised the communications group. Drawing on previous work done at AT&T's Bell Laboratories on visible speech (a sound spectrograph), which was associated with the vocoder (a 1930s device that reduced the amount of information needed to produce recognizable speech), Wiener and Jerry Wiesner, head of the lab, proposed to build a hearing glove to convert a vocoder-like signal to the sense of touch, rather than to the sense of sight. As pointed out by Mara Mills, the "team at MIT reinvented 'skin hearing,' based on the vocoder." The MIT researchers were not familiar with the decades-long history of using tactile sensations and gloves to aid hearing. But they did obtain permission from AT&T to continue their research after Leon Levine, the graduate student assigned to develop the hearing glove, learned that Bell Labs had covered the method in a 1937 patent.[30]

Levine designed and built an electronic device that converted spoken sounds into vibrations sensed by a person's fingertips (fig. 5). In the mechanism, code named Project Felix, a microphone converted sound waves into electrical signals, which were broken up into five signals representing five octaves. These were amplified and converted into mechanical vibrations applied to each finger. The goal was to generate a unique pattern of vibrations for each phoneme. Once the bugs were worked out of the system, the laboratory, which had built a prototype based on Wiener's suggestions, intended to miniaturize it into a portable hearing glove. They intended the device to be used by deaf people to improve their speech by comparing the patterns of vibrations they created when speaking into the microphone to those created by nondeaf speakers. Presumably, the device would also act as a regular hearing aid to translate speech into sensory patterns.[31]

Unfortunately, there were many bugs to be worked out in a device that Wiener had prematurely described in a public lecture in early 1949 and in academic journals.[32] In late 1949, Wiener touted the device again, this time in a lecture he gave on sensory prosthesis at the American Mathematical Society. The address drew more media coverage than the talk given earlier in the year. The *New York Times*, an Associated Press newspaper story, and *Life* magazine heralded the new wonder coming from the lab of the famous founder of cybernetics, coverage that reinforced the negative stereotype of a "helpless deaf audience."[33]

The bugs lasted from the outset of the project to its demise. In the fall of

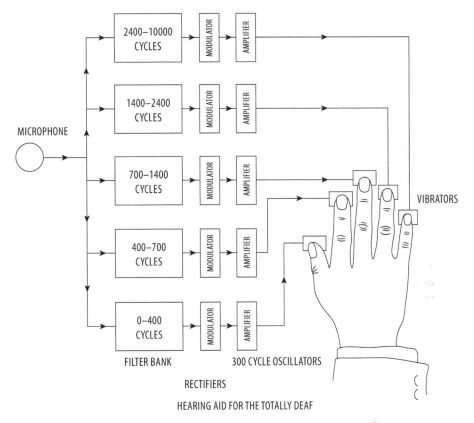

FIGURE 5. Designed and built by Norbert Wiener's cybernetics group at MIT in 1949, the hearing glove reproduced sound patterns as tactile sensations on the fingers of one hand. Wiener designed the device as a prosthesis and as a means to investigate the cybernetic principles of aural systems. From Norbert Wiener, *The Human Use of Human Beings: Cybernetics and Society* (Boston: Houghton Mifflin, 1950). Reprinted by permission of Houghton Mifflin Harcourt Publishing Company.

1949, the MIT lab reported that the five-channel system "failed to differentiate the phonemes adequately," but that a seven-channel unit gave a "unique pattern for each phoneme."[34] Researchers studied electrical stimulation of the skin and thought it would work better than mechanical vibrations. In early 1950, they addressed these problems by testing copper electrodes to electrically stimulate a subject's forearm on a seven-channel unit. In the spring, they used a different type of microphone and added a random-noise generator to tune the device, all to no avail. The researchers reported in the summer that "Felix has operational shortcomings. Whenever the subject's ability to receive words varied substantially from one test to another, we could not ascertain to what degree this was the fault of the subject or of the equipment." Digitizing the amplitude of the signals did not help.[35]

Because of the media hype, MIT and Wiener were kept busy explaining to impatient parents of deaf children, and also to Helen Keller, who had tried out the device in the lab, that it was still in the experimental stage. Keller wrote Wiener, "I can never be too grateful when I reflect that you have said the experiments you are trying out for the deaf are the first constructive application of cybernetics to human beings."[36] Wiener stopped working on Project Felix in 1952, after he severed relations with Jerry Wiesner, and the project languished.[37]

The press took a more active role in promoting Wiener's autonomous mechanism, the moth/bedbug. In the spring of 1949, Wiener convinced the Research Laboratory of Electronics to assign graduate students the task of building a machine to illustrate the principles of cybernetics for a forthcoming article in *Life* magazine. *Life* paid for its construction, and its editors negotiated with Wiener about what functions the device should perform and its general design. Wiener presented two alternatives for a mobile device, one with an umbilical cord connected to the apparatus, and one without a cord. Wiener wrote to Edward Kern, an editor at *Life*, "I think it highly desirable that you make up your mind what Life really wants and that you come up here prepared to make some definite decisions.[38]

The lab built a photo-cell mechanism mounted on a three-wheeled cart that could be configured as a light-seeking "moth" or a light-avoiding "bedbug" (fig. 6). No umbilical cord was used, making it more like a robot. The machine used two feedback paths (voluntary and postural) to model two types of tremors. It simulated an intention tremor—which occurs when a goal such as reaching for a glass of water is attempted—by overloading voluntary feedback. It simulated a Parkinsonian tremor—which occurs at rest but subsides when performing a specific function—by overloading postural feedback. Simulating a Parkinsonian tremor illustrated an analogy between maladies in humans and machines that Wiener, Rosenblueth, and Bigelow had identified in their 1943 paper on teleology. Wiener wrote to Kern that two European colleagues had seen the moth/bedbug and "were fascinated by it. One of them told me that it was very lucky for me that the burning of sorcerers was not in vogue at present."[39]

Although *Life* paid for the work, its senior science editor decided not to publish photographs of the moth/bedbug. The editor objected that the device illustrated the *analogy* between humans and machines by modeling the

FIGURE 6. Another "communication machine" designed and built by Norbert Wiener's group at MIT in 1949 was the moth/bedbug. An early light-sensing robot, it illustrated cybernetic principles by simulating Parkinsonian tremors. From Getty Images.

nervous system, rather than showing the *human characteristics* of computers, which was *Life*'s objective. The magazine wanted a device that performed a common human behavior that would be easily recognizable by readers, and it put the cybernetics story on hold.[40]

The science editors put their efforts, instead, into a cover story on computers for *Time* magazine, which was owned by the parent company Time-Life. Wiener was interviewed for the story, which publicized his and Warren McCulloch's work on cybernetics, Claude Shannon's on chess-playing machines, John von Neumann's on the computer at Princeton, and Howard Aiken's on the relay computers at Harvard, which were funded by the navy. Published in January 1950, the article marks the first time a computer, the

Harvard Mark III, appeared on the cover of *Time* magazine. Boris Artzy-basheff, a well-known cartoonist who specialized in anthropomorphism, drew the cover's Cold War illustration of the Mark III as a naval officer. Originally, Kern had wanted the cartoonist to depict the moth/bedbug for *Life*, but Wiener objected to an anthropomorphic drawing and preferred photographs of the actual device.[41] After the *Time* cover story appeared, Kern wrote to Wiener, "Plans for doing a Life article on Cybernetics have fallen flat for the moment because of Time's story last month." *Life* did not publish an article on the moth/bedbug, but the device received a modicum of publicity when Wiener included a diagram and a description of it in the first edition of *The Human Use of Human Beings* (1950).[42]

The interactions among Wiener, *Life* magazine, and MIT's Research Laboratory of Electronics reveal an extreme form of blurring the boundaries between genuine science and popularization. Funding for laboratory research came not from the university, industry, nor the federal government, as was typical in Cold War America, but from a popular magazine, which decided in the end whether or not to publicize Wiener's type of research and, thus, how to interpret cybernetics to the public.

Throughout these negotiations, Wiener stood by his position that the moth/bedbug was a research tool to study human physiology, not a robot. That view was shared by the two main builders of cybernetic mechanisms in England, both of whom attended the Macy conferences on cybernetics. W. Ross Ashby constructed the stationary homeostat to model how random neural networks adapted to changes in the environment. Although the homeostat achieved its goal in unpredictable ways, it worked toward the predetermined end of ultrastability. In contrast, W. Grey Walter built two mobile "tortoises" in 1949 to model physiological and psychological behaviors whose ends were not predetermined. Constructed of war-surplus electronic parts, the battery-powered, wheeled, dome-shaped devices (giving them their tortoise shape) used photo-cells to sense the presence of light and contact rings to sense the presence of physical objects. Having a small lightbulb on their "foreheads," they were attracted to moderate light and repelled by bright light. As they navigated their environment of physical objects and lights, the tortoises exhibited such behaviors as tropism, self-recognition (while facing a mirror), and mutual recognition (while facing each other). Grey Walter named them ELSIE (Electro-mechanical robot, Light Sensitive with Internal and External Stability) and ELMER (ELectro-MEchanical Robot). Later tortoises had small memory circuits and could thus "learn" how to interact with their environment.[43]

The humanlike tortoises, ELSIE and ELMER, and the mouselike Theseus proved to be irresistible to the American press, unlike Ashby's stationary homeostat. Waldemar Kaempffert, a science writer for the *New York Times* agreed with Grey Walter that the unpredictable movements of the tortoises exhibited a form of "free will." John Pfeiffer titled his *Popular Science* story on Theseus, "This Mouse Is Smarter Than You Are."[44]

A discordant note was struck by journalist Serge Fliegers, who interviewed Howard Aiken at Harvard and Wiener at MIT for a 1953 article in the *American Magazine*. Drawing on a distinction made by the 1950 *Time* magazine story on computers, Fliegers created the narrative of a "Battle of the Brains" between those who believed that computers could think (Wiener and Ashby) and those who dismissed the idea because computers simply followed instructions (Aiken). Fliegers called the latter group the "anti-Cybernetics faction," and identified Wiener as the "central figure of the new cult of 'Cybernetics.'" Tensions did exist between Aiken and Wiener. In 1947, newspapers had reported that Wiener refused to attend a military-funded conference on computers that Aiken had organized at Harvard because Wiener had publicly said that year that he would no longer accept military funding for his research (see below). But the metaphor of a "battle" between Aiken and Wiener, over whether computers could think or not, simplifies a complex situation. In fact, Fliegers noted Wiener's ambivalence on the subject. Because Fliegers admired Wiener's polymath abilities—telling Wiener's secretary that her boss was "definitely the lion at a LIFE party to which I took him"—he probably emphasized the conflict with Aiken to dramatize a rather conventional story about electronic brains.[45]

An actual dispute between Wiener and a computer designer remained private. In early 1951, George Stibitz, who invented relay computers at Bell Labs during World War II, complained to *Life* magazine about a 1950 photo caption that said Wiener had "successfully applied cybernetic theory to problems of radar and naval gunnery and the 'electronic brain' calculating machines." Stibitz acknowledged Wiener's wartime work on antiaircraft systems but bristled at the implication that cybernetics had led to the invention of the electronic computer. Stibitz wrote to Wiener that he and Aiken "had designed and had built computers with most of the important features of the present 'Giant Brains' long before you took an interest in the subject." Stibitz thought such confusions resulted when the press used the term *cybernetics* in two senses: to refer to the science of control and information processing; and to Wiener's own contributions in this area. Journalists thus made an illogical deduction of the form: Wiener invented cybernetics, computers were a branch of cybernetics, thus Wiener invented computers. Wiener replied by repeating the position he had stated in *Cybernetics*: in 1940, he wrote a memo to Vannevar Bush, head of the research and development effort in the war, about how to use digital computers to solve partial differential equations, methods that Bush did not pursue.[46]

The publicity for cybernetics shot up again when *The Human Use of Human Beings*, directed toward a broad audience, appeared in 1950. Wiener discussed the project in 1949 with Robert Morison at the Rockefeller Foundation, which funded his cybernetics research with Arturo Rosenblueth. Morison was not keen on the idea. He wrote in his diary, which was circulated to other foundation project officers, that Wiener "has decided that it would bore him merely to rehash the former book [Cybernetics] and he showed us a list of chapter headings which suggest that he has in mind a treatise on the entire subject of communication. He insists that he could write such a book in two months in the summer." Morison thought Wiener was writing too many books. Wiener had not discussed the project with Rosenblueth, a friend of Morison's, whom Wiener knew "takes a very dim view of any popularization of science whatsoever." Morison concluded, "it seems clear that W. really does need the money which he hopes that a popular book might bring him, but he is extremely conscientious and would not want to do anything which his fellow scientists might feel was infradig [undignified]." Morison privately hoped that Wiener would drop the project. "I doubt if he could stand the sort of criticism he would be likely to get if he writes a book in two months on the vexed subject of communication."[47]

Wiener ignored Morison's advice and signed a contract with Houghton Mifflin for a book tentatively titled the "Communication State." He planned to write the book over the summer at his farm in New Hampshire, dictating it to his secretary, Margot Zemurray, and drawing on previous essays. He promised Rosenblueth that he would only work on page proofs during his research stay in Mexico, unlike the last time, when he wrote the entire manuscript of *Cybernetics* in an office off of Rosenblueth's laboratory.[48]

True to his word, Wiener finished the book manuscript in August 1949 and revised it in the fall.[49] Houghton Mifflin gave it the title *The Human Use of Human Beings: Cybernetics and Society*. Editor Paul Brooks said a title such as "Pandora" or "Cassandra," also proposed by Wiener, was "absolutely out of the question. It would, in the opinion of everyone here, kill the book dead." Brooks derived the title from a sentence in the introduction, in which Wiener criticized the dehumanizing effects of fascism, big business, big government, and assembly-line work: "I wish to devote this book to a protest against this inhuman use of human beings." Wiener was "delighted" with the new title.[50]

Published in the fall of 1950, *The Human Use of Human Beings* was not simply a popular version of *Cybernetics* written for a lay audience. Wiener described the entropic definition of information without using mathematics, but he went beyond that to add a semantic dimension to the scientific concept of information (see chapter 4). More broadly, he promoted cybernetics as a universal science by extending it to social issues. He defined cybernetics as the "study of messages" and argued that "society can only be understood through a study of the messages and the communication facilities which belong to it." To make his case, Wiener introduced the basic analogy of cybernetics, that because both humans and communications machines consisted of effectors, sensors, "brains," and information-feedback paths, they could both be studied by the principles of control and communications engineering. In this manner, society itself could be analyzed as the communication of messages between humans, between humans and machines, and between machines and machines. As in *Cybernetics*, Wiener addressed social issues by warning of adverse social implications, and by relating the scientific concept of information to language, communication, and society. But now, he dealt with those topics at length. In a nonmathematical manner, he used the principles of information and feedback to correlate an organism's structure with its ability to learn, to analyze language in terms of information theory, to illustrate the nonmateriality of messages and information, and to describe new communication machines such as the moth/bedbug and the hearing glove. He addressed wider social issues by showing how adopting a cybernetic viewpoint could improve the writing of laws, reveal problems with the Cold War demands for secrecy in science, and point out the dangers and opportunities of the "Second Industrial Revolution."[51]

Most reviewers did not know what to make of the book. Journalist Stuart Chase, a venerable critic of capitalism, called it a "brilliant and disorderly book."[52] A reviewer in *Scientific American* satirically captured its idiosyncrasies by saying that the "reader is apt to be dazed, as well as dazzled, by a book which in a brief 200 pages discusses entropy, Mexican frescoes, the Industrial Revolution, Parkinson's disease, the patent system, the logarithmic scale in [thermodynamic] order-disorder relations," and a host of other seemingly unrelated subjects.[53] Because of the popularity of *Cybernetics*, a wider range of magazines and newspapers covered *The Human Use of Human Beings*, from the high-brow *Atlantic Monthly* to the low-brow *Parade Magazine*. The *New Yorker* noted, "Everyone who reads even *Quick* [a pocket "news weekly for busy readers"] has heard of Dr. Wiener." As with the media coverage of *Cybernetics*, reviewers emphasized the coming second industrial revolution created by robot brains, rather than information or feedback. Titles such as "Mind in Matter," "Automatic Age," and "The

Cybernetic Way of Life" reinforced this meaning of cybernetics.[54] Many commentators repeated Wiener's warnings of the harmful consequences of automation. The *Christian Science Monitor* said Wiener was a "Jeremiah with the taste and learning of a Renaissance humanist, the free-ranging intellectual gusto of a William James. He is a mathematician who sees red when he finds men reduced to soulless digits, a machine-maker who rebels against treating men as machines."[55]

Following the book's publication, Wiener actively promoted cybernetics in the media, more so than any other cybernetician in the United States or Britain. Personal reasons included the monetary one mentioned by Morison and the deep psychological insecurities that arose from being raised as a child prodigy by his father.[56] In the fall of 1950, he gave interviews on the book for the Mary Margaret McBride radio program and for programs on two other radio stations in New York City. He signed up with a Brooklyn lecture bureau to give three talks at the substantial sum of one hundred dollars per lecture.[57] Late in the year, *Life* magazine featured a plan by Wiener and two MIT colleagues, political scientist Karl Deutsch and historian Giorgio de Santillana, on how the United States could survive a nuclear attack by building railway lines and highways around cities to decentralize industry and disperse the population. Offered during the height of the Cold War, as Communist China entered the Korean conflict, the proposal was "cybernetic" in the sense that it viewed the city as a communications system.[58]

Wiener hit the publicity trail again with the publication in 1953 of *Ex-Prodigy*, the first volume of his autobiography. Simon and Schuster lined up interviews with New York City book review editors, a second stint on the McBride radio show, a slot on Edward R. Murrow's radio show to read his statement for the Cold War "This I Believe" series, and an appearance on NBC's *Today* television show.[59] Wiener also wrote a short article on cybernetics for the annual volume of the *Encyclopedia Americana* in 1952 and then a longer article for its 1955 edition.[60]

In 1953, at the urging of Jason Epstein, an editor at Doubleday, Wiener substantially revised *The Human Use of Human Beings* for a second edition, in order to broaden its appeal. The structure of the book remained the same, but Wiener removed technical passages, shortened chapters, and tied his cybernetic vision to a philosophy of life based on entropy. That concept explained how humans, societies, and communications machines used information feedback systems to create life, to exist. Fed on information (negative entropy), these local entities could temporarily delay the inevitable increase of entropy, which would lead to the earth's heat death. At the suggestion of Epstein, he introduced the categories of Augustinian and Manichean to explain two views of nature. Wiener saw the statistical aspect of nature as Augustinian (as an organic incompleteness), rather than as Manichean (as a malicious opposition), and thus as knowable through mathematics, because nature would not trick the observer in a Manichean way. He also employed the terms to distinguish the Augustinian scientific attitude of the individualistic prewar era from the Manichean militarized big science of the Cold War.[61] The hard work paid off. Total sales of *The Human Use of Human Beings* climbed to more than fifty thousand copies by the end of 1956, while the total sales of *Cybernetics* plateaued at thirty-three thousand at the end of 1959.[62]

Writing, lecturing, and giving interviews exhausted Wiener. He begged off writing two long articles on "cybernetics" and "information theory" for the 1957 edition of the prestigious *Encyclopedia Britannica*, citing other literary commitments (he was finishing the second volume of his autobiography). He recommended, instead, his former student and competitor, Claude Shannon, who had written a positive review of *Cybernetics*. Shannon wrote both articles, defining cybernetics as Wiener had, as the "science of control and communication processes in both animals and machines." On the question of the boundary between cybernetics and his own area of information theory, Shannon said that cybernetics "overlaps" such fields as neurophysiology, computers, and information theory. Having witnessed the debates on creating a common language of cybernetics at three of the Macy meetings, Shannon thought cybernetics was "universal" in the limited sense that it sought to "find features common to these diverse disciplines."[63]

## A Cold War Science, Science Fiction, or a Scientific Fad?

The social meaning of cybernetics had a triple valence in the 1950s. When the media portrayed cybernetics as the science of robot brains, it linked the discipline to civilian applications, such as automatic factories; military uses, such as guided missiles; and scientific "fads," such as dianetics. Wiener promoted cybernetics as a civilian science and tried to control its image by distancing it from the military and from fringe groups. To him, it was a means to bridge engineering and biology for peaceful purposes, not a Cold War science, science fiction, nor a scientific fad.

The case for viewing cybernetics as a Cold War science was strong, because the height of its media coverage, from late 1948 to early 1951, occurred during the fervor of the early Cold War. The blockade of Berlin by the Soviets and the American airlift of supplies into Berlin began in the summer of 1948 and continued into 1949. Tensions increased further in 1949 when NATO was established in the spring; the Soviet Union unexpectedly exploded its first atomic bomb in August, several years ahead of the predicted timetable; and China was "lost" to the Communists in the fall. The Cold War intensified in 1950 when President Harry Truman announced the hydrogen-bomb project, State Department official Alger Hiss was convicted of espionage, and Senator Joseph McCarthy made his infamous speech in West Virginia, claiming to hold in his hand a list of communists working in the State Department. The Korean War started that summer and the United States entered it in the fall. In response, Truman submitted a defense budget of $50 billion in early 1951, up from $13 billion for the previous year. Research in science and technology—especially in atomic energy, guided missiles, electronics, computers, and communications—skyrocketed in universities and government laboratories.[64]

The media covered these topics extensively during the Cold War. Newspapers and magazines typically portrayed electronic computers, a major subject in *Cybernetics*, as giant scientific calculators built for the military—symbolized by the Harvard Mark III depicted as a naval officer on the 1950 cover of *Time* magazine.[65] Cybernetics shared a World War II–pedigree with computers and many other sciences and technologies. In those hothouse years for research and development after the war, *Scientific American* ran articles on the invention and development of the transistor, Wiener's cybernetics, von Neumann's and Morgenstern's game theory, electronic computers, Shannon's information theory, Shannon's chess-playing automata, Grey Walters's tortoises, and operations research.[66]

Yet robot brains and automatic factories, rather than the military uses of cybernetics, dominated media coverage in the 1950s. By depicting computers as robots—with sensors, effectors, and memory—newspapers and magazines emphasized Wiener's view of cybernetics as a civilian enterprise, one associated with biology, psychology, medicine, and the mechanization of labor.[67] Although the new science produced life-enhancing prosthetics as well as automatic factories and the specter of technological unemployment, these were civilian hopes and fears, not military ones.

Wiener influenced the civilian image of cybernetics by publicizing his antipathy toward military research, a stance covered extensively by newspapers and magazines. In December 1946, he refused, on moral grounds, to send an engineer working on a guided-missile project at the Boeing Aircraft company a copy of his declassified wartime report on antiaircraft prediction. In the letter, published by the *Atlantic Monthly* in January 1947, Wiener said that the "policy of the government itself during and after the war, say in the bombing of Hiroshima and Nagasaki, has made it clear that to provide scientific information is not a necessarily innocent act, and may entail the gravest of consequences." In the present case, the "practical use of guided missiles can only be to kill foreign civilians indiscriminately, and it furthers no protection whatsoever to civilians in this country."[68] That same month, the *New York Times* reported that Wiener refused to speak at the computing conference organized by Howard Aiken at Harvard because of the military use of computers. He told the sympathetic *Bulletin of Atomic Scientists* in 1948 that government funding had degraded "the position of the scientist as an independent worker and thinker to that of a morally ir-responsible stooge in a science-factory."[69] Wiener made similar comments in private. He wrote to Warren McCulloch that the Harvard meeting "is under Navy auspices and that is enough to damn a job for me even though the job is less evil than poisons," an allusion to McCulloch's own research on biological warfare. "I am also giving up all work on the computing ma-chine because it is too closely associated with the guided missiles project." In 1950, he asked the Air Force Cambridge Research Laboratories not to put his name on their mailing list. "As you know, I do not work on militarily sponsored research."[70]

But how could Wiener, who had made his opposition to military research so public, work at the same time on two cybernetics projects at MIT's Re-search Laboratory of Electronics—where most research was funded by the Department of Defense's Joint Services Electronics Program—one of which, the hearing glove, was funded solely by the military?[71] How could he ask that his name not be placed on a military lab's mailing list, when it had appeared on a published laboratory progress report for the hearing glove, which he touted in *The Human Use of Human Beings*?

To understand Wiener's apparently inconsistent actions, consider how he distinguished between acceptable and unacceptable military research. Wiener wrote in the *Bulletin of Atomic Scientists*, "I refused to furnish that information [to the Boeing engineer] on the basis that I did not wish to participate in any way in a military program [guided missiles] of which I did not approve."[72] In a press release for *Cybernetics*, Wiener was quoted as saying he would "not work on any project that might mean the ultimate death of innocent people." In the book itself, he lamented bringing cyber-netics, the science of command and control, into the "world of Belsen and Hiroshima." The best cyberneticians could do was to educate the public about the "trend and bearing of the present work, and to confine our per-sonal efforts to those fields, such as physiology and psychology, most remote from war and exploitation."[73]

In late 1951, three years after the publication of *Cybernetics*, Wiener became discouraged with the growing use of cybernetics by the military. Wiener wrote to Grey Walter that he was "very much disturbed by the way in which cybernetics has been largely taken over by workers in the field of controlled missiles. When I get back to the States [from a research trip to Mexico], I shall have to reconsider radically what fields I shall preempt for myself for further research and what fields I shall abandon."[74]

Prosthesis was an acceptable area of research for Wiener, an area he had first discussed in *Cybernetics*.[75] The intense fighting during the early days of the Korean War in the summer of 1950 prompted him to clarify his research position at MIT. He wrote the head of the MIT's mathematics department, "I would like any association that it may be necessary to have with war work to be on the medical side. I am thinking of participation in the technology of prosthesis and automatic medication."[76] This justification covered both the hearing glove, a prosthesis, and the moth/bedbug, which simulated a physiological disorder. Even though the military had funded protheses since the nineteenth century, Wiener apparently viewed the field as an area "most remote from war and exploitation." He did not write about the military's growing interest in the medical and biological sides of cybernetics.

In the 1950s, Wiener questioned the Cold War political consensus, what Paul Edwards has called a "closed-world" discourse that supported the building of computers and cyborgs.[77] Wiener objected to the State Department's policy of containment when it interfered with the internationalism of science, and he called for more communication with the Soviet Union to ease Cold War tensions.[78] He told his publisher that Vannevar Bush's recent book, the Cold War manifesto *Modern Arms and Free Men* (1949), was the "absolute opposite and counterpart of my [new] book." In *The Human Use of Human Beings* Wiener pointed out the baneful effects of classifying scientific research and criticized his colleagues John von Neumann and Claude Shannon for being cold warriors. Shannon had pointed out that his research on a computer program to play chess could be used to evaluate the best military moves. "Let no man think that he is talking lightly," Wiener said. "The great book of von Neumann and Morgenstern on the Theory of Games has made a profound impression on the world, and not least in Washington. When Mr. Shannon speaks of the development of military tactics, he is not talking moonshine, but is discussing a most imminent and dangerous contingency."[79]

In the era of McCarthyism, Wiener defended Dirk Struik, his colleague and friend in MIT's mathematics department, from charges of trying to overthrow the government by teaching Marxism at a nearby college. Concerned about Wiener's agitation in the Struik case, MIT president James Killian assured the Rockefeller Foundation, which funded Wiener's research, that Wiener "has not himself been involved in any Communist or Communist front activities." Nevertheless, the FBI monitored Wiener's antimilitary statements and his long-standing friendship with the British biologist J. B. S. Haldane, a prominent communist.[80]

Wiener aided the Cold War effort to some extent in the 1950s. The proposed plan to protect American cities from nuclear attack fits in that category, even though it was a defensive measure. In 1953, he consulted for the Office of Naval Research, which supported mostly basic research, and lectured on cybernetics before the Industrial College of the Armed Forces.[81] Yet the contrast between Wiener and Warren McCulloch, the second-leading figure in cybernetics, was stark. Lily Kay has noted that McCulloch "was a militant anti-communist and was deeply committed to the cold war." He did research on biological warfare, held research grants from the navy, and consulted for NASA and all three of the armed forces, mostly on the medical and biological applications of cybernetics.[82] In the 1960s, he and Heinz von Foerster, former chief editor of the proceedings of the Macy conferences on cybernetics, helped establish a new military-supported area of cybernetics called "bionics."

The opposing attitudes toward the Cold War held by McCulloch and Wiener led to some uneasy tensions between the two men. In the letter telling McCulloch about his decision not to attend the computer conference at Harvard in 1947, Wiener extended his antimilitary policy to the work his daughter might do in McCulloch's laboratory, saying, "I would not want Barbara to be employed in any job having even the most distant relationship with biological warfare."[83] At the Macy conference on cybernetics in 1949, McCulloch said neither John von Neumann nor Julian Bigelow could attend the meeting because they "are at present overrun by civil servants inspecting the apparatus which they are at the moment soldering," the military-funded electronic-computer project at the Institute for Advanced Study. Wiener asked, "Civil servants or civil masters?" McCulloch replied sharply, "I spent the day yesterday at Princeton. I know what they are up against."[84] The exchange did not make it into the published proceedings of the conference. These tensions were probably a factor in Wiener's breaking-up with McCulloch in the winter of 1951–1952.

At the same time that Wiener was trying to keep the civilian meaning of cybernetics paramount in the public eye, he was fending off interpretations of it as science fiction. Although many scientists and engineers were fans of science fiction—including Wiener, who had tried his hand at the genre—others objected when the media portrayed cybernetics in sensational terms, as the science of robots. Wiener contributed to the problem, himself, in *The Human Use of Human Beings*, when he described the prospect of telegraphing a human by turning its body into a code on one end of the line, transmitting the coded signal, and then reconstituting the body from the code on the other end of the line. Wiener said he engaged in this fantasy, "not because I want to write a science fiction story . . . but because it may help us understand that the fundamental idea of communication is that of the transmission of messages, and that the bodily transmission of matter and messages is only one conceivable way of attaining that end." Scientific ends thus sanctioned the story for Wiener, but he criticized how others linked cybernetics to fantasy. In his 1956 autobiography, Wiener said that he had watched over the introduction of cybernetics "very carefully through a period where what I intended as a serious contribution to science was interpreted by a considerable public as science fiction and as sensationalism."[85]

That charge was a real concern to some social scientists. Management specialist John Diebold wrote in 1952, "Writers such as Norbert Wiener, by emphasizing the similarity of automatic control systems and the nervous systems of humans and animals, have made the world of science fiction seem indeed to be upon us, with a race of human-like robots already in the making. No interpretation of the facts could be more perverse—or disturbing." When psychologist George Miller reviewed the Macy conference proceedings, he complained, "Since the boundary between cybernetics and science fiction has never been overly sharp, those of us seriously interested in the social and psychological applications of this kind of thinking have often wished for some standard of scientific respectability in this young discipline." He did not think the rambling Macy proceedings helped in that regard.[86]

To make matters worse, science fiction writers identified cybernetics as the science of robots. Advertisements for Isaac Asimov's *I, Robot* (1950) and Kurt Vonnegut's *Player Piano* (1952) appeared alongside ads for books by Wiener, Berkeley, and Diebold in a special 1952 issue of *Scientific American* dedicated to automatic control systems.[87] Asimov did not mention cybernetics in *I, Robot*; all but one of the book's nine short stories were published before *Cybernetics* appeared.[88] But later science fictions writers took notice of cybernetics. Bernard Wolfe's disturbing 1952 novel, *Limbo*—about a future United States in which voluntary amputees are fitted with atomic-powered prosthetics, becoming cyborgs—drew inspiration from cybernetics. Wiener is a hero to the pro-prosthetics faction in the novel, whose narrator and characters speak of "cybernetic rape," "cybernetic might," "cybernetic brotherhood," "Wienerity," and so forth.[89] Author Erik Fennel took a dimmer view of Wiener's work and wrote to *Time* magazine to complain about its glowing review of *Cybernetics*:

> A plague upon Professor Norbert Wiener for his treatise on *Cybernetics*, and a king's-size murrain upon you for publicizing it!
>
> I make my living writing science-fiction—(Do you smoke opium or just get stinko to dream up that wild, fantastic, impossible stuff?' my friends ask)—and characters like Wiener are lousing up the racket. . . .
>
> Yes, the pincers of [modern] technology squeeze inexorably upon the poor science-fiction writer.
>
> Two of my pet themes have been machines (or robots) replacing humans with a civilization of their own, and machines that go crazy and raise assorted hell for/with their creators.
>
> But from now on the fans will bat my ears down with letters starting, 'You're nuts. Wiener says, page x, line x, that ———'.
>
> The era of carefree flirting with the psychiatry of mechanisms has departed.
>
> I wish a psychotic robot afflicted with the electronic variant of hydrophobia would bite Dr. Wiener. Hard.[90]

Wiener replied in kind, "Your plague received and contents noted." Then he turned to the serious question Fennel had raised about the relationship between science and science fiction. "You people really are between the devil and the deep sea. If your story isn't sound, it doesn't have any kick and if it is sound, it naturally is in a field that is going to be appropriated by the scientist." Because Wiener struggled to publish science fiction and mystery stories, he may have taken satisfaction in lecturing a successful writer from the standpoint of a scientist.[91]

The association with science fiction cut both ways for cybernetics. Clyde Beck, a fan and early literary critic of science fiction, argued that science fiction readers were responsible for the great popularity of *Cybernetics*.[92] But John Burchard, chair of MIT's Technology Press, which copublished *Cybernetics* with John Wiley and a French firm, thought that Wiley's proposed advertising for the book was undignified. Burchard complained, "The 'Science Fiction' flavor, to my mind, will attract the wrong kind of buyer, and by its nature, the book will disappoint those who buy it under such a misapprehension."[93]

Kurt Vonnegut put cybernetics front and center in the futuristic *Player Piano*, his first novel. In the book, Paul Proteus, head of an automatic factory at General Electric, gave the text of an important speech to his secretary, Dr. Katharine Finch, to type. "That's very good, what you said about the Second

Industrial Revolution," Finch said. Proteus replied that it was "Old, old stuff." Norbert Wiener had "said all that back in the nineteen-forties. It's fresh to you because you're too young to know anything but the way things are now." As the novel unfolds, inequalities become unbearable, and workers destroy the GE plant, the legacy of a new world order in which computers run factories and make all managerial, economic, and political decisions (a scenario Wiener had worried about in *The Human Use of Human Beings*). When asked by the publisher for an endorsement of *Player Piano*, Wiener said he was "complimented by the references to myself and cybernetics," but he thought it was mediocre science fiction and not up to the caliber of work by his heroes Jules Verne and H. G. Wells. "In short, I feel that it was inevitable that your book be written, and it will probably be written by different authors four or five times over with varying degrees of unoriginality."[94]

Defending cybernetics against charges that it was a scientific fad proved to be more contentious. The *Nation* magazine remarked in 1950, "Cybernetics should not be confused with words like technocracy and dianetics which involve prescriptions for the ills of mankind. Wiener has invented the word to describe a new field of scientific study, rather than a theory of behavior." The *Saturday Review of Literature* predicted that cybernetics "is not likely to parallel the rapid rise and fall of [engineer] Howard Scott's technocracy [movement of the 1930s] and we should do well to study carefully modern developments in 'communication and control in the man and the machine.'"[95] The proponents of technocracy, dianetics, and general semantics disagreed with these sentiments and saw strong parallels between cybernetics and their movements.

In 1950, Donald Bruce, the editor of *Technocracy Digest*, wrote to Wiener in response to a newspaper story on cybernetics that appeared in the *Vancouver Sun*. "We Technocrats have been keeping a close watch on any of your statements." They had been quoting the newspaper story in their public lectures and would print it in their magazine. Titled, "Machines May Ruin Us, if Reds Don't," the story quoted Wiener's warning that a period of technological unemployment more severe than that in the Great Depression would be brought about by automatic factories if these were not responsibly controlled by society. The story resonated with the technocracy movement's belief, publicized by Howard Scott in the 1930s, that only the politico-economic system of technocracy—which was based on energy units rather than on the price system, and run by engineers rather than politicians—could control modern machinery, thus avoiding technological unemployment and ensuring a life of leisure and abundance for all.[96]

In 1951, CHQ, Technocracy, Inc., criticized Wiener for visualizing the possibility of an "automatic social mechanism," a computerized feedback system that would make all political and economic decisions. "But the vision seems to frighten him, or at least worry him, for he cannot convince himself that it will be used to serve mankind beneficially." Furthermore, "Technocracy would like to remind Norbert Wiener and all other Americans that a technological social mechanism, working on the feedback principle, was presented to the American people nearly 20 years ago," the Technate of North America. It would provide the "know-what" and well as the "know-how" called for by Wiener to ensure the socially responsible use of the new technology.[97] Wiener responded by distancing himself and cybernetics from technocracy. "As technocracy not only involves a theory of life but a certain amount of propaganda organizations, I feel very hesitant to identify my name explicitly with it. I don't want to bring down the thunderbolts against you that are falling, and I think very justifiably, on the heads of the dianetics people. You must understand that in the confusion of fan mail, and possibly hostile mail, I must watch my step very closely." Wiener further explained his position in 1954, writing to *Technocracy Digest* that as a scientist, he had to remain impartial and not associate with "any group which attempts to make a claim to represent the influence of my ideas. In other words, my social function is to create, and not to propagandize."[98]

Wiener had much more trouble distancing cybernetics from dianetics and general semantics, both of which were labeled as scientific "fads" at the time.[99] L. Ron Hubbard, the engineer turned science fiction writer and founder of dianetics, the basis for scientology, published *Dianetics: The Modern Science of Mental Health* in 1950. Hubbard announced the "new science of the mind" in the April 1950 issue of *Astounding Science Fiction*, to which he was a frequent contributor. The theory claimed that the mind was composed of three divisions: the conscious "analytical mind" (which functioned like a computer with memory); the subconscious "reactive mind" (which operated like a low-level computer with memory at the cellular level); and the "somatic mind" (which put the commands from the analytical or reactive mind into physical effect). Mental illnesses (psychoses, neuroses, compulsions, etc.) and psychosomatic illnesses (asthma, allergies, arthritis, etc.) were caused not by faults in the computing mechanisms, which were "perfect," but by *engrams* (cellular memories of painful physical or emotional experiences) being sent as posthypnotic-like suggestions from the reactive mind to the analytical mind. Dianetics claimed to cure these illnesses through talk therapy, by auditors recovering engrams from patients and erasing them (refiling them in inaccessible sections of analytical memory).[100]

Dianetics superficially resembled cybernetics, in that both theories compared the mind to the electronic computer, and "neuroses" in humans to those in computers. Physicist Yvette Gittleson observed in *American Scientist* that Hubbard drew on cybernetic ideas without acknowledging them. "Cybernetics is the big new idea of the times, and in my opinion Hubbard (who never mentions the word) has got cybernetics, and got it bad; this is to say, he has got it wrong." Whereas Wiener talked about the *parallels* between minds and computers, Hubbard *literally* thought the mind was a computer that does not make mistakes. Gittleson referred to that way of thinking, common in newspapers, as a "mal-emphasis in regard to the cybernetic approach" because machines did not "hold the key to why we behave like human beings." She thought that it was "perhaps inevitable that the productive thinking which generated the cybernetic point of view should beget some incidental monstrosities amidst the voluminous literature accumulating in and about the field. It is to be hoped, however, that such ambitious misapplications as dianetics will be infrequent."[101]

Wiener could not have agreed more. He heard about dianetics from a former student of his at MIT, John Campbell, who was editor of *Astounding Science Fiction*. As one of Hubbard's earliest patients, Campbell wrote to Wiener about Hubbard's forthcoming book and his magazine's promotion of it. Campbell thought Wiener would "be greatly interested" in dianetics "as suggesting a new direction of development of the work from the cybernetics side," and that "further study of dianetics will be of immense aid in your projects." To the contrary, Wiener drew a sharp boundary between cybernetics and dianetics. He implored the dianetics organization, through his lawyer, to stop using his name in its promotional literature and told one correspondent who had noticed a similarity between dianetics and cybernetics, "I have no connection [with] dianetics nor do I approve the fanfare of claims which Mr. Hubbard makes. If he has used any part of my ideas, he has done so at his own responsibility, and I do not consider myself involved in any way in his ideas."[102] He complained to a social scientist that literary men did not understand the new machine age and were "ready to succumb to any moral quack," as seen in the "great sales of *Dianetics*, of Velikovsky's *Worlds in Collision*, and the recent spoof on the flying saucers."[103] In 1951, Wiener asked his lawyer to stop the Dianetics foundation from listing him as one of its associate members. Wiener considered the action to be "detrimental to my standing as an honest scientist."[104]

In sharp contrast, two other leaders of cybernetics and information theory thought highly of L. Ron Hubbard. In August 1949, Claude Shannon wrote to Warren McCulloch "in behalf of a friend, Mr. L. Ron Hubbard," asking McCulloch to meet with Hubbard to judge his work from the point of view of psychiatry: "If you read Science Fiction as avidly as I do you'll recognize him as one of the best writers in that field. Hubbard is also an expert hypnotist and has been doing some very interesting work lately in using a modified hypnotic technique for therapeutic purposes."[105]

McCulloch did not see Hubbard because of a trip to England he took that fall, but he did correspond with Hubbard. In December, Hubbard thanked Shannon for aiding him with his research on the mind and promised to send Shannon a copy of *Dianetics* as soon as it came out. Although Hubbard did not mention cybernetics in the book, he had said in an earlier article that dianetics was a "member of that class of sciences to which belong General Semantics and Cybernetics and, as a matter of fact, forms a bridge between the two," a view he repeated in the letter to Shannon. Hubbard told Shannon that his theory of the mind dealt with function rather than structure. He hoped that dianetics, as an engineering science, would lead to an understanding of the structure of the mind.[106]

Dianetics soon came under siege. The American Psychological Association denounced it in the fall of 1950 as bordering on pseudoscience, and the New Jersey State Board of Medical Examiners instituted proceedings against the Hubbard Dianetic Research Foundation for practicing medicine without a license in 1951. The foundation went bankrupt in 1952. Yet McCulloch cooperated with dianetics that year by giving permission for the organization to reprint his paper, "The Brain as a Computing Machine."[107]

Hubbard was not alone in perceiving a link between cybernetics and general semantics, a mental health therapy based on a synthesis of neurology, epistemology, and linguistics, created by Alfred Korzybski before the war. In 1949, Korzybski wrote to Wiener that after reading *Cybernetics*, he felt that his own books, which dealt with the relationships between neurology and mental functions, "were not mere idiocy on my part. I owe you a great debt and I am grateful to you for putting my own work on such solid grounds [in mathematics and the natural sciences] as you have done." Anatol Rapoport and Alfonso Shimbel, researchers in mathematical biology at the University of Chicago, argued that cybernetics and information theory supported the tenets of general semantics.[108]

Wiener would have none of it, though he had ties to Rapoport's and Shimbel's colleague at Chicago, Nicholas Rashevsky. After some pointed exchanges with the Institute for General Semantics, Wiener replied to a correspondent, who wondered why he had not mentioned Korzybski in *Cybernetics*, since they had such similar ideas. "You will find that Alfred Korzybski is quite as unwilling to be identified with me as I am with him. If I were to have many years contact with Southern California, I might become even as

Korzybski, but to be perfectly frank I find the theatricalness more than a little transpontine."[109] Several avant-garde musicians, writers, and artists also appropriated cybernetics as a liberating and guiding force in their work, which probably did not please Wiener either.[110]

## Cybernetics in the Sciences, Engineering, and the Humanities

In his books, articles, and private correspondence, Wiener drew sharp boundaries separating cybernetics from science fiction and scientific "fads" so that cybernetics would be taken seriously by scientists.[111] That project succeeded to a remarkable degree in the 1950s. By claiming that the human-machine analogy and the language of cybernetics were universal, Wiener and the interdisciplinary Macy group helped spread the cybernetic viewpoint throughout the sciences, engineering, and the humanities in the United States in the 1950s. Historians have emphasized two aspects of these universal claims. Geof Bowker argues that cyberneticians used rhetorical strategies such as legitimacy exchange (claiming expertise by linking one's research with another field) to establish cybernetics as a universal discipline that applied to all fields. Peter Galison labels cybernetics as an interdiscipline and argues that it was at the heart of the postwar transformation of the European Unity of Science movement in the United States, what he calls the "Americanization of Unity."[112] These arguments help explain the wide appeal of cybernetics in academia but not the contestations about the meaning of an interdisciplinary science that claimed to be universal.

Many natural scientists in the United States were excited about cybernetics. Physicist Churchill Eisenhart at the National Bureau of Standards praised *Cybernetics* for heralding a "new discipline" in an effusive review for *Science*, the premiere scientific publication in the country. In *Physics Today*, John von Neumann praised the "great virtuosity" of *Cybernetics* for describing the analogy between computers and biological systems. Although von Neumann criticized Wiener's claims of priority to mathematically equate information with entropy, he hoped his review conveyed "some feeling of the book's brilliancy, as well as of its bounds." For physicist Léon Brillouin at IBM, cybernetics offered a theory of information as negative entropy that helped explain why living organisms did not violate the second law of thermodynamics.[113]

Evelyn Fox-Keller and Lily Kay have shown that life scientists, especially molecular and developmental biologists, readily embraced cybernetics in the 1950s because of its promise to mathematically model living organisms and break the genetic code.[114] Henry Quastler at the University of Illinois, who attended the last two Macy conferences on cybernetics, stressed the information-theory aspect of applying cybernetics to biology.[115] Biophysicist Walter Rosenblith praised *Cybernetics* in a review published in the prestigious *Annals of the American Academy of Political and Social Science*. He regularly attended Wiener's dinner seminar on scientific method and later worked with him at MIT's Research Laboratory of Electronics. He lauded Wiener for attempting "to bring about a new synthesis" at a time of increasing specialization in science, by suggesting the "new trinity of matter, energy, and information."[116] The movement of cybernetics into ecology is epitomized by the work of Howard Odum at the University of North Carolina, who created ecological system models based on cybernetic principles.[117] The laboratory research of Warren McCulloch and his assistants at MIT provide the best examples of cybernetic neurophysiology.[118]

Researchers at AT&T's Bell Laboratories also held a high opinion of cybernetics. Henrik Bode, a prominent theorist of electrical circuits and Claude Shannon's boss at the labs, gave talks on cybernetics at local sections of the American Statistical Association and the American Society for Quality Control in 1949. That same year, Shannon reviewed *Cybernetics* for an electrical engineering journal, saying "Professor Wiener, who has contributed much to communication theory, is to be congratulated for writing an excellent introduction to a new and challenging branch of science."[119] He also joined one of the cybernetics clubs sprouting up in the United States. Before Shannon attended his first Macy conference on cybernetics, he and a colleague, engineer John Pierce at Bell Labs, attended meetings of "The Cybernetics Group" in New York City in 1949, when the subject was communication theory.[120] Control system engineers were more skeptical, but Hsue Shen Tsien at the Jet Propulsion laboratory at the California Institute of Technology titled his 1953 textbook on control engineering *Engineering Cybernetics*.[121]

Social scientists took a while to warm up to cybernetics. In 1949, the *American Sociological Review* thought the book *Cybernetics* and its doctrine were impressive. "Yet in the end it is doubtful if social scientists have much to learn here." Wiener's argument that the social sciences could learn from communications and control engineering might be true. But Wiener's "admitted unfamiliarity with the social sciences, and his dependence on analogical reasoning, will leave many social scientists skeptical." The book was "too technical," and Wiener was "too naive" about the social sciences.[122] Several researchers demonstrated the value of cybernetics to the social sciences, including two men who had learned their cybernetics from Wiener: political scientist Karl Deutsch, a colleague of Wiener's at MIT; and anthropologist Gregory Bateson, a mainstay of the Macy conferences on cybernetics and a confidant of Wiener's (see chapter 5).

Several humanists paid tribute to the new field of cybernetics in their role as public intellectuals. Dixon Wecter, a historian at Berkeley who wrote for upper-class magazines and newspapers, commended Wiener's self-regulating theory of communication as a means to understand culture and society. Wecter was one of the few social commentators to stress this aspect of cybernetics in the media. Lewis Mumford, the renowned architectural critic and historian of technology, received a complimentary copy of *The Human Use of Human Beings* from the publisher and effusively praised it to Wiener. "The combination of intelligence, human insight, and courage you display in that book sets a high level for the rest of us. I feel as if I should thank you, not just personally, but on behalf of the human race!"[123] A young Marshall McLuhan at the University of Toronto wrote to Wiener to praise *Cybernetics* and *The Human Use of Human Beings* as a way to introduce Wiener to his own first effort in the humanistic study of technology. In *The Mechanical Bride: The Folklore of Industrial Man* (1951), an innovative study of advertising, McLuhan thought Mumford wrongly assumed that the "organic is the opposite of the mechanical." To rebut this idea, McLuhan cited Wiener's assertion that "since all organic characteristics can now be mechanically produced [through electronic computers], the old rivalry between mechanism and vitalism is finished."[124] In his masterwork, *America as a Civilization* (1957), journalist and educator Max Lerner quoted Wiener's idea that "communication is a dialogue between people united against the common enemy, whether we call it entropy or chaos." It was one of the few popular references to Wiener's extensive discussion of entropy in his books.[125]

### Criticism, Ambiguity, and Multiple Interpretations

The enthusiasm for cybernetics in the American academy was not universal in the 1950s. Indeed, the field was dogged by persistent criticism, ambiguity, and multiple interpretations, which most histories of cybernetics in the United Stated ignore.[126] Alphonse Chapanis, a researcher in human-factors engineering, voiced a common criticism when he asked in a 1949 review of *Cybernetics*, "Is there anything more to cybernetics than an analogy and an interesting intellectual argument? Does it help us understand behavior any better? Does it open the way to new and exciting research on the physiological mechanisms of behavior? In the reviewer's opinion, the answer to all these questions is 'no.'"[127] Others criticized Wiener's coinage of a new term to encompass existing fields. In 1953, information theorist Robert

Fano, a colleague of Wiener's at MIT's Research Laboratory of Electronics, questioned the need for the term *cybernetics*. He wrote to a colleague in Britain: "I, for one, have some doubts about the wisdom of giving a new name to a collection of well-established fields. I know, for instance, that some people in the servomechanism field do not particularly care to be called 'Cyberneticians.'"[128]

Part of the problem scientists and engineers had with cybernetics was the ambiguity about its meaning. Mathematician Joseph Doob wrote to Wiener in the summer of 1948 that the "title of your new book is intriguing. Did you make up the word [cybernetics] or does it really have a meaning?" After receiving a copy of the book, Doob paternalistically wrote, "I am glad my wife is a doctor; she was invaluable in discussing your nonmathematical remarks." Esther Potter, director of the Dewey Decimal System at the Library of Congress, was equally puzzled. "We have read and reread reviews and explanations of the content of your book and have even tried to understand the book itself, only to become more uncertain as to what field it falls into." Books could only be assigned one number in the Dewey system. Should *Cybernetics* be catalogued under psychology, computing devices, or mathematics?[129]

British information theorist Donald MacKay commented on this ambiguity in a report he sent to the Rockefeller Foundation about his fellowship year in the United States in 1951. MacKay included a diagram to illustrate the "field of study covered under this Fellowship," which drew lines between boxes labeled "probability and statistical mechanics, mathematical logic, information theory, computing theory, physiology, psychology, and stability and control." The diagram, MacKay said, "outlines some of the interconnections between subjects sometimes compounded—and confused—under the name 'Cybernetics'; but its purpose is as much one of disentanglement as of unification."[130]

Warren Weaver at the Rockefeller Foundation, who funded the research of Rosenblueth and Wiener that led to the writing of *Cybernetics*, read the book when it appeared and wrote to Wiener that it contained "fascinating and important material," but he was puzzled by its philosophy. "I have read the first chapter three times, and think I really understand it now." Weaver then turned to the article on teleology by Rosenblueth, Wiener, and Bigelow, a founding paper of cybernetics, for clarification. "I want to read the article," he wrote to Wiener in a second letter, "but so far I have not succeeded in getting beyond the first four paragraphs. Perhaps if I were just a little brave and patient, and went on with the rest of it I would not need to ask the following questions" about the tautological nature of the argument.[131]

Many scientists had difficulty understanding the book's mathematics. British biologist J. B. S. Haldane, a founder of the (mathematical) field of population genetics and a friend of Wiener's since the 1930s, had trouble understanding the "mathematical part of the book, and fear I never may master it, as I doubt if I have time to learn about group characters [set theory], and the like."[132] Others objected to Wiener's predictions about the automatic factory. Sociologist Daniel Bell, who was the labor editor for *Fortune* magazine in the 1950s, noted in a 1956 essay that Wiener "has pictured a dismal world of unattended factories turning out mountains of goods which a jobless population will be unable to buy. Such projections are silly." The percentage of workers affected would be small, and, as in the past, new technology would destroy some jobs, but create others.[133]

The reception by philosophers was mixed. As we have seen, Filmer S. C. Northrop, a founding member of the Macy conferences on cybernetics, thought research on neural nets and teleology provided a physiological basis for relating cultural factors to biological factors in social institutions. Russell Ackoff at Wayne State University lavishly praised *Cybernetics* for opening a "whole new direction of thought which may enable many [scientists and philosophers] to fruitfully reformulate their own problems."[134]

Ackoff was a member of the American Unity of Science movement, which embraced cybernetics, but the relationship was a rocky one. In Cambridge, the Institute for the Unity of Science, headed by physicist and philosopher Philip Frank and funded by a grant from Warren Weaver at the Rockefeller Foundation, formed a study group on "Cybernetics and Communication" in late 1950. Initially, Gerald Holton, the Harvard historian of science and physicist who, as secretary of the institute, organized the study group, told Wiener that it would probably be less technical than Wiener's dinner seminar on scientific method. Peter Galison notes that cybernetics and communications "remained one of the most active topics at the Institute for several years." The cybernetics study group, which was cosponsored by the American Academy of Arts and Sciences, met from 1950 to at least 1953. Wiener gave a paper, "Cybernetics," to the Academy in early 1950 and attended many meetings of the study group.[135]

But Wiener had also criticized a pillar of the institute when he reviewed Frank's book, *Modern Science and Its Philosophy*, in 1949. Wiener praised the general tenets of Frank's logical positivism—the intellectual basis of the unity of science project—but he criticized Otto Neurath's political aims in organizing the Unity of Science movement in Europe. "Even those quite ready to admit the need of progress in scientific unity," wrote Wiener, "may find this movement somewhat distasteful on account of the excess baggage of organization and propaganda it contains."[136]

In practice, the cybernetics study group deviated from the principles of the

Unity of Science movement. The chair of its steering committee was none other than Robert Fano, who later told his British colleague that U.S. researchers in servomechanisms did not like to be called cyberneticians. In late 1952, Fano reorganized the study group so that speakers would not feel compelled to use cybernetic language and theories in their papers, which were mostly on applications. Fano thought that the interdisciplinary discussions had become "at times somewhat amateurish, in spite of the fact that the participants were far from amateurs in their own field." The steering committee valued cross-fertilization between disciplines, but the "same general problem does not imply identity of approach or of method of analysis."[137] So much for the methodological unity of science in this unity of science group!

As a whole, the Institute for the Unity of Science thought cybernetics had more promise. In 1949, it included Karl Deutsch's seminal article on cybernetic models in the social sciences in a special section of the journal *Synthese* devoted to publications sponsored by the institute. Then at MIT, Deutsch had attended the group's seminar in Cambridge.[138] Semiotician Charles Morris at the University of Chicago, who had criticized cybernetics at a Macy conference, was a leader of the Unity of Science movement in the United States and vice-president of the institute. In a 1951 article published under the auspices of the institute, Morris was more appreciative and pointed to Wiener's cybernetics and Shannon's theory of information as a scientific basis for semiotics. The institute continued to support the new discipline in 1953 by publishing a bibliography of cybernetics that focused on its applications in the social sciences.[139]

Philosophers of science, many of whom had close links to the American Unity of Science movement, held mixed views about the ability of cybernetics to unify the sciences. Wiener was known in this relatively new field. He was a founding member of the editorial board of *Philosophy of Science*, and he had published three articles in that journal, including the teleology paper with Rosenblueth and Bigelow.[140] It must have pained Wiener, then, when Ernest Nagel, a prominent logical empiricist at Columbia and a vice president of the Institute for the Unity of Science, panned *Cybernetics* in the prestigious *Journal of Philosophy* in 1949. "Despite what enthusiastic publishers and publicists may claim for cybernetics," Nagel wrote, "in its present stage of development it is far from being a universally applicable body of theory or a complete philosophy."[141] On the other hand, two other members of the Unity of Science movement, Russell Ackoff and C. West Churchman, defended Rosenblueth's and Wiener's concept of purposive behavior from criticism by philosopher Richard Taylor at Brown. They were concerned, however, that researchers, especially social scientists, would rush to apply the new ideas without understanding the criteria of purposefulness recently established by Rosenblueth and Wiener and thus endanger the ability of cybernetics to integrate the sciences.[142]

The rocky relationship between the American Unity of Science movement and cybernetics held implications for the future. The movement's rhetoric of unity provided an academic rallying call for the broad-based interdisciplinarity desired by Wiener and other cyberneticians, and the Institute of the Unity of Science provided short-term support through conferences and publications. But contention among leading lights in both movements dampened support from a sympathetic academic constituency.

## The Scientific Legacy of the Cybernetics Boom

The enormous interdisciplinary range of cybernetics and the resulting multiple interpretations of the field—in the United States, Europe, and the Soviet Union—challenge Bowker's claim that cybernetics was a universal discipline.[143] In many respects, prominent cyberneticians in the United States in the 1950s did interpret their field as "universal" in the manner analyzed by Bowker. They saw cybernetics as a universal discipline or metascience that provided—through the principles of control and communications engineering—a universal, analogical method that could analyze all complex systems, from the level of the cell to that of society. The universal language of cybernetics, expressed in terms of feedback, control, information, and homeostasis, enabled researches to apply these concepts to the broad range of fields that adopted cybernetics in the 1950s. And several cyberneticians used the rhetorical strategy of "legitimacy exchange," identified by Bowker, to link their field with established disciplines and garner research grants.

Although cyberneticians agreed on the universal character and general principles of their field when they engaged in the *metadiscourse* of cybernetics analyzed by Bowker, they disagreed on many points, even on how to interpret their field. The widespread interest in cybernetics led to multiple meanings of the term, which existed below the metadiscourse of cybernetics as a universal discipline. At the local level of their own research, workers tended to interpret cybernetics from the point of view of their discipline and social concerns, a point noted by a sociological study of cybernetics conducted in the early 1970s.[144] Ironically, most cyberneticians were specialists, and they specialized in a wide range of fields, including automatic control, computers and automata, information theory, biology, neurophysiology, prosthetics, the social sciences, and the philosophy of science. Dedicated cyberneticians such as Warren McCulloch spoke of cybernetics as a universal discipline when they wrote about the philosophy of cybernetics. But his definition of cybernetics as an empirical epistemology differed substantially from that of another dedicated cybernetician, W. Ross Ashby, who defined cybernetics as the "study of systems that are open to energy but closed to information and control—systems that are 'information-tight.' "[145]

Wiener was the exception in doing research in all of the areas listed above, but he did not define cybernetics as a universal science. The closest he came to stating that position was to define cybernetics in 1952 as the "complex of sciences dealing with communication and control in the living organism and in the machine," and to refer to the "cybernetic nexus of disciplines." That same year, he stated, "Cybernetics is a name which has been invented to cover those aspects in which the theory of communication in instruments, the theory of control apparatus, and the theory of the nervous system and other modes of communication and control within the body, and the theory of social control resemble one another and justify the use of parallel methods."[146] Wiener thus did not think of cybernetics as universal in the sense of being a unitary metascience.

In sum, the enthusiasm for cybernetics in the media reinforced the meaning of cybernetics as the science of robots and automatic factories, while the enthusiasm for it in academic circles created ambiguity and multiple meanings that sustained the research of a multitude of specialists. Furthermore, the cybernetics boom did not lead to the founding of robust research or educational institutions, which Wiener had dreamed about in the late 1940s. Several prominent social scientists drew on cybernetics to carry out influential research in the fields of political science, anthropology, and psychology, but they did not establish cybernetics research centers or departments. Instead, cybernetics found an American home in two military-funded, interdisciplinary research laboratories in the 1950s. MIT's Research Laboratory of Electronics, funded by the Joint Services Electronics Program, established cybernetics research groups in communications (supporting the work of Wiener and his protégé Yuk-Wing Lee), biophysics (headed by Walter Rosenblith), and neurophysiology (headed by Warren McCulloch).[147] In the late 1950s, Heinz von Foerster established the Biological Computer Laboratory at the University of Illinois. Funded by the Office of Naval Research and the air force, the lab worked mainly on artificial neural nets.[148] As we will see in later chapters, the scientific fate of cybernetics was largely in the hands of these laboratories and some social scientists in the tumultuous 1960s and 1970s.

# CHAPTER FOUR

# The Information Bandwagon

I N DECEMBER 1955, Louis A. de Rosa, the chair of the Institute of Radio Engineers' Professional Group on Information Theory (PGIT) published a provocative editorial in the group's transactions. Although cyberneticians considered the theory of information introduced by Norbert Wiener and Claude Shannon to be an essential part of cybernetics, communication theorists established two organizations—the London Symposium on Information Theory in 1950, and the PGIT in the United States in 1951—to mark off a separate discipline under the name of *information theory*. A London organizer even called it a "new branch of science."[1] De Rosa wrote his editorial, titled "In Which Fields Do We Graze?" in response to the wide range of topics discussed at the third London symposium, held in 1955. The meeting's interdisciplinary flavor was enhanced by the presence of three prominent members of the Macy Foundation conferences on cybernetics: Warren McCulloch, Margaret Mead, and Walter Pitts.[2] De Rosa, who worked at the Federal Telecommunication Laboratories in New Jersey, observed that the "application of Information Theory to fields other than radio and wired communications has been so rapid that oftentimes the bounds within which the Professional Group interests lie are questioned. Should an attempt be made to extend our interests to such fields as management, biology, psychology, and linguistic theory, or should the concentration be strictly in the direction of communication by radio or wire?" De Rosa had heard arguments on both sides of the question and asked for the opinions of members on the issue.[3]

Shannon and Wiener, neither of whom attended the third London symposium, wrote the first two (and only) editorials to respond to de Rosa, probably by invitation. The acknowledged founder of information theory in the United States, Shannon warned of the dangers of broadly applying information theory. In a widely cited editorial, "The Bandwagon," published in March 1956, he drew a firm boundary around the new field:

> Information theory has, in the last few years, become something of a scientific bandwagon. Starting as a technical tool for the communications engineer, it has received an extraordinary amount of publicity in the popular press as well as the scientific press. In part, this has been due to connections with such fashionable fields as computing machines, cybernetics, and automation. . . . Applications are being made to biology, psychology, linguistics, fundamental physics, economics, the theory of organization, and many others. . . . Although this wave of popularity is certainly pleasant and exciting for those of us working in the field, it carries at the same time an element of danger. . . . It will be all too easy for our somewhat artificial prosperity to collapse overnight when it is realized that the use of a few exciting words like *information, entropy, redundancy*, does not solve all our problems. . . . I personally believe that many of the concepts of information theory will prove useful in these other fields . . . but the establishing of such applications is not a trivial matter of translating words to a new domain. . . . Research rather than exposition is the keynote. . . . Only by maintaining a thoroughly scientific attitude can we achieve real progress in communication theory and consolidate our present position.[4]

Not surprisingly, Wiener called for more interdisciplinarity in his editorial, "What Is Information Theory?," published in June. A world-renowned mathematician, Wiener had a high reputation in the Professional Group on Information Theory. His statistical theory of the prediction and filtering of signals anchored a major research area in the group, and he was the banquet speaker at the PGIT's annual symposium held at MIT in 1954.[5] In his editorial, Wiener noted that he and Shannon shared equal credit for creating the entropic theory of information. He agreed with Shannon's editorial that the theory was "beginning to suffer from the indiscriminate way in which it has been taken as a solution of all informational problems, a sort of magic key." Wiener pleaded that "Information Theory go back of its slogans and return to the point of view from which it originated," Wiener's own statistical concept of communication, which he had based on time series, a branch of statistical theory. "What I am urging is a return to the concepts of this theory in its entirety rather than the exaltation of one particular concept of this group, the concept of the measure of information into the single dominant idea of all." Wiener hoped that the group's transactions would extend "its hospitality to papers which, while they bear on communication theory, cross its boundaries, and have a scope covering the related statistical theories."

The danger did not lie in the improper application of information theory, but in "overspecialization." Wiener further hoped "that these Transactions may steadily set their face against this comminution of the intellect."[6]

The Professional Group on Information Theory also published three letters from communications engineers who responded to de Rosa's editorial. All of them praised the value of applying information theory outside of communications engineering.[7] The editorials and letters reveal some of the battles fought over information theory in its first decade as it moved out of the control of Shannon. In those years, scientists, engineers, and mathematicians hotly debated the meaning of *information*, which concepts were to be included in *information theory*, the mathematical rigor of Shannon's theorems, and how to apply them outside of communications engineering. Carried out amidst the postwar fervor about cybernetics and electronic brains in the popular press, and at such professional settings as the Professional Group on Information Theory and the London Symposium on Information Theory, these debates drew disciplinary boundaries around information theory, what sociologists of science call *boundary work*. Communication theorists drew sharp boundaries separating the English version of information theory from the American version, and information theory from cybernetics. Other researchers did not respect these boundaries and interpreted information theory flexibly when they applied it to all manner of fields, from physics and biology to the social sciences.[8]

## The English School of Information Theory

In the early Cold War, the term *information theory*, which Shannon had used in his 1948 paper, "A Mathematical Theory of Communication," referred to a wide range of research that both preceded and followed Shannon's classic paper. Theories of information were proposed in the 1930s by British statistician Ronald Fisher in classical statistics; in 1946 by physicist Denis Gabor at Imperial College, London, in waveform analysis; independently in 1948 by Shannon and Wiener in communication theory; and in 1950 by British physicist Donald MacKay, who crafted a comprehensive theory that embraced the work of Fisher, Gabor, Shannon, and Wiener.

American communication theorists tended to use the phrase *information theory* in a narrower sense than did their British counterparts in the 1950s. Yehoshua Bar-Hillel, the Israeli philosopher of science who attended a Macy conference on cybernetics, observed in 1955, "Whereas the term 'Theory of Information' came to be used in the United States, at least as from 1948 (probably due to the impact of Wiener's *Cybernetics*), as a certain not too well defined subscience of Communication Theory, the British usage of this term moved away from Communication and brought it into close contact with general scientific methodology." Bar-Hillel referred to the British group as the "European school of information theoreticians, MacKay, Gabor, [Colin] Cherry, and others."[9] In 1957, Cherry noted that British researchers distinguished between the "theory of communication" and a broader discipline. "This wider field, which has been studied in particular by MacKay, Gabor, and [French physicist Léon] Brillouin [who had moved to the United States], as an aspect of scientific method, is referred to, at least in Britain, as *information theory*, a term which is unfortunately used elsewhere [e.g., in the United States] synonymously with communication theory. Again, the French sometimes refer to communication theory as *cybernetics*. It is all very confusing!"[10]

The British information theorists held faculty positions at two London institutions. Gabor and Cherry were in the electrical engineering department at the Imperial College of Science and Technology—Gabor as a reader in electronics, Cherry as a reader in telecommunications. MacKay was a lecturer at Kings College, London, where he was finishing his Ph.D. in physics. The tightly knit group formed what Warren McCulloch later called the "English School of Information Theory," as opposed to the "American School," which focused on Shannon's work and was centered at Bell Labs and MIT.[11]

Gabor said as much in early 1951 when he corresponded with colleagues about his research for a paper, "Light and Information." (In 1971 Gabor won the Nobel Prize in Physics for inventing holography.) Gabor wrote to Warren Weaver at the Rockefeller Foundation, who had popularized Shannon's theory, that he was "trying to drive Information Theory forward on a track somewhat different from the ones outlined by you, towards physics." He told Lord Cherwell, a physicist at Oxford, that the "American School, /Wiener and Shannon,/ have introduced Communication Theory from the statistical side. But for a start at least, MacKay and I, who have been chiefly working in this field in this country, have found this [physical] approach easier."[12]

The British usage of the term *information theory* dates to the first London Symposium on Information Theory, which Cherry, Gabor, and MacKay helped to arrange in 1950. Initially, the sponsors organized the symposium around Shannon's work and his availability to attend the conference. The heads of the symposium's two sponsoring agencies—Willis Jackson, chair of the Electrical Engineering Department at Imperial College, and F. S. Barton, director of communications development at the Ministry of Supply, a British military agency—negotiated with Shannon's boss, Mervin Kelly, head of Bell Labs, who was in England at the time, for Shannon to attend the conference. Kelly wrote home in March that the British "hope to build a Communication Theory Conference around Shannon. His work has made a great impression here."[13] Shannon agreed to attend the symposium, to be held in the fall, and to give three lectures: one on the fundamentals of his theory of information, one on coding problems, and one to summarize the three-day conference. In April, the organizing committee—dominated by Jackson, MacKay, Gabor, and Cherry—changed the name of the event from the "Symposium on Communications Theory" to the "Symposium on Information Theory."[14]

The organizers, however, did not change the name of the symposium to highlight Shannon's theory, as Kelly had supposed. They adopted the name *Information Theory* to refer to a broader concept of information than that held by Shannon. The organizers performed this inclusive form of boundary work in a variety of ways in the papers they presented at the symposium. Cherry wrote a lengthy history of information theory that gave it a prewar pedigree, a time-honored way of justifying the intellectual basis of an emerging field. After setting his subject in the historical context of communication theory, Cherry identified two approaches to information theory: the work of Fisher, Gabor, and MacKay in Britain in scientific measurement; and that of W. S. Percival, in England, and Ralph Hartley, Wiener, Shannon, and Robert Fano in the United States in communications engineering. Cherry explained privately that he had tried to "present particularly the work of European (especially British) origin and its connection with the modern work in [the] U.S.A."[15] Gabor noted in his introductory remarks at the symposium that the "concept of Information has wider technical applications than in the field of communication engineering. Science in general is a system of collecting and connecting information about nature, a part of which is not even statistically predictable. Communication theory, though largely independent in origin, thus fits logically into a larger physico-philosophical framework, which has been given the title of 'Information Theory.' "[16]

MacKay then described his synthesis of information theory. As noted in chapter 2, MacKay broadly defined information theory as the making of representations. MacKay synthesized the work of Gabor and Fisher under the heading of "scientific information," which measured information in "logons" and "metrons," and the work of Wiener and Shannon under the heading of "selective information," which measured information in "bits." At the request of the symposium's organizing committee, MacKay and another organizer, neurologist John Bates (who had organized the Ratio Club), prepared two glossaries for the meeting. MacKay produced one on information theory; Bates, one on neurophysiology. The latter reflected a cybernetics thread at the symposium, at which members of the Ratio Club, including W. Grey Walter, but not MacKay, gave papers on automata and neurophysiology.[17] Before the symposium, the organizers decided that the "Chairman on [the] first day should call attention to [MacKay's] Glossary, and define clearly the meaning of 'Theory of Information.' " MacKay offered to revise his glossary in response to criticisms that it focused too much on his own work. But the glossary retained his terminology and concepts, so much so that it served as the foundation of MacKay's theory of information for years to come.[18]

The symposium's organizers acknowledged MacKay's theory in their papers. In his historical review, Cherry said that Gabor's logons "are now called units of 'structural information' " and gave a short section on MacKay's theory of scientific information. Gabor went further and revised his introductory talk to include a passage suggested by MacKay: the sentence quoted above that communication theory was part of the "larger physico-philosophical framework" of information theory.[19] Willis Jackson summed up the consensus of the organizers in his preliminary remarks to the symposium, which were reprinted in *Nature*, the premier scientific journal in Britain: "This recent work proves to have a significance beyond the sphere of electrical communications. A new branch of science is emerging which reveals and clarifies connexions between previously largely unrelated fields of research concerned with different aspects of the processes in which living organisms—in particular man—collect, classify, convert and transmit information." Wiener could not have said it better. Jackson went beyond his preliminary remarks to add that MacKay's paper "gave an appropriate frame for the unification of apparently divergent definitions of 'amount of information.' . . . It may be hoped that the terminology which he presented in an extensive glossary will be used by future contributors to the subject."[20]

That did not occur, however, despite the efforts of such promoters as Willis Jackson in Britain and Warren McCulloch in the United States. In 1951, McCulloch introduced MacKay by letter to two colleagues whom MacKay would visit during his fellowship year in the United States, funded by the Rockefeller Foundation, by saying they would find MacKay's "notions of information more powerful and, in some ways, more exact than any I have had the good fortune to encounter." Yet McCulloch did not use MacKay's terminology in his own work.[21] Neither did the vast majority of participants at the next two London symposia on information theory, held in 1952 and 1955. The proceedings of the 1952 symposium show that only two presenters referred to the terms MacKay had introduced two years earlier. David Bell, a professor of electrical engineering at the University of Birmingham, had wanted to employ the phrase *structural information* to analyze the information contained in English words, but he could not do so because "that term has already been used by MacKay for something quite different in another branch of information theory." Unsurprisingly, MacKay used his own terminology in both of the papers he presented at the symposium: one on types of modulation systems; and one on human information generators. Also, in the discussion of a paper by Yehoshua Bar-Hillel and Rudolf Carnap, MacKay suggested that his theory of information could clarify some problems with their "semantic theory of information."[22]

One of the reasons that more participants at this symposium did not refer to MacKay's theory may have been that the organizers limited the conference to communications engineering, which encompassed the work of Gabor, Wiener, and Shannon, but formed only one part of MacKay's synthesis. The symposium was called the "Applications of Communication Theory," instead of "Symposium on Information Theory."[23] MacKay was not on the organizing committee, because he was in the United States on the Rockefeller fellowship, and he was unhappy with the changes. He wrote to McCulloch, "I suppose they're entitled to have a strictly bread-&-butter meeting if they want one, but it seems a pity in many ways."[24] In his introductory remarks to the symposium, Gabor referred to "communication theory," rather than "information theory," and said it consisted of two branches: "signal analysis" (Gabor's work) and the "statistical theory of communication" (the work of Wiener and Shannon).[25] The second branch dominated the symposium. Furthermore, U.S. researchers at the meeting commonly referred to Shannon's theory as "information theory," while those outside the United States used that term more broadly.[26] In one discussion of this difference at the meeting, Bar-Hillel agreed with MacKay and deplored the "replacement of the unambiguous term 'statistical theory of communication' by 'theory of information,' but this replacement is an historical fact, at least in the United States."[27] MacKay may have thought that the U.S. usage preempted his terminology.

Matters improved for MacKay at the third symposium in 1955. Its organizers, led by Colin Cherry, returned to the broad scope of the first meeting and called it, once again, the London Symposium on Information Theory. MacKay, three other presenters, and two discussants referred to either Mac-Kay's general theory of information or used parts of it, such as logon and metron content. These papers and discussions dealt with the topics of statistics, linguistics, and psychology, rather than coding (the focus of Shannon's theory), indicating the wider scope of MacKay's theory. Nearly all of those who used MacKay's terminology came from Britain or Continental Europe. The exception was radiologist Henry Quastler at the University of Illinois, who had attended a Macy conferences on cybernetics and had turned from studying information processes in biology to those in psychology.[28]

MacKay's theory also suffered from severe criticism. In 1951, an anonymous referee of Cherry's historical paper on information theory, which he had submitted for publication, thought the space given to MacKay's theory should be reduced. MacKay's views "have never been properly developed, said the referee, "and are not generally accepted or even understood by the majority of the experts themselves. Unless a clear distinction is made, readers will certainly be confused by two differing concepts of information, one of them thoroughly established in the field of communication engineering [Shannon's], the other not [MacKay's]."[29]

Even Denis Gabor had some doubts. In 1951, he wrote to philosopher Bertrand Russell at Cambridge University that the "Theory of Communications, which has started in the field of electrical engineering, and in whose beginnings I had some part, has grown into a somewhat ill-defined but more general structure, called Information Theory, which has made contact with a good many branches of science. We had even some trouble in refraining D. M. MacKay, one of its young and ardent supporters, from breaking into Epistemology."[30] In 1953, Gabor encouraged his brother André, an economist with whom he was writing a joint paper on the mathematical theory of freedom, to reapply for a grant from Warren Weaver at the Rockefeller Foundation for this project. Gabor pointed out that Weaver had granted MacKay "a year's study in America without any strings, after [he wrote] his first, rather confused paper ["Quantal Aspects of Scientific Information"] which would have been even more confused if I had not spent days putting it right."[31] Gabor alluded to these criticisms while praising MacKay's theory in a popular article he published on "information theory" in 1953. The "still somewhat blurred outlines of this new science became visible" at the first London symposium, Gabor maintained. "It was chiefly D. M. MacKay who staked out the territory, and showed how the concepts of communication theory could be used for a general classification of scientific information, and how they could fertilize other sciences." Gabor then engaged in an exclusive form of boundary work by saying, "Information Theory is not a new name for epistemology, and is not philosophy. . . . Though Information Theory has a philosophical fringe, its real value is in its hard core of mathematics, and of electronic models."[32]

But even in that realm, the English school of information theory did not wholly adopt MacKay's theory. Gabor used only some of MacKay's terminology, such as *metron content*, and some terms inspired by MacKay, such as *structural communication theory*, in his technical papers.[33] Cherry included MacKay's theory of scientific information in his history of information theory, published in several versions from 1951 to 1957. Cherry said it marked the way the term *information theory* was used in Britain. He also used MacKay's terminology of *selective information content* and *selective information rate* to refer to the theory developed by Wiener and Shannon, and he referred readers to MacKay's glossary for a more complete treatment of terms.[34] But Cherry did not utilize MacKay's synthesis of the scientific and statistical theories of information. Instead he drew a boundary between the theory of scientific information and communication theory. In his book *On Human Communication* (1957), Cherry stated, "Questions of extracting information from Nature and of using this information to change our models or representations lie outside communication theory—for an observer looking down a microscope, or reading instruments, is not to be equated with a listener on a telephone receiving spoken messages. Mother Nature does not communicate to us with signs or language. A *communication channel* should be distinguished from a *channel of observation*."[35]

Despite their differences, Gabor, MacKay, and Cherry agreed that the English school of information theory comprised the measurement of information in scientific experiments and in communications systems. But they conducted research on different aspects of that synthesis. Gabor confined his work to physical questions, focusing on signal analysis, the relationship between entropy and information, and the application of information theory to the theory of light. MacKay ventured much further afield, despite Gabor's attempts to restrain him. He developed an even broader theory of meaning than he had presented at the Macy conference on cybernetics in 1951 as part of his expanded theory of information in the mid-1950s. During his research on automata theory, which became part of the new field of "artificial intelligence," MacKay operationally defined meaning as how the human brain selectively interpreted a message in its "adaptive-response-space."[36] MacKay related this claim to Shannon's measure of information at the third London symposium in 1955.[37]

Cherry dealt with meaning by drawing on the semiotics of Charles Morris to speak of information in terms of syntactics (relation between signs—the realm of the "Wiener-Shannon statistical theory"), semantics (relation between signs and designata), and pragmatics (relation between signs and their users). In *On Human Communication*, written for a lay audience, Cherry extensively discussed Bar-Hillel's and Carnap's mathematical theory of "semantic information" that Bar-Hillel had presented at the Macy conference in 1953, which dealt with the meaning contained in a statement rather than the communication of meaning. Cherry did not discuss MacKay's theory, which dealt with the meaning interpreted by the recipient in human communication. Cherry said that the theory by Bar-Hillel and Carnap was the "only investigation of which your author is aware, into the possibility of actually applying a measure to semantic information," an implicit critique of MacKay.[38]

All three founders of the English school of information theory made research trips to the United States in the early 1950s, where *information theory* typically referred to Shannon's mathematical theory of communication and its subsequent development. MacKay made Warren McCulloch's laboratory in Illinois the home base for his extensive travels during his fellowship year in the United States in 1951, and he sparred with Shannon at the Macy conference on cybernetics that year. Gabor worked at MIT's Research Laboratory of Electronics (RLE) during his visit to the United States in 1951, which was also funded by the Rockefeller Foundation. He gave several talks on his version of information theory that summer, including one on physics and information theory at Bell Labs, and one on "Light and Information" at the General Electric Research Laboratory in Schenectady, New York. At MIT, he presented six lectures on "Communication Theory," and one on "Information Theory and Scientific Method" at the Institute for the Unity of Science in Boston.[39]

Cherry also worked as a visiting research associate at the RLE during his six-month stay in the United States in 1952. He gave a talk on the "Unity of the Communication Sciences" at a conference on speech analysis held at MIT in the summer of 1952 and became well acquainted with Jerry Wiesner, head of the RLE, and three researchers in the lab's communication section: Robert Fano, Yehoshua Bar-Hillel, and linguist Roman Jakobson. After returning to England, Cherry debated the scope of information theory with Fano, professor of electrical engineering at MIT and a follower of Shannon. Bar-Hillel, a visiting research associate who worked on the mechanical translation of languages, commented on a draft of Cherry's *On Human Communication*. It was the first book published in the MIT Press series Studies in Communication, coedited by Jakobson, who was a professor of Slavic languages at Harvard and an affiliate at the RLE. Cherry became something of a linguist himself. He coauthored a paper with Jakobson on using information theory to analyze phonemes.[40]

During their research visits to the United States, neither MacKay, Gabor, nor Cherry were converted to the dominant American usage that interpreted the phrase *information theory* as Shannon's theory. Their interactions with American colleagues in the 1950s did not shake their adherence to the broad interpretation that characterized the English school of information theory.

## The American School of Information Theory

While the British spoke of English and American brands of information theory, most researchers in the United States did not give the term *information theory* a nationalistic label. As an umbrella term, it encompassed both Shannon's and Wiener's (statistical) theories of communication, but it was often reduced to Shannon's theory of coding.

The boundary work between British and U. S. information theorists began during the first London symposium in 1950. Ironically, Shannon himself was rather ecumenical about the matter. In a survey of the field, he noted that the "word 'information' has been given many different meanings by various writers in the general field of information theory. It is likely that at least a number of these will prove sufficiently useful in certain applications to deserve further study and permanent recognition. It is hardly to be expected that a single concept of information would satisfactorily account for the numerous possible applications of this general field" (a position he repeated at the Macy conference in 1951; see chapter 2). But when Shannon presented a "new approach to information theory" to handle multiple sources, it still rested on his entropic definition of information.[41] Robert Fano was just as inclusive at the second London symposium in 1952. In his concluding remarks, Fano said, "'Information Theory' has aroused the interest of people in a rather large variety of fields, and has brought them together in a potentially fruitful association. . . . There is room for a great variety of interests and points of view, as evidenced by the discussion at this symposium."[42]

In early 1953, Fano explained his remarks to Colin Cherry, who was editing them for the proceedings of the 1952 symposium. "Roughly speaking," Fano wrote, "there have been four main areas of work in the general field of what you call 'Information Theory.' The first area is the one to which you refer as 'Communication Theory,' which originated from the work of Shannon, and which I have been calling 'Information Theory.'" The second and third areas were the "Wave Form Analysis" of Denis Gabor and "Classical Statistics," which dealt with the "design of experiments," an allusion to Ronald Fisher's definition of information. The fourth area comprised "miscellaneous philosophical speculations on broad communication problems and on other related problems which cannot yet be formulated in a precise enough manner to be attacked by mathematical means," probably an allusion to MacKay. "Confusion of these four broad areas of work because of ambiguous terminology," Fano continued, "is responsible in my view for many of the misunderstandings between the two sides of the Atlantic and also on *each* side of the Atlantic." Fano did not recommend standardizing one usage of *information theory*. Everyone should "make an honest effort to make clear what we are talking about in spite of difficulties with terminology." He had "been occasionally at fault in this regard," and "did not realize that certain terms such as 'Information Theory' meant to other people something quite different from what they meant to me."[43] One example is a 1950 article in which Fano spoke of the "Wiener-Shannon theory of communication, also called 'Information Theory.' "[44] But Fano seems to have remained tone-deaf in this regard. Cherry told Bar-Hillel later in 1953 that he had "recently sent a passionate plea to Fano *not* to call his new book 'Information Theory'—but he cannot see what I am talking about. He will cause further confusion, with this title" because the proposed book focused on Shannon's theory.[45]

As Cherry, Bar-Hillel, and others observed, Fano's usage of the term *information theory* was prevalent in the United States in the 1950s. This was especially true at MIT's Research Laboratory of Electronics and at AT&T's Bell Laboratories.[46] To David Slepian at Bell Labs, the variety of topics covered at the third London symposium in 1955 was amazing. In an internal report, he wrote, "There were papers on taxonomy, codes, theory of hearing, game theory, translating machines, language, neural networks, psychology, delta modulation, philosophy of science, etc. . . . Most of the Europeans to whom I spoke did not find the inclusion of these various topics under the heading Information Theory at all startling. The term 'Information Theory' or 'The Information Theory' is in common use there and has a very much wider meaning than it seems to have here. The best definition I was able to get as to what constitutes 'The Information Theory' was 'the sort of material on this program.' "[47]

A different form of boundary work at the 1955 symposium dealt with gender. In 1955, Jerry Wiesner, who was the American liaison for the symposium, telephoned Warren Weaver about the prospect of the Rockefeller Foundation's funding two U.S. delegates to travel to the meeting. The Office of Naval Research was providing military travel for the rest of the U.S. delegation. Weaver wrote in his diary at the foundation, "JW has handled the transportation of most of the U.S. delegation, but there are two cases that cannot be handled by military transport: *Margaret Mead* (since she is a woman), and *Calvin Mooers* of the Zator Company (concerned with information retrieval). JW says CM is doubtlessly something of a nut, but the

British have asked for him. . . . These two persons could go for $500 apiece. Could the RF handle?" Wiesner told the Ford Foundation, from whom he tried to get funds, that Mead "presented a problem because the Navy is reluctant to use MATS [Military Air Transport Service] transportation for women."[48] Mooers, who received a master's degree in electrical engineering from MIT in 1948 for a thesis on coding problems, was ineligible for MATS because he did not have a military contract and funded his research from the proceeds of his company.[49] Mead, of course, was not new to information theory. She attended all of the Macy conferences on cybernetics and co-edited its proceedings—including the debate between Shannon and MacKay at the 1951 meeting. Wiesner was able to secure travel funds for Mooers, but Mead had to attend at her own expense.[50]

The Americans drew boundaries around the meaning of the term *information theory* within a new organization, the Institute of Radio Engineers' Professional Group on Information Theory, which was founded in 1951. The first chairs of the group, from 1951 to 1957, worked in the Cold War electronics industry located around New York City and Washington, DC, not at MIT and Bell Labs. Nathan Marchand worked for Sylvania Electric in New Jersey; William Tuller for the Melpar avionics company in Arlington, Virginia; Louis de Rosa for the Federal Telecommunications Laboratory in New Jersey; and Michael di Toro, Jr., for the Guided Missile Division of the Fairchild company on Long Island.[51]

Tuller took an active part in getting under way the group's transactions and its annual symposia in a manner that displayed a broad view of information theory, which Shannon later criticized in his bandwagon editorial. Tuller had completed his Ph.D. on the information rate in communications systems at MIT in 1948 under communication theorist Ernst Guilleman while working as a research associate at the Research Laboratory of Electronics. He thought his (independent) work on a theory of information rivaled that of Shannon.[52] As chair of the Professional Group on Information Theory in 1953, Tuller negotiated with Willis Jackson at Imperial College for permission to publish the proceedings of the first London Symposium on Information Theory in its entirety as the first volume of the group's transactions, which appeared in February 1953.[53] The second volume, published that November, consisted of a lengthy bibliography by F. Louis H. M. Stumpers, a Dutch research physicist who worked on communication theory as a visiting research associate at the RLE. Stumpers noted, "In this new field the boundaries are not well defined. I have kept to the wide view of Wiener's books, the first London Congress [Symposium on Information Theory], and the Macy Foundation Conferences." The approach is evident in Stumpers's title, "A Bibliography of Information Theory (Communication Theory—Cybernetics)," and such nonengineering headings as "Experimental psychology," "Linguistics," "Biological Applications," and "Group Communication."[54]

The wider view is also evident at the PGIT's first two annual symposia held at MIT. Tuller worked with Fano at MIT to organize the 1954 symposium, but tragically died in an airplane crash near Ireland (ironically near Shannon airport) a week before the symposium began in September.[55] One-third of the symposium's sixteen papers were on automata and the social sciences (the structure of organizations and human information processing), rather than on communications engineering. Peter Elias, Fano's successor as chair of the organizing committee for the 1956 symposium, expanded on that trend. Almost one-half of this symposium's nineteen papers covered the topics of automata and the social sciences. These included a paper on the grammatical transformation of the English language by linguist Noam Chomsky at MIT, on human memory and information storage by psychologist George Miller at Harvard, and an early paper on artificial intelligence by social scientists Herbert Simon and Allen Newell at the Carnegie Institute of Technology (now Carnegie-Mellon University). All three papers were later seen as foundational in their respective areas. U.S. cognitive scientists have followed Miller in dating the origins of their field to this meeting, one hosted by an engineering society. The interdisciplinarity of the MIT symposia mirrors the interdisciplinary research done in the communications section of MIT's Research Laboratory of Electronics, a group that took its inspiration from Wiener's cybernetics. In fact, Chomsky was affiliated with the RLE at this time.[56] The volume containing the 1956 symposium papers was published six months after Shannon's bandwagon editorial appeared in the same journal. The fact that Shannon had invited Simon and Newell to give their paper at the meeting indicates that Shannon had more of an interest in applying his theory outside of communications engineering than his bandwagon editorial suggests.

Yet the broad view of information theory present at the MIT symposia is not evident when we consider the scope of the papers published in the early years of the PGIT transactions. Of the nearly 150 technical papers published between 1954 and 1958,[57] including those from the 1954 and 1956 MIT symposia—about 60 percent deal with the prediction, filtering, detection, and analysis of signals, 25 percent with Shannon's theory and coding, and about 15 percent with topics outside of communications engineering. Because the latter group of papers invariably applied Shannon's theory of information, Shannon is cited or mentioned in about 40 percent of the total number of papers. Wiener is only cited or mentioned in about 25 percent of the papers because of the multiple ways to analyze the statistical prediction, filtering, detection, and analysis of signals.[58] These papers illustrate the difference between Shannon's and Wiener's approaches to the statistical theory of communication. While Shannon encoded signals to transmit them efficiently in the presence of noise, Wiener filtered signals from noise to predict the future states of signals.[59] In 1955, Peter Elias at MIT notably combined the work of Wiener and Shannon when he used Wiener's prediction theory to develop new type of codes in Shannon's theory, a method he called "predictive coding."[60]

The number of papers outside of communications engineering published by the Professional Group on Information Theory dropped dramatically after the 1956 MIT symposia. Paradoxically, this seems to have been due to an editorial that Elias, the organizer of that very meeting, published in the transactions in December 1958. In "Two Famous Papers," Elias used satire to reinforce the boundary work started by Shannon's 1956 bandwagon editorial. Elias described two problematical types of papers written by information theorists. The first one produced results in prediction and filtering that Wiener and his followers had "obtained years before." The second was a shallow form of interdisciplinarity, which Elias satirized by saying it had a title along the lines of "Information Theory, Photosynthesis and Religion." Its author, usually an engineer or a physicist, discussed the "surprisingly close relationship between the vocabulary and conceptual framework of information theory and that of psychology (or genetics, or linguistics, or psychiatry, or business organization). . . . Having placed the discipline of psychology for the first time on a sound scientific base, the author modestly leaves the filing in of the outline to psychologists," instead of collaborating extensively with a psychologist, as he should have done. Elias suggested "that we stop writing" those types of papers and work on more fruitful areas.[61] My survey of the PGIT transactions indicates that if Elias was referring to papers published in that journal, it was to those that had been given at the symposium he himself had organized![62]

Elias's editorial, like Shannon's, received some pushback. In June 1959, Robert Price at MIT's Lincoln Laboratory, wrote an editorial that pleaded for a "cross-fertilization between pure and applied research in our field." Recognizing that this might open him up to the kind of criticism leveled by Elias's editorial, Price alluded to it by saying, "Lest someone expect a sequel to appear under the title of 'Information Theory, Photosynthesis, and Religion,' I want to emphasize that truly basic studies are thorough and well-grounded. Their rigor is not sapped by casual overextension into a conglomeration of neighboring fields."[63]

The scope of the papers that appeared in the PGIT transactions did narrow following the publication of Elias's editorial in 1958. But the journal continued to focus on two areas in the statistical theory of communication: the theory of information and coding, drawn from Shannon; and the prediction, filtering, and detection of signals, drawn from Wiener and general probability theory. The group's leaders recognized the importance of both areas in the 1950s and 1960s. In his 1959 editorial, Price identified four subdisciplines within the purview of PGIT: "filtering theory, information theory, detection theory, and the analysis of signal statistics." A report commissioned for the International Scientific Radio Union, on progress in information theory in the United States from 1957 to 1960, listed a similar set of areas: information theory ("Shannon's theory and theory of coding"), random processes, pattern recognition, detection, and prediction and filtering ("Wiener's theory and its extensions"). The report for 1960 to 1963 included those fields, as well as "artificial intelligence" and "human information processing," which were present at the MIT symposia in 1954 and 1956.[64]

The term *information theory* thus had two valences within the Professional Group on Information Theory during its first decade. In the title of the group, it was as an umbrella term that referred to the broad areas of information processing discussed at the first London Symposium on Information Theory, from which the organizers of the PGIT probably took its name. Second, as the title of a subdiscipline within the PGIT, *information theory* was a synonym for Shannon's theory of information.

Elias made a similar observation in 1959 when he noted three usages of the term *information theory*. The narrowest meaning denoted a "class of problems concerning the generation, storage, transmission, and processing of information, in which a particular measure of information is used," that derived from Shannon. This area was often called *coding theory*. A second, broader meaning included, in addition to Shannon's theory, "any analysis of communication problems, including statistical problems of the detection of signals in the presence of noise, that make no use of an information measure." This area included Wiener's theory of prediction and filtering and Joseph Doob's work on stochastic processes. The third and broadest meaning was a "synonym for the term 'cybernetics' introduced by Wiener to denote, in addition to the areas listed in the foregoing, the theory of servomechanisms, the theory of automata, and the application of these and related disciplines to the study of communication, control, and other kinds of behaviour in organisms and machines." The third meaning was used by the London and MIT symposia on information theory, as well as by the journal *Information and Control* (1958), whose founding editors were Elias, Colin Cherry, and Léon Brillouin. Elias noted that the title of the PGIT encompassed the second meaning of *information theory*, the boundary his 1958 editorial tried to enforce. The group attempted to clarify matters in the 1960s by establishing an associate editorship of its transactions for "Shannon theory."[65]

Communication theory textbooks directed to graduate students also mark the distinction between Shannon's theory and Wiener's theory. One of the earliest texts, *Information Theory* (1953) by Stanford Goldman, a professor of electrical engineering at Syracuse University, presents the work of both Shannon and Wiener, but later ones focus on either one or the other. Fano's *Transmission of Information: A Statistical Theory of Communications* (1961) deals exclusively with Shannon's theory of information, while Yuk-Wing Lee's *Statistical Theory of Communication* (1960), written by Wiener's former Ph.D. student and long-time colleague at MIT, deals with Wiener's theory. Although both Fano and Lee give joint credit to Shannon and Wiener for coming up with the entropic theory of information, they observed that modern communication theory had developed along two main lines: the application of Wiener's theory of correlation to problems of the prediction, filtering, and detection of signals in the presence of noise; and the application of Shannon's theory of information to problems of how to code signals to transmit information in the presence of noise.[66] Textbooks published in the 1960s with "information theory" in their titles typically focused on Shannon's theory of information.[67]

### Mathematicians and Shannon's Theory

The biggest scientific challenge to the theory of information proposed by Wiener and Shannon came from mathematicians who questioned its relationship to classical statistics and the rigor of Shannon's proofs of his theorems. Joseph Doob, a prominent mathematical colleague of Wiener's who specialized in probability theory at the University of Illinois, criticized Wiener and Shannon on both issues. In 1948, Doob wrote to Wiener about the definition of information in *Cybernetics*. "I admit I found your 'Information' concept a bit mysterious. That integral is the amount of information relevant to doing what? [Statistician Ronald] Fisher's amount of information for example . . . is the amount of information relevant to estimating the parameter from a given sample. I was not clear on the significance of your logarithmic integral." Wiener replied that the concept "need not be so mysterious. It is essentially a matter of counting effective independent choices. . . . For a given [communications] problem, I am interested in the number of coded choices that get through when we use the original code more effectively."[68] Wiener developed this approach while working with Walter Pitts after the war, not from his wartime research on prediction theory. Doob was more familiar with the latter work. He read Wiener's Yellow Peril report during the war, and Wiener contributed material on prediction theory for a book that Doob published on probability theory in 1953.[69]

British mathematicians soon took up the debate about the relationship between the theories of information proposed by Fisher, Wiener, and Shannon. In 1951, G. A. Barnard, a British statistician who had been at the first London Symposium on Information Theory, delivered a paper in this regard before the prestigious Royal Statistical Society in London. "To most mathematicians," Barnard said, "the idea of 'quantity of information' will be associated with the name of R. A. Fisher," whose definition of information he would discuss later in the paper. "But for the present it is desirable to forget it altogether," while he discussed "another notion of 'amount of information,'" developed by Shannon. He then described a connection between a modified version of Shannon's theory and another sense in which Fisher had used the term "amount of information," to refer to a statistic relating the information in a sample to a given parameter. During the discussion of the paper, another British statistician, M. S. Bartlett, who had analyzed the relationship between Shannon's and Fisher's theory at the first London symposium, criticized Barnard for modifying both Shannon's and Fisher's theories by ignoring average values. Wiener, who was in Europe lecturing at the Collège de France, briefly commented on Barnard's paper by discussing the communications aspect of information theory. He stressed that information could only be lost in transmission and that the receiver needed to be taken into account. Wiener may have limited his remarks, at least in print, because Bartlett had criticized him at the London symposium for claiming in *Cybernetics* that his theory of information could replace Fisher's.[70]

Of equal if not more concern to mathematicians was the rigor with which Shannon had proved his theorems in information theory. Doob questioned the proofs in an uncharitable, often-quoted review of Shannon's classic 1948 paper. "The discussion is suggestive throughout, rather than mathematical, and it is not always clear that the author's mathematical intentions are honorable."[71] The stinging remarks were fresh in Shannon's mind forty years later during an interview. "I didn't like his review. He didn't read my work carefully. You can write mathematics line by line, with each tiny inference indicated, or you can assume the reader understands what you are talking about. That's what I did. I was confident I was correct—both intuitively and rigorously. I knew exactly what I was doing, but maybe it takes people a little brighter to understand it. You can always find new proofs of things, better proofs, shorter proofs."[72]

Brockway McMillan, a colleague of Shannon's at Bell Labs who received his Ph.D. in mathematics under Wiener at MIT in 1939, began that work in earnest in the early 1950s. In 1953, he published an early paper that placed the basic theorems of information theory (regarding source, channel capacity, coding) on a more precise mathematical basis. McMillan did not leave Wiener out of the picture, pointing out that "this discipline has come specifically to the attention of mathematicians and mathematical statisticians almost exclusively through the book of N. Wiener and the paper of C. E. Shannon."[73] Work of this sort continued throughout the 1950s. By the end of the decade, as claimed by David Slepian, who worked under McMillan at Bell Labs, the "foundations of Information Theory were reset on a rigorous basis that could leave no skeptics." The translation of the *Mathematical Foundations of Information Theory*, by Soviet mathematician A. I. Khinchin in 1957, and the publication of *Foundations of Information Theory*, by Amiel Feinstein, who worked at MIT's Research Laboratory of Electronics, in 1958 marked the culmination of that endeavor in the 1950s.[74]

The work succeeded so well that Doob did not complain about the lack of rigor in information theory in an editorial he wrote for the transactions of the Professional Group on Information Theory in early 1959. But he still found plenty to criticize. "In spite of all the suggestive work by Wiener, Shannon, and their successors," Doob wrote, "the main thing that strikes an outsider is that there are so few theoretical results." The work of purging and purifying the theorems of information theory made them "more and more attenuated and inapplicable as their hypotheses become more restrictive. . . . Can it be that the existence of a mathematical basis is irrelevant, and that the basic principle is the very idea that there is a context in which the word 'information' is accepted by general agreement and used in an intuitive way, and that no more is needed?" Denis Gabor replied indirectly to Doob in a guest editorial that fall which reviewed the accomplishments of Shannon's theory. "For one thing," noted Gabor, "Information Theory has now become mathematically respectable. Pure mathematicians like Doob will no longer query 'whether the mathematical extensions [intentions] of the author (Shannon) are always strictly honorable.' The work of McMillan, Feinstein, Khinchin and of Shannon himself has made Information Theory rigorous—and almost unreadable to engineers!"[75]

## The Popularization of Information Theory in the United States

If a mathematically rigorous theory of information was almost unreadable to engineers, even a mathematically light version was tough sledding for lay audiences. Nevertheless, journalists, popular science writers, scientists, and engineers introduced Shannon's theory to general readers soon after it appeared in the *Bell System Technical Journal* in 1948.

One method was to associate the theory with an information discourse that arose to explain the newly invented electronic computers to the public. In *Giant Brains: Or Machines That Think* (1949), for example, Edmund Berkeley spoke extensively about "storing information" and the "handling of information" by computers. An insurance actuary who wrote articles on symbolic logic, had operated the Mark I computer at Harvard during the war, and had helped found the Association for Computing Machinery in 1947, Berkeley used Shannon's term *bit* (short for binary digit) to discuss "quantity of information" and "unit of information." But he did not discuss the theory of information, even though he cited *Cybernetics*.[76] Many writers used the term "information" in this manner, to describe what computers processed; they did not link it to such theoretical concepts as "quantity of information." The same held true for accounts that did not mention cybernetics. In a 1952 article on "office robots" that predicted what would later be called the "paperless office," *Fortune* magazine said, "many engineers now feel that a more accurate name for these machines is 'electronic information processors,' for computation is only a part of their real or potential talents."[77] An exception was the *Christian Science Monitor*, which described "amount of information" as the choice involved in selecting a message. It alluded to cybernetics by featuring the autocorrelator signal analyzer developed by Yuk-Wing Lee, Robert Fano, and Jerry Wiesner at MIT's Research Laboratory of Electronics.[78]

A few scientists attempted to describe information theory more fully in the popular-science press.[79] Warren Weaver wrote the earliest account, which appeared in *Scientific American* in July 1949, the same magazine that had published an article by Wiener on cybernetics. Weaver wrote his article to promote the forthcoming book, *The Mathematical Theory of Communication* (1949), which he coauthored with Shannon. In the article, Weaver mentioned Wiener's work on information theory, but focused on making Shannon's more extensive theory understandable to a lay audience and to social scientists in particular. An applied mathematician, Weaver described in a nonmathematical manner the tenets of Shannon's theory: information as a measure of uncertainty and choice in sending messages; the equivalence between information and entropy; and Shannon's central theorem that if the capacity of a noisy channel is equal to or greater than the entropy of the source, a coding scheme can be devised to send information "over the channel with as little error as possible." Weaver covered other technical topics in Shannon's theory, including redundancy, the structural probability of languages, and the digital sampling theorem. He reprinted Shannon's diagram of a general communications system, consisting of an information source, transmitter, noise source, receiver, and destination.[80] It was the first appearance outside of technical journals of what became an icon in Shannon's theory of information. The book contained a more mathematical version of Weaver's article and reprinted Shannon's 1948 paper from the Bell journal. Weaver succeeded in translating Shannon's technical paper into a form understandable by nonmathematical readers.

A firm believer in the popularization of science, Weaver became the expositor of Shannon almost by accident. Impressed with Shannon's wartime research, which he had funded as head of the government's research effort in gun-control systems, Weaver had read the first part of Shannon's paper in the *Bell System Technical Journal* when it came out in the summer of 1948. Weaver discussed the theory in-person with Shannon that fall. He initially thought the theory might help solve the problem of using the computer to translate Russian-language scientific papers into English during the Cold War.[81] After extravagantly praising Shannon's work to Chester Barnard, the head of the Rockefeller Foundation, Weaver wrote a lengthy, less-mathematical treatment of the theory for Barnard in early 1949. A promoter of interdisciplinarity, Weaver thought the theory applied beyond the limitations imposed on it by Shannon and considered submitting a briefer version of his paper to *Scientific American* if Shannon approved.[82]

The project moved forward rapidly in the spring of 1949 after Weaver visited physicist Louis Ridenour, dean of graduate studies at the University of Illinois, whom Weaver had known on radar antiaircraft projects during the war, to see if the Rockefeller Foundation should fund a biological sciences program at the university.[83] After the visit, Weaver sent a copy of his paper to Ridenour, who, on the advice of Wilbur Schramm, head of the new Institute of Communications Research at the university, recommended that the University of Illinois Press publish a small book containing the papers of Weaver and Shannon.[84] Ridenour was interested in Shannon's theory because the university was constructing a large digital computer and thought the book would fit well with the press's new series of lectures by computer builders. Schramm saw it as the basis for a general theory of communication.[85] After negotiating with Ridenour, Schramm, Shannon, the head of Bell

Labs, and editors at the University of Illinois Press and *Scientific American,* Weaver agreed to write a briefer and simpler version of his paper as an article and a longer, more mathematical piece for the book.[86]

Although many writers have mistakenly cited Weaver as a coauthor of Shannon's theory of information,[87] Weaver had no doubts on this score. He told Ridenour, "No one could realize more keenly than I do that my own contribution to this book is infinitesimal as compared with Shannon's." Weaver wrote to Vannevar Bush that his contribution was "interpreting" Shannon's work and explaining its larger implications outside of engineering.[88] Consequently, the text of a proposed dust cover for *The Mathematical Theory of Communication* read, "This book is largely intended to bring to the attention of social scientists who are concerned with one or another aspect of the broad communications problem, and to students of language, the existence of a theory which promises to have important applications far outside the engineering field for which it was created." Weaver's essay "presents a highly persuasive argument that the theory has important applications to the whole problem of communications in society." Shannon did not suggest changes to this wording. In fact, Weaver had told Ridenour earlier in the year that Shannon "thinks the ideas in Parts I and III of the paper [containing Weaver's speculations] are sound and important."[89] That position was a far cry from the bandwagon editorial Shannon wrote seven years later in 1956.

Although newspapers and magazines did not cover information theory nearly as much as cybernetics, a few science writers did popularize it in the 1950s. Francis Bello, technology editor for *Fortune* magazine, published the most extensive account in 1953. Not one to be modest, Bello compared information theory to the great scientific achievements of the twentieth century. "Within the last five years a new theory has appeared that seems to bear some of the same hallmarks of greatness [as relativity and quantum theory.] The new theory, still almost unknown to the general public, goes under either one of two names: communication theory or information theory. Whether or not it will ultimately rank with the enduring great is a question now being resolved in a score of major laboratories here and abroad." Bello had little doubt that the question would be resolved in the favor of information theory. "It may be no exaggeration to say that man's progress in peace, and security in war, depend more on the fruitful applications of information theory than on physical demonstrations, either in bombs or in power plants, that Einstein's famous equation works." For evidence of this claim, Bello pointed to the use of information theory on the "Distant Early Warning Line" of radar stations along the Canadian border built to detect incoming Russian planes loaded with atomic bombs.[90]

Bello popularized the theory more fully than Weaver had done in *Scientific American* five years earlier. Bello explained the details of the theory (amount of information, channel capacity, noise, relationship to entropy, redundancy, and coding) in a nonmathematical manner. He enlivened his exposition by discussing several applications, including the digitization of telephone and television signals at Bell Labs and experimental research on neurophysiology at MIT's Research Laboratory of Electronics. He noted pitfalls of applying the theory in psychology, unsuccessful efforts to fulfill Weaver's hope that it would form the basis for a new theory of meaning, and cautioned in the meantime against the temptation to ascribe meaning to the scientific term "information."[91]

The researchers whom Bello interviewed—Shannon at Bell Labs, and Wiener and Warren McCulloch at MIT—tried to shape the article. Shannon praised a draft of the article "as a bang-up job of scientific reporting, comparable to your transistor article [in *Fortune*]." He and his colleagues at Bell Labs suggested some technical changes, and Shannon unsuccessfully tried to convince Bello to tone down the hyperbole. "Much as I wish it were so," Shannon wrote, "communication theory is certainly not in the same league with relativity and quantum mechanics. The first two paragraphs should be rewritten with a much more modest and realistic view of the importance of the theory." In regard to who should get credit for creating the theory, Shannon suggested adding the following sentence, which is similar to that in his 1948 paper: "And it was M.I.T.'s great mathematician, Norbert Wiener, who, with his work on prediction and cybernetics, laid the cornerstone of information theory." Bello rewrote Shannon's sentence using the metaphor of exploration. Wiener received "credit for discovering the new continent and grasping its dimensions," Shannon for "mapping the new territory and charting some breath-taking peaks." In response to Shannon's complaint that more MIT researchers were named than those at Bell Labs, Bello reversed the emphasis and said much more about research at Bell Labs than at MIT, even to the point of not mentioning McCulloch.[92]

Bello apologized to McCulloch that in compressing the article, he could not cover the neurological research in his lab more fully.[93] As noted in chapter 1, Wiener argued at length that Bello should present his work on the statistical theory of communication as forming the basis for information theory, and he described how his findings on measuring information paralleled those of Shannon. Jerry Wiesner and Robert Fano at MIT had pointed out to Bello that his first draft did not give adequate credit to Wiener, for which he apologized.[94] The final version acknowledged Wiener's statistical theory of communication, but not his entropic definition of information, which left the impression that this had been developed by Shannon alone.

Other early attempts to popularize information theory were not as detailed. In 1954, *Science Digest* claimed that satellite television and improved telephony were "some of the advances to be expected from a new-born but fast-sprouting science that goes by the name of Information Theory. It is a research field that is absorbing more and more first-rate brains." Its "giants" included Wiener, the "father of cybernetics," and Shannon, "whose book, *A Mathematical Theory of Communication*, was the technological bombshell of 1948 [*sic*]." The article dealt in a simplified and nonmathematical manner with coding, redundancy, and amount of information, using examples from research on the digitization of radio and TV signals.[95] *Business Week* gave a much fuller account in 1955. Its lengthy article, "Information: Now It's the Realm," closely followed Bello's article in the rival magazine *Fortune* by giving credit to Wiener and Shannon for creating a statistical theory of information, describing some technical details, including Shannon's communication schema and information as a measure of choice, and by comparing the theory favorably to Einstein's relativity theory. But the article also discussed eventual business applications, saying the theory formed the basis for computerized data processing and that psychologists used it to understand how humans handled information. The author maintained that information theory could thus help managers of large companies make better use of office automation by understanding how the combinations of humans and machines could best deal with information overload, a growing concern in this period.[96]

Shannon did not popularize information theory nearly as much as Wiener popularized cybernetics, perhaps because he had criticized the bandwagon frenzy of his colleagues. But, as we have seen, he did his share of popularization. He allowed Weaver's article to appear alongside his in the *Mathematical Theory of Communication*, commented on Bello's article in *Fortune*, and wrote articles on cybernetics and information theory for the *Encyclopedia Britannica* in 1957, upon a referral from Wiener. He also published an article in *Scientific American* on how to program a computer to play chess.[97] In the *Britannica* article on information theory, Shannon mathematically described it at the level of Weaver's account and mentioned applications in linguistics and in cryptography, to which he had contributed. By limiting the range of applications, he adhered to the spirit of the bandwagon editorial. Shannon said the theory was "known as communication theory, or, in its broader applications, information theory."[98] That distinction seems to have been a new one for him. Although he titled his 1948 paper "A Mathematical Theory of Communication," Shannon had used the term *information theory* as a synonym for *communication theory* in that paper and throughout the late 1940s and 1950s—in titles of talks, in private letters, in a book review of *Cybernetics*, as a title of one paper, and as the name of a seminar he taught at MIT starting in 1956.[99] Even though he referred to "information theory" as a field in the bandwagon editorial, perhaps it was that dispute which stimulated him to make the distinction in the *Britannica* article—to draw a boundary between a narrow "mathematical theory of communication" and a broader "information theory"?[100] In any event, the boundary work failed because the formulation did not catch on, mainly because his followers had been a mainstay in the Professional Group on Information Theory since the early 1950s.

The media did not publicize Shannon nearly as much as they did Wiener. Many more stories appeared in newspapers and magazines about Wiener during the "cybernetics craze" of the 1950s and following the publication of his two-volume autobiography, *Ex-Prodigy* (1953) and *I Am a Mathematician* (1956). The press portrayed the two men as opposites. Wiener was celebrated as a mathematical genius who spoke several languages, cared about the social implications of science, and made "sensational" claims about cybernetics. Shannon was seen as a modest genius, a wiry figure like Abe Lincoln, who quietly did important research for the corporate giant AT&T.[101] Instead of trying to explain Shannon's mathematical theories, newspapers and magazines focused on his inventions: the electromechanical mouse Theseus (which Bell Labs publicized), and programming computers to play chess. A few stories tried to jazz-up his image by quoting his views that automata could reproduce themselves, and by noting that he rode a unicycle, was a "jazz addict," a fan of science fiction, and worked late into the night. An article in *Vogue* magazine managed to sandwich information theory and artificial intelligence between descriptions of the toys and automata that filled Shannon's home.[102]

The contrast between a modest Shannon and an immodest Wiener is common in private correspondence. In 1939, Vannevar Bush wrote to a Harvard professor that Shannon, then Bush's Ph.D. student, was a "very shy and retiring sort of individual, exceedingly modest," one who needed to be handled with "great care."[103] We have seen that Shannon deferred to Wiener in regard to who should receive credit for originating information theory. Shannon told Bello to be more "modest" in comparing it to other scientific theories. On the other hand, Weaver wrote about Wiener's immodesty in his diary at the Rockefeller Foundation, which was circulated to other grant officers. He even tested the limits of Wiener's immodesty when Wiener stopped by the foundation in 1950 to report on the five-year grant that he and Rosenblueth had received from the foundation.

W [Wiener] drops in, in a quite characteristic way, to "report" concerning his activities and to ask whether or not we think he is coming up to the expectations which we had made when we made the grant! W is becoming just pathologically excited by the general interest which his work on cybernetics has aroused, and all of this popular acclaim has him acting precisely like a very small boy who has just won first prize. . . . In reply to W's direct question: "Well, boss, how am I doing?", WW [Warren Weaver] decides to try an experiment. He assures W, in progressively more extravagant language, that we are simply overwhelmed by the success of this grant, and by the blinding brilliance of W's performance. WW thought that it might be fun to see just how much of this sort of thing W could take before he began to see the joke. In a strict sense, therefore, the experiment was a failure! For W never did see the joke. He ate up all of these remarks with perfectly obvious and increasing relish.[104]

The immodest claims of cybernetics (and of Wiener by association) is a theme in John R. Pierce's writings on popular science. A colleague of Shannon's at Bell Labs who had done pioneering work in satellite communications, Pierce drew on his experience writing science fiction (under the pen name of J. J. Coupling) to popularize science in a series of books. His first book, *Electrons, Waves, and Messages* (1956), contained two chapters on information theory and warned against the improper use of the theory outside of engineering.[105]

In *Symbols, Signals, and Noise* (1961), in which he called himself a "Shannon worshiper," Pierce did even more boundary work. He gave examples of the dangers of misapplying information theory and distanced it from cybernetics. Pierce used elementary mathematics and insightful examples to explain the basic elements of Shannon's theory, including the relationship between entropy and information and the efficient coding of signals in noisy channels. The book remains the most widely read English-language account of information theory to date. Pierce, who had attended meetings of an early cybernetics group with Shannon, devoted one chapter to cybernetics. He described Wiener's work on the smoothing, prediction, and filtering of signals, rather than on the broad sweep of cybernetics, noting that it included almost all of modern science and technology. Pierce satirically criticized the universal claims of cybernetics by saying, "So far, in this country the word *cybernetics* has been used most extensively in the press and in popular and semiliterary, if not semiliterate, magazines. I cannot compete with these in discussing the grander aspects of cybernetics. Perhaps Wiener has done that best himself in *I Am a Mathematician*. Even the more narrowly technical content of the fields ordinarily associated with the word cybernetics is so extensive that I certainly would never try to explain it all in one book, even a larger book than this." Pierce observed that few scientists in the United States called themselves cyberneticists. But "even if a man acknowledged being a cyberneticist, that wouldn't give us much of a clue concerning his field of competence, unless he was a universal genius. Certainly, it would not necessarily indicate that he had much knowledge of information theory."[106] As we have seen, British engineer Colin Cherry issued similar warnings about misapplying information theory in his popular book, *On Human Communication* (1957), but he treated cybernetics with respect.

Pierce ridiculed the pretensions of cybernetics to a lay audience in a manner similar to the way Peter Elias had used satire to reign in an expert community of information theorists a few years earlier. Pierce was concerned that the enthusiasm for information theory as a "glamor science" would cause the field to lose its scientific status (as was happening to cybernetics; see chapter 7).[107] The fate of cybernetics was thus a cautionary tale for information theory. Pierce did exclusionary boundary work in *Symbols, Signals, and Noise* by discussing what he considered to be the proper applications of information theory—that is, mathematical and experimental ones—in linguistics, music, and psychology, to the extent of doing some research in the latter field. Pierce acknowledged that an information theorist had criticized him for exploring these applications at all. But he thought it was important to address the issue, in a popular book, by stating the theory's value in communications engineering and countering the belief that the theory was valuable chiefly because of its connection with wider fields, including cybernetics.[108]

### Navigating the Informational Turn

Pierce felt some urgency to spread this message because information theory was being applied as widely as cybernetics had been in the 1950s. Enthusiastic researchers, believing that Shannon's theory provided a quantifiable basis to investigate any type of communication, applied it vigorously to a wide range of fields. These included physics, statistics, artificial intelligence, biology, physiology, psychology, linguistics, economics, sociology, anthropology, communication studies, library and information science, and music.[109]

The spread of information theory was studied in 1957 by Randall Dahling, a master's student in communication at Stanford University. Dahling found that the number of articles in scientific journals and the popular press that cited the theory of information developed by Shannon and Wiener increased exponentially from five articles per year in 1948 to thirty-two per year in 1951. Among the earliest fields to pick up the new theory were electronics, psychiatry, and psychology in 1949; engineering, biology, physiology, and linguistics in 1950; library science, education, and statistics in 1951; followed by social science in 1952 and journalism in 1955. Dahling found that Shannon and Wiener were equally cited in these publications.[110] I interpret that citation practice to mean that these authors, many of whom were at Bell Labs and MIT, recognized that Wiener's work formed the statistical basis for Shannon's theory, as Shannon himself acknowledged in his 1948 paper. The pattern noted by Dahling shifted during the development of each field. Citations of Shannon dominated in psychology, engineering, and linguistics; citations of Wiener dominated in biology, physiology, and psychiatry. Although the Macy conferences on cybernetics brought Wiener's and Shannon's work to the attention of researchers in these areas, Dahling points out the importance of three research centers that diffused the theory of information: Bell Labs, MIT, and the University of Illinois.[111]

In turning to information theory, researchers typically drew on specific aspects of Shannon's theory—such as coding schemes, information measurement, and entropy relationships—rather than on the theory as a whole, with mixed results in biology, the social sciences, and physics. Researchers in molecular biology spent several fruitless years trying to use information theory to break the genetic "code," before realizing that DNA was not a linguistic code, and therefore not able to be analyzed by Shannon's theory. Evolutionary biologist E. O. Wilson had more success quantifying the amount of information in the odor trails insects use for communication.[112] Information scientists, such as Calvin Mooers, devised information-theoretic codes for information-retrieval systems.[113] Physiologists and experimental psychologists turned to Shannon to quantify the information capacity, in bits, of human hearing, vision, memory, and behavioral responses to stimuli. A report commissioned by the Professional Group on Information Theory in 1963 called this field "human information processing." Linguists such as Roman Jakobson and Benoit Mandelbrot used the information measure to quantify the structural elements of languages. Sociologists and anthropologists employed a nonquantitative, semantic version of information theory, an endeavor encouraged by Warren Weaver's popularization of Shannon, and by Wiener's book, *The Human Use of Human Beings* (1950).[114] But American economists criticized the intrusion of Shannon's theory into neoclassical economics on the grounds that it did not provide a value for information in market transactions and assumed that the perfect transmission of information was impossible.[115] In regard to thermodynamics, physicists such as Léon Brillouin and Denis Gabor tried to relate information theory to entropy in a physical manner, rather than the purely mathematical manner favored by Shannon. Norbert Wiener corresponded with Gabor and Jerome Rothstein in this regard and collaborated briefly with Gabor on the theory of optical information.[116]

Communications engineers were slow to apply information theory because they were initially skeptical about the design value of such a highly mathematical field. In the 1960s, information theorists established two MIT-related spinoff companies (Codex and Linkabit) that applied algebraic and probabilistic coding schemes, developed from Shannon's theory, to enable reliable and low-power communications to deep space. The first success was with the solar-orbiting Pioneer 9 spacecraft, launched in 1968. It used a sequential coding scheme that Codex derived from the research of Fano and Elias. Linkabit employed the Viterbi algorithm on the Voyager 1 spacecraft in 1977. In these applications, the Gaussian white-noise characteristics of the real channel of deep space matched well the theoretical channel used by information theorists to develop new codes.[117] In the early 1990s, engineers utilized Huffman coding, a compression scheme derived from Shannon's theory of information decades before, to invent the MP3 format, the basis for almost all of the music files downloaded over the Internet today.[118]

The application of information theory outside of communications engineering held many difficulties. Henry Quastler, who organized conferences on its applications in biology and psychology, summed up the matter in 1955. "There is something frustrating and elusive about information theory. At first glance, it seems to be the answer to one's problems whatever these problems may be. At second glance it turns out that it doesn't work out as smoothly or as easily as anticipated. . . . So nowadays one is not safe in using information theory without loudly proclaiming that he knows what he is doing and that he is quite aware that this method is not going to alleviate all worries. Even then, he is bound to get his quota of stern warnings against unfounded assumptions he has allegedly made."[119] The warnings can be classed under three headings—namely, the way researchers dealt with semantics, entropy, and quantification—which I'll discuss in turn.

The most common injunction was not to interpret Shannon's theory semantically, as a theory of meaning. At the London symposium in 1952, Yehoshua Bar-Hillel and Rudolph Carnap said it was unfortunate that "impatient scientists in various fields applied the terminology and the theorems of statistical information theory to fields in which the term 'information' was used, presystematically, in a semantic sense, i.e., one involving contents or designata of symbols, or even in a pragmatic sense, i.e., one involving the users of these symbols." They thought their semantic theory of information overcame these difficulties. Psychologist George Miller noted in 1953, "Most of the careless claims for the importance of information theory arise from the overly free association to the word 'information.' This term occurs in the theory in a careful and particular way. It is not synonymous with 'meaning.' " But the very term *information* encouraged this interpretation. In 1955, Bar-Hillel thought the misunderstanding arose because Shannon's predecessor, Ralph Hartley, had argued that speaking of the capacity to transmit information implied a quantification of information. "However, it is psychologically almost impossible not to make the shift from the one sense of information, for which this argument is indeed plausible, i.e. information = signal sequence, to the other sense, information = what is expressed by the signal sequence, for which the argument loses all its persuasiveness."[120]

By all accounts, it was Warren Weaver at the Rockefeller Foundation who provided the intellectual sanction for a semantic interpretation. In his popular exposition of Shannon's work published in *The Mathematical Theory of Communication* (1949), which sold well (six thousand copies in its first four years), Weaver admitted that the theory "at first seems disappointing and bizarre—disappointing because it has nothing to do with meaning," and bizarre because it dealt with ensembles of messages and equated information with uncertainty. Nevertheless, Weaver claimed that Shannon's "analysis has so penetratingly cleared the air that one is now, perhaps for the first time, ready for a real theory of meaning." Weaver argued his case by describing three levels of analyzing the communication problem: (a) "How accurately can the symbols of communication be transmitted?"; (b) "How precisely do the transmitted symbols convey the desired meaning?"; and (c) "How effectively does the received meaning affect conduct in the desired way?" Although Shannon restricted his theory to level (a), Weaver claimed that it "actually is helpful and suggestive for the level B and C problems."[121] Many researchers who semantically applied Shannon's theory drew on this passage to justify their work.

Weaver thus did more than simplify information theory. His mathematical account was simpler than Shannon's, but he went beyond Shannon to point out the semantic and pragmatic implications of the theory, to which social scientists responded. A similar blurring of the boundaries between genuine science and popularization,[122] occurred in Wiener's *The Human Use of Human Beings* (1950), which presented a semantic interpretation of information theory.

In the book, Wiener first presented a nonsemantic definition of information as negative entropy, but then equated amount of information with the "amount of meaning" of a message to support his semantic use of the word *information* throughout the book. Statements such as, "It is quite clear that a haphazard sequence of symbols or a pattern which is purely haphazard can convey no information" show his distance from Shannon, who said that the meaning conveyed by symbols was irrelevant to his theory.[123] Perhaps because Wiener viewed information as a measure of order, rather than disorder, as Shannon did, he was inclined to use the term in a more everyday, semantic way to mean knowledge.

Wiener's concept of "amount of meaning," published in a popular book, was criticized in scientific journals by two experts in information theory: Bar-Hillel and Donald MacKay, both of whom proposed alternative semantic theories of information. At the Macy conference in 1951, MacKay suggested "that we ought not to talk, as Wiener does, about 'amount of meaning,' but that we ought to keep the concepts of information and meaning quite distinct."[124] Bar-Hillel agreed. In a philosophical critique of information theory, published in 1955, he thought the title "Theory of Transmission of Information" or, better yet the "Theory of Signal Transmission," was a more descriptive name for Shannon's theory than the "Theory of Information." That name "could not fail to stimulate the illusion that finally a radically new solution of some age-old puzzles connected with the concept of meaning had been found," a reference to Weaver's claim. "Wiener, himself, for instance, deliberately treats as synonyms 'amount of information' and 'amount of meaning.'" Wiener thus conflated the probability of a certain message being received with the probability of an event expressed by the message occurring, which Bar-Hillel thought was unreasonable.[125]

How to treat individual messages was a related semantical problem. The mathematical theory underlying Wiener's and Shannon's measure of information required analysts to consider a statistical ensemble of messages, not simply individual messages. In this instance, Weaver drew on his background in applied mathematics to warn readers about the problem. "Note that it is misleading to (although often convenient) to say that one or the other message conveys unit information. The concept of information applies not to the individual messages (as the concept of meaning would), but rather to the situation as a whole, the unit information indicating that in this situation one has an amount of freedom of choice, in selecting a message, which it is convenient to regard as a standard or unit amount." Many researchers ignored this injunction. As early as 1952 Denis Gabor complained about misinterpretations of the theory that calculated such quantities as the amount of information in a received telegram—a point Bar-Hillel also raised in his 1955 critique.[126]

The second issue in applying information theory concerned the relationship between entropy and information. While physicists extensively debated the physical aspects of this relationship, many social scientists interpreted it more broadly.[127] This drew the ire of Colin Cherry. By 1957, he had "heard of 'entropies' of languages, social systems, and economic systems and of its use in various method-starved studies. It is the kind of sweeping generality which people will clutch like a straw." Although Cherry thought some of the interpretations were valid, he stated that Shannon's concept of entropy "is essentially a mathematical concept and the rules of its application are clearly laid down."[128] Warren Weaver probably encouraged this interpretation as well, by saying "when one meets the concept of entropy in communication theory, he has a right to be rather excited—a right to suspect that one has hold of something that may turn out to be basic and important."[129]

The third issue—the tension between the quantitative and qualitative applications of the theory—was pervasive. During the Cold War, many scientists—such as biologists, psychologists, and economists—tried to emulate the success and status of physics in World War II by becoming more quantitative and thereby welcomed a mathematical theory of information.[130] Yet information theorists complained about how their mathematical concepts were being applied. Robert Fano told Cherry in 1953 that he had "been leading a private fight" in this regard. "I believe, for instance, that expressions for the amount of Information such as those which appear in Shannon's paper can only be used in problems where the transmitter and receiver are well-identifiable, and where one can assume the existence of an ensemble of messages with known statistical characteristics."[131] An equally serious problem from this point of view had to do with metaphorical applications of the theory, which used such technical terms as *amount of information, entropy, channel capacity*, and *noise* in a nontechnical manner. Shannon referred to this sentiment in his bandwagon editorial of 1956 when he said that "establishing such applications is not a trivial matter of translating words to a new domain, but rather the slow tedious process of hypothesis and experimental verification."[132] That prescription proved to be difficult to follow, especially for "impatient" social and behavioral scientists.

In applying information theory so broadly, scientists and engineers created a variety of strategies to deal with the issues of semantics, entropy, and quantification. They adopted, adapted, modified, resisted, and sometimes even opted not to use the theory developed by Claude Shannon and Norbert

Wiener. Not only did they spread the American version of information theory throughout nearly every discipline—from engineering and the natural sciences to the social sciences—they changed the theory in the process. Like users of the early automobile and telephone who reinterpreted and adapted them to fit the cultural practices of everyday life, scientists and engineers were "innovative users" of information theory.[133]

Robert Fano predicted as much at the second London symposium, in 1952, when he said that bringing together researchers from a variety of fields "will undoubtedly result in a mutually useful cross-fertilization, even if information theory should prove ultimately to be of little direct help outside the field in which it originated."[134] He was correct. Cross-fertilization did occur in the cybernetics moment, when information theory was seen as a "glamor science," and researchers eventually recognized that it was of "little direct help" in many disciplines outside of communications engineering, such as in biology and economics.

Despite these shortcomings, an information discourse survived in these and numerous other fields to redefine them in terms of the generation, transmission, and processing of information.[135] The informational turn outlasted its specific scientific results to define how we think about intelligent machines, living organisms, and society itself.

# CHAPTER FIVE

# Humans as Machines

HERBERT SIMON MUST HAVE FELT DISAPPOINTED when he read the letter from Norbert Wiener. A rising star in the social sciences, Simon had begun to apply cybernetics to management problems at the Carnegie Institute of Technology (now Carnegie Mellon University). The correspondence was initiated by his colleague at Carnegie, economist Charles Holt, who wrote to Wiener in early 1954: "Your pioneering work in the statistical approach to the design of filters, servos, etc., has been stimulating to Professor Simon and myself in connection with our work on decision rules for production and inventory control." Holt thought Wiener "may be interested in seeing some of your theory applied in a different context," and asked for his comments and criticism on three papers he had coauthored with Simon. About two months later, Wiener returned the papers and replied: "Any opinion I would pass on econometric matters would not be of much value, but if I may be permitted to say so, the general treatment you give does not seem to me to be particularly closely related to any technique of observation, and I get the causal impression of rather tenuous theory as well. Don't take this too seriously—I am not in a position to work my way through your paper."[1]

Holt and Simon were probably unaware that Wiener had given his standard response to social scientists who asked about applying cybernetics to their field. Since the first Macy conferences, Wiener had maintained that the mathematics of cybernetics was of little use to the social sciences because their data were spotty and they ignored the role of the observer in collecting the data.[2] Yet Simon and many other social scientists ignored Wiener's skepticism, as well as warnings about how to apply information theory, and adopted cybernetics and information theory with vigor. In fact, Simon would win a Nobel Prize in economics in the 1970s for his theory of bounded rationality, which drew on cybernetics.

This chapter discusses the work of a handful of prominent social scientists, including Simon, who enthusiastically took up cybernetics and information theory in the 1950s and early 1960s. These researchers worked within a larger movement to quantify the social sciences in Cold War America. Social scientists had emulated the physical sciences since the late nineteenth century by adopting laboratory experimentation in psychology, physical models in economics, and statistics in several fields. A few workers had also created mathematical rather than physical models of the economy in the 1920s and 1930s. But the effort to apply mathematical models, the hallmark of cybernetics and information theory, throughout the social sciences did not gain ground outside of economics in the United States until after World War II.[3]

The concept of a mathematical model was novel enough at the time to catch the attention of philosopher of science Max Black at Cornell University. Black observed in 1960 that the "expression [*mathematical model*] has become very popular among social scientists," and that it referred to a set of equations describing the behavior of individual humans or social groups under simplifying assumptions. He contrasted a mathematical model with the more common theoretical model, whose paradigm was nineteenth-century mathematical theories of the ether, which assumed it to be an imaginary fluid.[4] As early as 1951, economist Kenneth Arrow had argued that mathematical models should be extended from economics to the rest of the social sciences. Because mathematics was a language, Arrow, explained, "any generalization about social behavior can be formulated mathematically." Arrow surveyed a variety of mathematical models in the social sciences that quantified the behavior of individuals or groups. These included rational choice theory, John von Neumann's and Oskar Morgenstern's game theory, Nicholas Rashevsky's theory of human relations, George Zipf's principle of least effort as applied to the size of cities, and statistical models of inductive inference.[5] Arrow could have added the postwar sciences of operations research, economic input-output analysis, and linear programming.[6]

Mathematical modeling was a hallmark of the postwar "behavioral revolution" in the social sciences. Funded in large part by the Ford Foundation, the movement transformed sociology, anthropology, psychology, economics, psychiatry, and political science. The term *behavioral sciences* came into widespread use after the Ford Foundation named it as one of their program areas in 1949 and defined it in 1952 as an interdisciplinary study of the behavior of humans as individuals and as members of social groups. This goal was pursued by interdisciplinary groups at Harvard, Chicago, Yale, Michigan, and the Carnegie Institute of Technology, and survived the termination of the Ford Foundation's program in the behavioral sciences in 1957. Military research funding agencies, the Social Science Research Council, and the Center for Advanced Study in the Behavioral Sciences, established at Stanford University in 1954, vigorously promoted the field at a time of dramatic expansion of the social sciences in the United States. For polymath Herbert Simon, who advised the Ford Foundation on the planning of the center, the behavioral program would help unify the social sciences. In the words of Simon's biographer Hunter Crowther-Heyck, Simon thought the "criteria for a reintegrated social science were that it would be mathematical, that it would be behavioral and functional, and that it would be both empirically grounded and theoretically sophisticated. These views placed him in the van of the 'behavioral revolution' in social sciences. Indeed they defined it." The postwar behaviorialists sought to remake "social science in the image of modern physical science," and valued mathematical modeling in that regard.[7]

It is little wonder, then, that cybernetics and information theory—two postwar sciences loaded with mathematics and tied to engineering—were attractive to behavioral scientists. Not long after the interdisciplinary Macy conferences did their part to introduce cybernetics and information theory to social scientists, several notables in these fields—W. Ross Ashby, Claude Shannon, political scientist Karl Deutsch, and psychologist George Miller—were fellows at the newly established Center for Advanced Study in the Behavioral Sciences, from 1954 to 1959.[8] That Shannon chose to spend an academic year at the center, starting in 1957, shows that he had more sympathy with the application of information theory to the social sciences than his bandwagon editorial of the year before had implied.[9]

I focus on the work of six researchers in the behavioral sciences: Herbert Simon, George Miller, Karl Deutsch, Roman Jakobson, Talcott Parsons, and Gregory Bateson. They employed cybernetics and information theory in two ways to explain human behavior. One method used an "information calculus" to model human judgments and the structure of language in psychology and linguistics. The second method used cybernetic feedback circuits to model human decision making and social interactions in management science, psychology, sociology, political science, and anthropology. These social scientists created innovative ways to apply information theory as a measurement technique, to address Wiener's complaint that cybernetics would not work for their field, and to extend cybernetics and information theory into the realm of philosophy.

Of primary importance was the choice of scale, at what level to model humans as machines. Should it be the individual, the social group, society as a whole, or the larger ecosystem of which humans were a part? Cybernetics, as interpreted by Wiener, especially in *The Human Use of Human Beings*, sanctioned the entire range of modeling, while information theory, as interpreted by Shannon, was more restrictive. In large measure, the disciplinary background of researchers determined their scale of modeling. Experimental psychologists focused on individuals, while linguists, sociologists, management scientists, and anthropologists favored social groups. Ever the iconoclast, Bateson extended cybernetic modeling to the ecosystem and beyond.

## The Information Calculus

The entry on "Information Theory" in the second edition (1968) of the classic reference work, the *International Encyclopedia of the Social Sciences*, written by psychologist Irwin Pollack, indicates the extensive application of Shannon's theory to the behavioral sciences (in its broadest interpretation). Pollock opened by saying,

> The concepts and measures of the statistical theory of selective information (information theory), have been so thoroughly enmeshed with the whole of behavioral science that delineation of the exact contribution of the theory is nearly impossible. The very verbal descriptive fabric of the behavioral sciences has become thoroughly interlaced with informational concepts: individuals or groups are described as "information sources" or "receivers"; skilled performance is described as "information processing"; memory is described as "information storage"; nerves are described as "communication channels"; the patterning of neural impulses is described as "information coding"; the brain is described as "an information complex," etc. Indeed, the molecule, the cell, the organ, the individual, the group, the organization, and the society have all been examined from the point of view of a general systems theory which focuses upon the information-processing, rather than upon the energetic characteristics of each system.[10]

Pollack did not describe the nonmathematical and semantic uses of Shannon's theory, which had been criticized by information theorists. He focused instead on a precise mathematical usage called *information measurement* by psychologists and *information calculus* by Israeli philosopher of science Yehoshua Bar-Hillel. Bar-Hillel observed in 1956 that "in most cases, it is rather the *information calculus* that is applied [in physics and the social sciences] and not at all its communication engineering interpretation."[11] Although Pollack described a wide range of scale at which humans were modeled, the informational calculus was mostly applied to individuals and social groups. I focus on the invention and application of the information calculus in two related fields: psychology and linguistics.

As noted by Pollack, the experimental psychologists who pioneered this technique in the early 1950s used Shannon's entropy equations to calculate indices of stimulus-response behavior, rather than the parameters of human communication. The information calculus allowed them to measure how humans responded to a stimulus and the ability of humans to transmit information in a stimulus-response situation—the main methods of applying information theory to psychology in the 1950s. Both methods sharply diverged from the original intention of Shannon's theory: to analyze the transmission of information from a human sender to a human receiver.

Psychologist George Miller did much to create and popularize this technique, which he called *information measurement*. Miller began his career during World War II as a graduate student at the Psycho-Acoustic Laboratory at Harvard, then moved in 1951 to MIT's Research Laboratory of Electronics to work on human-machine problems for the SAGE air-defense system with psychologist J. C. R. Licklider, who attended a Macy conference on cybernetics. Miller returned to the Harvard psychology department in 1955. A close student of the work of Shannon and Wiener, Miller utilized information measurement in experimental psychology and psycholinguistics, and employed cybernetics to develop a computer metaphor of mind. Both endeavors made cognitive psychology into an information science.[12]

Miller and Frederick Frick, also at Harvard, presented the method of information measurement in a 1949 paper, which is generally recognized as the first application of information theory to psychology.[13] They introduced an information calculus to improve the quantification of long-standing research problems in experimental psychology. Miller and Frick were not interested in the transmission of *information*; in fact, they did not employ the term as a scientific concept in the paper. Instead, they used Shannon's entropy equations to calculate an index that measured the statistical patterns in a sequence of responses generated by subjects during an experiment, for example, the right and wrong guesses in a test of extrasensory perception. Miller and Frick argued that their index, the "index of behavioral stereotypy," measured in "bits," analyzed well the statistical structure of experimental responses, and thus provided the "kind of predictability necessary for a science of behavior." The method would enable researchers to describe all learning experiments in an improved, uniform manner so they could be compared. In 1951, they employed this method in one of their own experiments to quantify the behavioral pattern of rats in an operant conditioning experiment, pressing a bar for food in a modified Skinner box. They calculated "average uncertainty in bits per response," but, again, did not speak about the transmission of information.[14]

In the 1950s, Miller helped popularize the second form of information measurement in experimental psychology: measuring the ability of humans to transmit information in a stimulus-response environment.[15] This method was also an adaptation of information theory. It did not treat humans as senders and receivers that transmitted information through a channel, as in Shannon's theory, but as communication channels themselves. In experiments on making absolute judgments about random stimuli (of tones and loudness in hearing, taste, visual distinctions, etc.), psychologists viewed human subjects as receiving "stimulus information" (sensory data) from an experimenter (the psychologist) and sending "response information" (the judgment) back to the experimenter. Psychologists viewed the subjects' ability to make these judgments as the property of an unknown "channel" in the Shannon model. They thus treated this human ability as a "black box," with inputs and outputs, a typical procedure in behaviorism. They used the mathematics of information theory (its entropy equations) to calculate the amount of stimulus information and the amount of response information, from which they could calculate the amount of information "transmitted" through the human channel. As the amount of stimulus information was increased, more errors of judgment occurred, and the amount of transmitted information decreased, leveling off at the subject's "channel capacity" for a task. In his influential paper, "The Magical Number Seven, Plus or Minus Two," Miller compared many such experiments performed with this method on absolute judgment and immediate memory. He found that, typically, a human's channel capacity for these tasks averaged around 2.5 bits, representing about 7 items. In this paper, Miller also proposed the idea that humans were able to operate at more than this channel capacity by "chunking" information into aggregates of data, instead of processing it bit by bit.[16]

Experimental psychologists enthusiastically employed both methods of information measurement in the 1950s, aided by digital and analog computers to do the tedious calculations. They modeled humans in this manner in the areas of reaction time, speech perception, language redundancy, absolute judgments, auditory dimensions, pattern perception, and motor skills.[17] In the long run, however, cognitive psychologists and historians, alike, agree that the importance of information theory lay not in these precise measurements, but in the introduction of an information discourse into psychology.

When experimenters used the technique of information measurement, they popularized concepts that helped establish information processing as the paradigm for the new fields of cognitive psychology and cognitive science. Ulric Neisser, whose book *Cognitive Psychology* (1967) did much to define the new discipline as a blend of information processing and experimental psychology, recalled, "It was not to be the mathematics of information but just the concept of information, as an entity in its own right, that would shortly transform psychology" after the 1956 conference at MIT discussed in chapter 4.[18] Ironically, the concept came from the practice of information measurement, rather than utilizing Shannon's theory to model human communication.

Linguists employed the information calculus to quantify their field in a similar manner. Although several information theorists had tried their hand at analyzing language—including, as we have seen, such prominent figures as Shannon, Wiener, and Bar-Hillel—the main figure to employ the information calculus in linguistics was a linguist, Roman Jakobson, aided by an information theorist, Colin Cherry. An émigré Russian who taught initially at Columbia University, Jakobson attended a Macy conference on cybernetics in 1948, then worked at MIT's Research Laboratory of Electronics after moving to Harvard in 1949. Firmly enmeshed in the cybernetic-information theory nexus between MIT and Harvard, Jakobson led the effort to remake linguistics as a structuralist information science in the Cold War.[19]

Jakobson latched onto cybernetics and information theory at an early date. He wrote to Wiener in February 1949, shortly after *Cybernetics* was published, "It is indeed a book which is epoch-making. At every step I was again and again surprised at the extreme parallelism between the problems of modern linguistic analysis and the fascinating problems which you discuss. The linguistic pattern fits excellently into the structure you analyze and it is becoming still clearer how great are the outlooks for a consistent cooperation between modern linguistics and the exact sciences."[20] Jakobson discovered a more fully worked-out and accessible form of information theory than Wiener had presented in *Cybernetics* when Warren Weaver at the Rockefeller Foundation sent him late that year a copy of his and Shannon's recently published book, *The Mathematical Theory of Communication*, which contained Shannon's seminal article on information theory and Weaver's popularization of it. In 1950, while visiting European linguists on a trip funded by the Rockefeller Foundation, Jakobson enthusiastically wrote to Weaver, "I frankly pointed out [to the linguists] your and Shannon's *Mathematical Theory of Communication* as the most important among the recent American publications in the science of language." He also praised the book in a letter to linguist Charles Fahs, director of the Humanities Division at the Rockefeller Foundation, to whom he was applying for a large grant to analyze the structure of the Russian language, a Cold War priority of Weaver's. "I fully agree," wrote Jakobson, "with W. Weaver that [with Shannon's theory] 'one is now, perhaps for the first time, ready for a real theory of meaning,' and of communication in general." The foundation awarded Jakobson a five-year, fifty-thousand-dollar grant to study the logical structure of Russian at Harvard that year.[21]

The grant led to several publications, including an article in *Language* that Jakobson coauthored with Morris Halle, his protégé at MIT, and Colin Cherry, the prominent British information theorist who was on leave at MIT. The paper used the information calculus to logically describe the phonetic structure of the Russian language. This was an initial step in Jakobson's larger program to establish a logical description of languages in which phonemes were the key to understanding language. Jakobson and his coauthors analyzed the forty-two phonemes of spoken urban Russian in terms of eleven binary oppositions, the yes-no states of Jakobson's "distinctive features" of phonemes, such as grave or acute, voiced or not voiced. Assuming that phonemes occurred independently of each other and removing redundant features, they used Shannon's entropy equations to calculate that each phoneme conveyed 4.78 bits of information. That is, they calculated a "measure of the 'information' conveyed when the speaker selects any particular phoneme." They recognized that the "simple analysis that we have made so far must be regarded as a somewhat artificial though quite efficient description of the language in its simplest aspect." They did a more realistic analysis by dropping the assumption that phonemes occurred independently and employed Shannon's conditional-probability equations to calculate the information conveyed by the Russian language per digram (group of two phonemes) and per trigram (group of three phonemes).[22]

Like Miller, Jakobson and his colleagues departed from Shannon's theory of communication. They did not measure the information of a group of messages transmitted in a written language as a selection of all available messages, as Shannon did. Instead, they used a form of information calculus to measure the binary choices (such as grave or acute) necessary to identify the phonemes in a particular language on the basis of their distinctive features. They put the term *information* in scare quotes so that the readers of *Language* would not interpret that term semantically. Like the experimental psychologists, Jakobson and his colleagues used information theory to measure human choices, in this case choices of how to vocalize speech, not to measure the information content of messages. Shannon's theory provided a sophisticated, statistical method of doing so—thus the interdisciplinary collaboration with Cherry. The fact that the term *information* usually had a semantic meaning in linguistics and psychology tended to confuse matters, but not for practitioners such as Jakobson and Miller, who had a firm grasp of the nonsemantic aspect of Shannon's theory.[23]

## Cybernetic Circuits

Those who promoted cybernetics to solve the problems of how to quantify the social sciences and accurately model human behavior disregarded Norbert Wiener's view that the mathematics of cybernetics could not be applied to the social sciences. In effect, they demonstrated how cybernetic circuits could quantify human behavior in a realistic mathematical model. They touted cybernetic modeling because it was based on an established mathematical theory from control systems engineering and because its information-feedback loops described actual communication pathways in which humans adapted to their environment. For these social scientists, cybernetics was no mere analogy. More so than in the information calculus, these researchers adopted feedback circuits in a variety of ways to model human behavior. While almost all of them modeled at the scale of the social group, they differed on how to use mathematics, how central cybernetics should be to their work, and the scope of behavior to model.[24]

Karl Deutsch, an early advocate of cybernetics and information theory, adopted those techniques as his primary methods of modeling political processes. Of all the social scientists who engaged with cybernetics, Deutsch was the closest to Norbert Wiener, even though he did not attend the Macy conferences. They were colleagues for more than a dozen years at MIT, from 1941 until 1957, when Deutsch moved to a prestigious position in political science at Yale. A member of Wiener's dinner seminar on scientific method, Deutsch cited conversations with Wiener and Walter Pitts in his early papers on cybernetics and the social sciences.[25] He worked with Wiener on an urban dispersal plan, publicized in *Life* magazine in 1950, to enable large cities, viewed as communications systems, to survive atomic warfare. The following year, when Wiener was in Europe, Deutsch handled a dispute with Bell Lab's computer designer George Stibitz, who had charged that a photo caption in a *Life* magazine story gave Wiener too much credit for inventing digital computers. After Deutsch moved to Yale, he and Wiener coauthored an article on the "lonely nationalism" of Rudyard Kipling.[26] Deutsch addressed Wiener's well-known skepticism about applying cybernetics to the social sciences by including a long extract from a letter from Wiener in his book, *The Nerves of Government* (1963). Wiener had written to Deutsch that even though social systems had more complex communication processes than did machines, both cases abided by the "same grammar." Therefore, the analysis of social systems could benefit from the technical advances of communication and information theory.[27]

Deutsch praised the merits of cybernetics at an early date. Shortly after Wiener's book *Cybernetics* appeared in 1948, Deutsch published articles advocating that cybernetic models of self-adapting communications networks should replace previous models favored by social scientists. Noncybernetic mechanistic models were mathematically rigorous, but they were deterministic and could not represent growth and evolution. Organismic models could handle that type of complexity, but they were "incapable of extensive rearrangement" of their parts by the analyst because of the wholeness of the system. Historical process models lacked "inner structure and quantitative predictability." In contrast, cybernetics provided information-feedback structures and operational terms that could model the behavior of organizations and other large groups. According to Deutsch, this approach could even model such attributes as purpose, learning, will, consciousness, and social cohesion.[28]

Deutsch drew on these concepts in his first and most influential book, *Nationalism and Social Communication* (1953), to argue that the extent of nationalism could be quantified by assuming that the communication defining a nation flowed through cultural channels of information. Information exchanges defined a culture, economic exchanges defined a society. Because Deutsch did not have extensive data on the amount of information flowing in mass communications, he used such stand-in measures as the growth of urban ethnic populations to graph the assimilation of different groups in a country. He also drew on the feedback concept to model the self-determination and self-autonomy of a people. In an appendix, Deutsch presented a formal mathematical model, based on differential equations, of the relationship between assimilated populations and the development of nationalism, which he had developed with the help of MIT economist Robert Solow.[29]

In *The Nerves of Government*, Deutsch extended cybernetic modeling to illuminate general problems in political science. One example is a verbal model he created to analyze the performance of governments, which used concepts from servomechanism theory such as the effect of the "load" of information on the decision process, the "lag" time necessary to respond to a challenge, the "gain" (speed and size) of the response, and the "lead" time it took to predict new problems. The performance of a government depended on the interplay among these factors, which could indicate the effectiveness of political systems in general or a government's performance of a specific function.[30] In addition to verbal models, Deutsch presented a diagram of information flows in foreign policy decisions. Although not worked out in the book, the model represents his cybernetic vision of a self-modifying communications network. The foreign policy decision-making system receives *Information* about the outer world from foreign and domestic inputs; processes that information through several *Screens* of attention and policies based on *Selective Memory* and recall; and makes *Tentative Decisions*, which are fed back through the system to produce *Final Decisions*. These are sent as outputs to foreign and domestic entities, and are also fed back as inputs to the entire system.[31]

Understandably, Deutsch did not attempt to put this formalized, complex, nonlinear, multiple-loop, information-feedback system into mathematical form. Although *The Nerves of Government* leaves the impression that Deutsch had turned away from mathematical modeling, which went against the grain of trends in behavioral science, one prominent modeler saw value in his method. Anatol Rapoport, professor of mathematical biology at the University of Michigan, wrote, "Deutsch's solid achievement is in his persistent attempts to replace generalities by operational definitions," which will lead to "clarification, insight, and impetus to further development" in political science.[32] Deutsch saw his task as that of justifying the validity of the cybernetics model and of deriving operational conditions for applying it to political science. Future researchers could then develop specific mathematical models to represent complex social systems.

Deutsch had enough confidence in his modeling expertise to critique Herbert Simon, one of the foremost mathematical modelers in the social sciences and a founder of the postwar behaviorist movement. Deutsch praised Simon for besting contemporaries whose simplified models did not match the sophistication of their mathematics, and for modeling different types of rationality. "Unfortunately," Deutsch continued, "Simon has not applied his powerful and suggestive models to large-scale political and social processes, although he has applied his style of thinking with excellent effect to problems of private and public administration, the business firm, and the study of organizations."[33]

This is a good account of the extent of Simon's modeling. Remarkably, however, Deutsch did not comment on Simon's debt to cybernetics, even though he cited papers in which Simon made that debt clear in developing his major contribution to the social sciences, the theory of bounded rationality. In that concept, the basis of his Nobel Prize in economics, humans satisfice rather than optimize to solve problems. Faced with limited information and calculating ability, humans nevertheless make decisions rationally in an area bounded by these limitations. Crowther-Heyck has demonstrated that Simon drew on three intellectual threads to develop this new model of humans, which combined the sciences of decision and choice: "(1) his own previous work on administrative decision-making, (2) contemporary cybernetics and servo-mechanism theory, especially the work of W. Ross Ashby; and (3) Gestalt psychology." Simon recalled that he "stayed up all night to finish W. Ross Ashby's *Design for a Brain*," when it appeared in 1952.[34] Simon developed his adaptive modeling technique in the early years of a long, distinguished career at the Graduate School of Industrial Administration at the Carnegie Institute of Technology. There, he contributed to a variety of fields—political science, economics, cognitive science, and computer science—and modeled a range of human behavior—from individual rationality in organizations to simulating human decision making by computers.

Cybernetics makes an early appearance in an article he published in *Econometrica* in 1952, most likely before he had read Ashby, in which he applied servomechanism theory to the problem of optimizing production control in manufacturing. The problem consisted of deciding how much of an item to produce to minimize the cost of production and keep an optimum amount of inventory, given a rate of customer orders. Saying his method came under the rubric of "Wiener's general program of cybernetics," Simon gave a tutorial for social scientists on the mathematics of servo theory. He acknowledged that experienced managers already knew intuitively the results of his analysis of the production problem, yet he argued that servo theory provided a rigorous methodology with which to derive management decision rules.[35] In a 1954 paper on constructing models in the social sciences, Simon again acknowledged Wiener, rather than Ashby, but he probably drew on Ashby's ideas to distinguish between models of optimization and models of adaptation in his theory of bounded rationality. Optimization referred to a "rational process in which the choice of a 'best' is central." Adaptation referred to a "rational process in which movement toward a 'better' is central." In the classic economic theory of the firm—which is similar to the production control problem—an entrepreneur could try to calculate how to maximize profit with respect to changes in raw material (optimization), or change the amount of raw material used over time to approach the condition of maximum profit (adaptation). Optimization required the entrepreneur to possess unrealistic prediction and calculation abilities, while adaptation represented actual management practices and could be modeled with servo theory, i.e., with cybernetics. After presenting the differential equations for the adaptation model, Simon said, "An engineer looking at this model would recognize in it something he is accustomed to call a 'servomechanism' or a 'closed-loop control system' ... in which the individual measures the error in his behavior, and adjusts the behavior seeking to eliminate the error. Norbert Wiener has argued persuasively that the servomechanism model may be a useful model for describing physiological, psychological, and sociological adaptive systems."[36]

In *Models of Man* (1955), a collection of articles, Simon made explicit the fact that servo theory modeled one method of human decision making in his theory of bounded rationality: a feedback control system that adapted to changes in its environment, seeking viability rather than optimization.[37] A decade later, Simon framed his popular book *Sciences of the Artificial* (1969) with the cybernetics of Wiener and Ashby, citing them both. He stated that one of the four characteristics that distinguished the artificial from the natural was that "artificial things can be characterized in terms of functions, goals, and adaptation."[38] Simon unified the sciences of choice and control in his theory of bounded rationality and also in the Logic Theorist, a computer program that he and his colleague Allen Newell created to simulate human reasoning in proving mathematical theorems (see chapter 6).

George Miller was so stimulated by the Logic Theorist (which Simon and Newell presented at the same 1956 meeting at MIT in which Miller presented his "Magical Number Seven" paper) that he designed an information-processing model for human decision making. In *Plans and the Structure of Behavior* (1960), Miller and his coauthors, psychologists Eugene Galanter and Karl Pribram, drew heavily on cybernetics, and very little on Miller's method of information measurement, to propose a new basic unit of psychology, the TOTE (Test-Operate-Test-Exit), to replace the reflex arc in that field. Modeled after a computer program and based on the "cybernetic hypothesis" that the "fundamental building block of the nervous system is the feedback loop," the TOTE unit produced a behavioral response by continually testing its state against a goal, by means of a feedback loop, then exited when finished. At the highest level of generalization, the arrows in the TOTE diagram represented control, not the flow of energy or information. The authors underscored their debt to cybernetics by saying that the foundational paper in that field, published by Wiener, Rosenblueth, and Bigelow in 1943, had made teleology scientifically respectable: "Today we can, almost as a matter of course, propose teleological arrangements such as the TOTE unit. ... The particular realization of the unit in tissue or in metal need not deter us, for we know that it can be accomplished in a variety of ways."[39]

My final example in this section is Harvard sociologist Talcott Parsons, whose structural-functional theory of society (a master social science) dominated sociology and anthropology in the Cold War.[40] Influenced by biochemist L. J. Henderson's ideas about systems and Walter Cannon's theory of homeostasis in the 1930s, Parsons established an elaborate theory of action, which comprised four subsystems: the organism at the physical level, the social system at the level of interaction between individuals, the cultural system at the symbolic level, and the personality system.[41] Cybernetics came into play to control the interactions of these levels. Parsons recalled, "Clarification of the problem of control, however, was immensely promoted by the emergence, at a most strategic time for me, of a new development in general science—namely, cybernetics in its close relation to information theory. It could now be plausibly argued that the basic form of control in [human] action systems was of the cybernetic type." This insight "later became a dominant theme in my thinking."[42] The timeliness came from his participation in the Macy conferences.

Cybernetics is explicit in Parsons's theorizing in the 1960s, in the concept of a "cybernetic hierarchy" of control. Drawing on Wiener's writings, Parsons applied the cybernetic insight that systems high in information but low in energy (like the household thermostat) could control systems low in information but high in energy (like the household furnace). For example, Parsons's high information cultural system could control the low-energy organism system. Although Simon was indebted to Parsons's behavioral-functional approach in mathematizing the social sciences, Parsons eschewed mathematics, and Simon disagreed with his interpretation of cybernetics.[43]

### Gregory Bateson: From Schismogensis to Immanent Mind

The career of Gregory Bateson, a mainstay of the Macy conferences on cybernetics, illustrates how radically cybernetics and information theory could be transformed when applied to subjects as diverse as New Guinea headhunters, schizophrenic patients, alcoholism, and religion. Bateson was unsparing in his praise for cybernetics and information theory. In the 1950s, he said that the idea of equating information with negative entropy, which he attributed to Norbert Wiener, "marks the greatest single shift in human thinking since the days of Plato and Aristotle, because it unites the natural and social sciences and finally resolves the problems of teleology and the body-mind dichotomy which Occidental thought has inherited from classical Athens"! In the 1960s, he called cybernetics and the Treaty of Versailles, which ended World War I but set the stage for World War II, the "two historic events of the twentieth century." It is no wonder, then, that in the foreword to his most famous book, *Steps to an Ecology of Mind* (1972), which is still read today, Bateson acknowledged that his intellectual debt to Norbert Wiener, Warren McCulloch, and other members of the Macy conferences "is evident in everything that I have written since World War II."[44]

In the introduction to the book, Bateson reveals why he considered cybernetics to be so crucial and the surprising fact that this outsider to mainstream social science saw himself as engaged in a reform effort similar to that of the "behavioral revolution." The two revelations are intertwined. In explaining how his work over the past thirty-five years had produced "benchmarks" (or "steps") with which to establish a new science of mind, an "ecology of mind," Bateson said his goal had been to create heuristic concepts to bridge observable behavioral data and scientific fundamentals. He divided the fundamentals into those related to substance and to form. Bateson thought that the behavioral scientists of the past fifty years bridged to the inappropriate part of the fundamentals—those related to substance, such as the laws of matter and energy—while he bridged to the appropriate part—those related to form, or pattern, which had been enriched by cybernetics and information theory. "This book," Bateson concluded, "is concerned with building a bridge between the facts of life and behavior and what we know today of the nature of pattern and order." The fact that Miller, Jakobson, Deutsch, Simon, Parsons, and many other social scientists had also turned to cybernetics and information theory to model human behavior, and to create a theory of mind in the case of Miller, seems not to have concerned Bateson, who did not cite them, even though he had met Jakobson and Parsons at the Macy conferences.[45] In that sense, he was an outsider to the behavioral revolution.

Trained as a biologist and then in the functionalist (sociological) British school of anthropology, Bateson turned to the mathematicians at the Macy conferences, especially Wiener, for help in reforming the social sciences.[46] Despite Wiener's skepticism of this endeavor, Bateson persisted and won Wiener's approval of his work. In 1951, Wiener wrote to Larry Frank, a founder of the Macy conferences, praising Bateson for doing "valiant work in attempting to bring psycho-analytic processes under the heading of cybernetics." Wiener noted, however, that Bateson's "work is and must be sketchy" because in psychology, the "problem of an adequate knowledge of the elementary processes is far from complete." The "vagueness and clichés" in psychoanalytic discourse made it "very difficult to find a substantial agreement in the application of this or that term," a criticism that had been made repeatedly at the Macy conferences.[47] There is little doubt that the emotionally insecure Wiener, who had also been psychoanalyzed, liked Bateson's praise of cybernetics (quoted above) in the book *Communication: The Social Matrix of Psychiatry* (1951), an ethnographic study of psychiatry in a San Francisco clinic that Bateson coauthored with psychiatrist Jurgen Ruesch. For his contribution to the book, Bateson developed a general theory of communication, based on the idea of information as negative entropy, to explain the epistemology and practices of psychiatry.[48] In 1952, Bateson visited Wiener to discuss how the notion of presenting logical paradoxes to computers, which had been debated at the Macy conferences, was analogous to doing psychotherapy with patients. Wiener replied that he had a "high opinion" of Bateson's work and would support his grant proposal to the Macy Foundation. Bateson received the grant and acknowledged that "it was our conversation at your house which gave me the central idea" for what became his influential double-bind theory of schizophrenia.[49]

Cybernetic circuits are evident in two other areas in which Bateson adapted cybernetics to his research: anthropological fieldwork in New Guinea and what he called "cybernetic epistemology." In 1958, Bateson added an epilogue to the second edition of *Naven* (1936), his path-breaking ethnography of the New Guinea Iatmul people, which employed cybernetics to revise how he treated his concept of schismogenesis in the first edition. Bateson had developed the concept to explain how schisms were generated between Itamul groups in two ways. In symmetrical schismogenesis, assertive behavior such as boasting by one group elicits an equally assertive behavior by the second group, which continues back and forth until a break occurs. In complementary schismogenesis, assertive behavior by one group elicits submissive behavior by the second group in a circular manner until a break occurs. He explained the stability of the society by the fact that its complementary schismogenesis coincidentally balanced its symmetrical schismogenesis. After being exposed to cybernetics, Bateson realized that both cases exhibited destructive (positive) feedback, and he looked for a self-corrective (negative) feedback circuit that kept the society stable. He found it in the transvestite *naven* ceremony, which typically celebrated the accomplishments of a young person. Bateson called it an "exaggerated caricature of a complementary social relationship between *wau* [e.g., mother's brother] and *laua* [e.g., sister's child]." In the first edition of the book, he had analyzed schismogenesis within the *naven* ceremony and did not relate it to the wider structure of Iatmul society. Now he realized, "In any instance of intense symmetrical rivalry between two clans we may expect an increased probability of symmetrical insult between members, and when the members of the pair happen to be related as *laua* and *wau*, we must expect a triggering of the complementary rituals [*navens*] which will function toward mending the threatened split in society."[50]

Bateson developed his new epistemology in the late 1960s while holding a series of peripatetic research appointments on the West Coast and in Hawaii, and published it in *Steps to an Ecology of Mind*. In the furtherest extension of his ideas—to the largest possible scale of modeling—he stated in 1970 that the "cybernetic epistemology which I have offered you would suggest a new approach. The individual mind is immanent but not only in the body. It is immanent also in pathways and messages outside the body; and there is a larger Mind of which the individual mind is only a subsystem. This larger Mind is comparable to God and is perhaps what some people mean by 'God,' but it is still immanent in the total interconnected social system and planetary ecology."[51] The appropriation of cybernetics to create a holistic concept of mind, formed by an information-feedback circuit extending from the organism to its environment and back again, inspired a large following in the 1970s, especially among the ecologically minded counterculture (see chapter 9).

Although Bateson's work won favor from Wiener, hardly any of his modeling was mathematical. In one exception, Bateson attempted to put his theory of symmetrical schismogenesis into mathematical form to analyze the "steady state" of Balinese society, using a set of differential equations that had been developed in the 1930s to analyze international arms races. The equations did not apply to complementary schismogenesis, but Bateson did reprint the 1949 paper that mentions this attempt in *Steps to an Ecology of Mind*. The lack of mathematics in Bateson's modeling put him in good company in the social sciences, with Talcott Parsons, for example.[52] But it also kept the already marginal Bateson on the margins of the new behaviorism, as even Deutsch had dome some mathematical modeling.

# CHAPTER SIX

# Machines as Human

I N *GOD AND GOLEM, INC.*, published in 1964, the year of his death, Norbert Wiener extended the reach of cybernetics beyond science, engineering, politics, labor, and other social concerns into the realm of religion. He introduced the topic by discussing the medieval Jewish legend of the Golem, an automaton servant that Rabbi Löw in Prague created out of clay and brought to life by reciting a religious incantation. In this morality tale of the dangers of human hubris, of imitating God by creating life, Rabbi Löw destroyed the Golem when it got out of control and ran amok. To Wiener, cybernetic machines were the "modern counterpart of the Golem of the Rabbi of Prague."[1] As the newest science dealing with the relationship between humans and machines, cybernetics provided an opportunity to write a modern-day commentary on the age-old question about the pitfalls of humans playing God. The central analogy of cybernetics—that both humans and machines could be studied using the same principles from control and communications engineering—thus raised fundamental questions about the relationship between science and religion.

*God and Golem* dealt with three areas of cybernetics that touched on religion. Machines that learned—like the IBM computer program that played an expert game of checkers—brought up the question of whether a creator should build a machine that he or she could not understand. Machines that reproduced themselves—which existed in the speculations of Wiener and physicist Denis Gabor, who showed mathematically that ideal machines with precisely known inputs and outputs could self-replicate—questioned the unique status of humans as created in God's image. Cyborgs and automatic machines—human prosthesis and computerized factories, for example—raised ethical concerns of fusing humans with machines and of technology throwing humans out of work. Wiener considered the latter issue to be "one of the great future problems which we must face" in the cybernetics age. Assigning proper roles to humans and intelligent machines would eliminate the specter of literal-minded automatons wreaking havoc across the land, the subject of such morality tales as the "The Monkey's Paw," by W. W. Jacobs, and Goethe's "The Sorcerer's Apprentice." Wiener feared this would occur when military planners computerized atomic warfare. He also affirmed his long-standing position that cybernetics could illuminate social issues, but severely criticized mathematical attempts to do so.[2]

Although written for a lay audience, *God and Golem* pointed to a research technique—cybernetic modeling—that scientists and engineers had used to apply cybernetics to their individual disciplines, leaving the impression that it was a universal discipline. At the height of the cybernetics moment in the United States, in the 1950s and the 1960s, these researchers believed that cybernetic models—composed of information-feedback loops—could solve major problems in the social sciences, in computer science, in military engineering, and in medicine. In addition to modeling the behavior of humans and social groups, discussed in the previous chapter, researchers used the digital computer to create artificial intelligence, designed weapons systems modeled after living organisms in the new field of bionics, and augmented humans by coupling them with feedback machines as cyborgs. In each of these areas—artificial intelligence, bionics, and cyborgs—scientists and engineers held an almost religious belief in the power of cybernetic models to solve the scientific, technological, and military problems of the Cold War.

### Artificial Intelligence

We have seen that cyberneticians invented numerous humanlike machines, such as Wiener's moth/bedbug, W. Ross Ashby's homeostat, and W. Grey Walter's tortoises, to model human physiology and behavior. This section explores the role of cybernetics and information theory in creating machines to exhibit a higher level of human intelligence by examining the Dartmouth Summer Research Project on Artificial Intelligence. Held at Dartmouth College in Hanover, New Hampshire, in the summer of 1956, and funded by the Rockefeller Foundation, the conference is generally regarded by historians and computer scientists as the "birthplace" of AI. The "new paradigm" of symbolic information processing was represented there by Allen Newell and Herbert Simon's Logic Theorist computer program, and its participants founded early AI programs at MIT, Stanford, and Carnegie-Mellon.[3] These institutions promoted the approach that came to be called "symbolic AI," rather than the earlier method of brain modeling, a neurophysiological approach associated with cybernetics and neural nets.

Of interest is the disciplinary boundary work done by researchers and the Rockefeller Foundation to separate cybernetics (brain modeling) from symbolic information processing during the editing of the volume *Automata Studies* (1956) by Claude Shannon and John McCarthy, a founder of symbolic AI, which initiated the conference, and during the planning, funding, and conduct of the Dartmouth conference. It should come as no surprise that in the cybernetic moment the practice of designing intelligent machines was related to the practice of modeling humans as machines in the social sciences. Indeed, Herbert Simon engaged in both endeavors, which soon came under the rubric of "cognitive science."[4]

In addition to Simon, several other participants at the Dartmouth conference had taken up cybernetics. Oliver Selfridge had proofread Wiener's *Cybernetics* as a graduate student at MIT, and Marvin Minsky, another founder of symbolic AI, had been "enamored" with Warren McCulloch's work on neural nets as an undergraduate at MIT.[5] Minsky later stated that, owing to advances in the digital computer in the mid-1950s, "cybernetics divided, in my view, into three chief avenues": "minimal Self-Organizing Systems" (represented by Ashby); the "Simulation of Human Thought" (represented by Newell and Simon); and "Artificial Intelligence" (represented by Minsky and McCarthy). Newell recalled that cybernetics split with AI over the issues of "symbolic versus continuous [analog] systems" and "psychology versus neurophysiology." The break was not abrupt. "Through the early 1960s," Newell said, "all the researchers concerned with mechanistic approaches to mental functions knew about each other's work and attended the same conferences. It was one big, somewhat chaotic, scientific happening."[6]

As we have seen, Claude Shannon had a strong interest in cybernetics and automata; he attended three Macy conferences on cybernetics and exhibited his maze-running mouse at one meeting. It is no wonder, then, that when Shannon and McCarthy invited authors to contribute to a volume on automata studies in 1953, they wrote to several researchers who had attended the Macy conferences on cybernetics. These included Wiener, Donald MacKay, Grey Walter, John von Neumann, and W. Ross Ashby.[7] They invited other prominent researchers in the theory of automata and neural nets, including Alan Turing; S. C. Kleene at the University of Wisconsin–Madison, who had written a paper on neural nets while at RAND in 1951; and Albert Uttley, who built conditional-probability devices for pattern recognition for the British military.[8] Marvin Minsky, who had also been hired to work with Shannon at Bell Labs in the summer of 1952,[9] agreed to submit something from his Ph.D. dissertation on neural nets and brain modeling. Wiener thought highly of Minsky's work, saying in a 1954 recommendation letter that it went "considerably further than the best work to date of such men as Ashby and Grey Walter."[10]

Turing, MacKay, Ashby, Grey Walter, and Uttley were members of the cybernetics Ratio Club in Britain. Turing had a strong interest in intelligent machines before he wrote a 1950 article on what became known as the "Turing test" for artificial intelligence. In 1946, when he was designing a digital computer at the National Physical Laboratory, Turing wrote to Ashby that he agreed with Ashby's approach of modeling the brain: "In working on the ACE [Automatic Computing Engine] I am more interested in the possibility of producing models of the action of the brain than in the practical applications to computing." In 1953, Turing referred directly to the cybernetic aspect of the automata-studies enterprise when he responded to the invitation from Shannon and McCarthy to contribute a paper to the volume. Turing had to beg off because he had "been working on something quite different for the last two years; viz. the mathematics of morphogenesis, though I expect to get back to cybernetics very shortly."[11]

McCarthy played a substantial role at the beginning of the project. Prospective authors corresponded with McCarthy in care of "Automata Studies, Fine Hall" at Princeton. McCarthy forwarded the papers and correspondence with his comments to Shannon at Bell Labs, even after McCarthy had moved to a temporary academic position at Stanford in 1954. The young McCarthy was not reticent in stating his views on how the papers fit into his vision of an emerging field. MacKay's paper, "The Epistemological Problem for Automata," discussed how automata could symbolically represent the external environment in a self-organizing, adaptive manner rather than in a prescribed way, and he proposed an analog-electronic circuit to illustrate the principle of negative feedback involved in his scheme. McCarthy told Shannon, "I think we should include it [the paper] with whatever changes he wishes to make. I don't think that the eventual solution to this problem will be an extension of his ideas, but I have no very precise grounds for this opinion." After criticizing MacKay for making questionable assumptions using the McCulloch-Pitts neural net model, McCarthy concluded, "Nevertheless, I think the paper has a number of good ideas and also represents a point of view which cannot be dismissed as unfruitful this early in the game."[12]

McCarthy had more trouble with MacKay's colleague W. Ross Ashby. Having achieved prominence in the field of automata with the publication of *Design for a Brain* in 1952, Ashby interpreted the invitation from Shannon and McCarthy as being an automatic acceptance for his paper, "Design of an Intelligence Amplifier." Extending his work on the analog homeostat, Ashby equated "intelligence" with the capacity to make appropriate selections. He presented the general principles for designing a "selection amplifier" that would speed-up the process and reach solutions not realized by the designer, resulting in an "amplification" of intelligence.[13] Ashby told McCarthy, "I am afraid I cannot consent to having it [his paper] 'refereed,' as it is on a subject to which I have devoted a great deal of work, and on which I can now speak with some authority. Frankly, I know of no referee to whose opinion I would defer (though I would, of course, be prepared to answer specific objections)." In contrast, MacKay welcomed referee reports on his paper. McCarthy and Shannon ended up accepting the papers of Ashby and MacKay without refereeing them.[14] The experience most likely did nothing to improve McCarthy's opinion of Ashby, the dean of the English cyberneticians.

Shannon and McCarthy divided the articles into three categories: Finite Automata, Turing Machines, and Synthesis of Automata. The categories resemble the subject headings in Shannon's published review of computers and automata in 1953, and his description of the "theory of automata" in 1958 as a recent, ill-defined "interdisciplinary science," which spanned symbolic logic and Turing machine theory, nonnumerical programming of large-scale computers, and neural nets.[15] Cyberneticians would not have quarreled with the definition.

All in all, McCarthy was disappointed with the papers. He wrote to Shannon in the summer of 1954, "The collection as a whole does not represent great progress but is certainly representative of current thought." McCarthy's expectations were high, and probably unrealistic. He thought the papers by the Ratio Club group (Ashby, MacKay, and Uttley) "contribute heuristic ideas of some value but none lead directly to a solution of the problem of thinking automata." Two papers on neural networks "solve problems of mainly formal interest." Another paper in this group introduced "concepts which may prove useful but does not get far off the ground." He marked "Same comment" for his own paper, and said Shannon's paper was a "short step in the direction of the theory of computing rate." Von Neumann's paper, "Probabilistic Logics and the Synthesis of Reliable Organisms from Unreliable Components," got the best mark: "An important result, perhaps slightly off the main track." Minsky's paper had not yet arrived for comment.[16] Shannon had a better opinion of the papers; earlier, for example, he had praised Ashby's homeostat in print "as a basis for learning machines and brain models."[17]

The tensions over editing *Automata Studies* surface in a letter McCarthy wrote to Shannon in late 1954. McCarthy had written a draft of a preface to the volume. "It consists mainly of a point of view on how the various lines of investigation represented by the included papers may contribute to the eventual design of intelligent machines." The sentence is crossed out by a wavy line in the original letter at the Claude E. Shannon Papers at the Library of Congress, perhaps by Shannon. McCarthy also put some pointed questions to Shannon:

1. Do you agree that the book should be regarded as a step in the direction of the design of intelligent automata? Or is this too presumptuous? Might it surprise the contributors?

2. What about the title 'Towards Intelligent Automata.' Is it too bombastic? Others?"[18]

Apparently Shannon thought McCarthy's interpretation of the volume was "too presumptuous" and the proposed title was "too bombastic," or something along those lines. The preface to the published volume deemphasized the goal of designing "intelligent automata," and spoke instead in terms of the "analytic and synthetic problems" of automata.[19] The title remained the staid one of "Automata Studies." Yet Shannon sympathized with McCarthy's goal, as indicated by his 1953 review of computers and automata. Privately, Shannon told his former high school science teacher in 1952, "My fondest dream is to someday build a machine that really thinks, learns, communicates with humans and manipulates its environment in a fairly sophisticated way."[20] And he did approve the publication of papers by the Ratio Club group, representing the British cybernetics movement, on design principles of intelligent automata. Perhaps what McCarthy viewed as conservatism was Shannon's desire to keep the book's title consistent with the sober tenor of a volume published in the Princeton University Press series, Annals of Mathematics Studies.

McCarthy was not adverse to mentioning cybernetics in the preface. In fact, he listed "Cybernetics" as one of six areas toward which the preface could, perhaps, make "brief bows" (the others were Mathematical logic, Statistics, Communication theory, Neurophysiology, and Telephone engineering).[21] Bows were made toward all of these areas or to their practitioners by name in the preface, except cybernetics.

The period between the completion of *Automata Studies* and the approach to the Rockefeller Foundation to fund a summer seminar on "artificial intelligence" was short. Upon moving to Dartmouth College in February

1955, to take up an assistant professorship in mathematics, McCarthy sent Shannon the final version of his paper for *Automata Studies* and said he was now free to do additional editing of the volume and to visit Shannon at Bell Labs.[22] In early April, McCarthy had an interview with Warren Weaver at the Rockefeller Foundation about the possibility of funding a six-week summer seminar at Dartmouth for about ten participants to work together on brain theory and automata. Weaver recorded in his program officer's diary that McCarthy was confident that Shannon and Nathan Rochester at IBM would attend and that their employers would pay their salaries while they were at Dartmouth. Weaver said he would think it over.[23]

Weaver came to have mixed feelings about the project and told McCarthy that Robert Morison, director of the Biological and Medical Research division of the Rockefeller Foundation, should handle the proposal, probably because it dealt with brain models.[24] Unable to attend the scheduled lunch in mid-June with Morison, McCarthy, and Shannon, whom the young McCarthy had enlisted to promote the project, Weaver expressed his opinion on the matter in a memo to Morison. Weaver felt it was "rather a 'personal' project, in the sense that McC. [McCarthy] and two or three other people would enjoy spending some time together, talking about various aspects of information theory," at someone else's expense. Weaver did not think the foundation should support it on that basis. Weaver had the highest regard for Shannon, whose theory of information Weaver had popularized, and thus worried about a conflict of interest. "I think it is on the whole rather better that I am not going to be at the luncheon. My personal connection with the subject, especially since Claude Shannon is now going to be present, tends to remove the discussion from the objective atmosphere in which we ought to approach it."[25]

Shannon's presence seems to have inclined Morison more favorably toward the proposal. Morison recorded in his diary that McCarthy did almost all of talking, with Shannon only answering questions put to him. "*McCarthy* strikes one as enthusiastic and probably quite able in mathematics," Morison wrote, "but young and a bit naive." Participants suggested for the seminar were McCarthy, Shannon, Rochester, Minsky, von Neumann, Oliver Selfridge, and psychologist Donald Hebb, well-known for his neural net theory of learning. Having funded the cybernetics research of Norbert Wiener and Arturo Rosenblueth in the mid-1940s, Morison apparently wanted to include someone from Wiener's group at MIT in the Dartmouth conference. Morison said that Selfridge, now a group leader at MIT's Lincoln Laboratory working on pattern recognition, "apparently strikes M. [McCarthy] and S. [Shannon] as the soundest and most amenable to group activity of the various younger men who have been associated with *Norbert Wiener*." Morison was alluding to the difficulties Wiener and Rosenblueth had in obtaining consistent work from the erratic, but brilliant, Walter Pitts during the Rockefeller-funded research. Minsky may also have recommended Selfridge, his group leader at Lincoln Lab and with whom he had organized a session on machine intelligence at the Association for Computing Machinery's Western Joint Computer Conference held in Los Angeles in spring 1955.[26]

Two other prominent cybernetics researchers were considered at the luncheon with Morison. Despite their difficulties working with Ashby during the *Automata Studies* venture, McCarthy and Shannon put forward Ashby as "the most attractive foreign name." MacKay, whose research on information theory and an analog version of machine intelligence had been funded by the foundation since 1951, did not fare so well. Even though MacKay's paper was accepted for *Automata Studies*, Morison recorded that "S. [Shannon] worries a good deal about *MacKay's* interest in theological metaphysics," which was evidently a concern that weighed against MacKay's participation in an informal six-week session. Morison suggested including two psychologists: Hans Lukas Teuber at New York University, who had coedited the proceedings of the Macy conferences on cybernetics, and cognitive psychologist Karl Pribram, who directed a laboratory in a mental hospital near Hartford, Connecticut. Morison thought both men had "enough breadth of education and curiosity to have informed themselves, at least fairly well, about cybernetics and mathematical speculations,"[27] one of the few direct references to "cybernetics" in regard to the Dartmouth conference.

Encouraged by the luncheon, McCarthy organized a committee of four to prepare a proposal to fund the conference. McCarthy, Shannon, Minsky, and Rochester, with whom McCarthy had worked at IBM in the summer of 1955, finished their research sections of the proposal by late August, and McCarthy submitted the proposal in early September 1955.[28] As historian Paul Edwards argues, the "still-nascent split between the computer-brain and computer-mind metaphors already appears clearly" in the proposal.[29] Yet the proposal includes both approaches as part of a common enterprise. The introduction, probably written by McCarthy, lists aspects of the "artificial intelligence problem" under the following headings: Automatic Computers, How Can a Computer Be Programmed to Use a Language, Neuron Nets, Theory of the Size of a Calculation, Self-Improvement, Abstractions, and Randomness and Creativity. The section on computer languages, McCarthy's emergent research area, was clearly in the camp of what came to be known as "symbolic AI." The section on neural nets mentioned the research of McCulloch and Pitts; Uttley, who had contributed two articles to *Automata Studies*; and two of the organizers, Minsky and Rochester.[30]

The research statements of the organizers reveal the "nascent split" in more detail. Three of the four organizers were on the brain-modeling side. Shannon proposed to work on one or both of two topics: the "application of information theory concepts to computing machines and brain models," and the "matched environment-brain model approach to automata." The first project extended his previous work on computing with unreliable components, which could apply to computers and brains, if viewed as computers. The second project would focus on mathematizing the model of the brain's "environment." Minsky hoped to extend his dissertation research and complete a model of a trial-and-error learning neural network that could be programmed on a computer. Rochester wanted to continue the research of his group at IBM on programming neural nets on a digital computer to exhibit originality when faced with a new problem by inserting productive randomness into this type of brain modeling. Only McCarthy leaned toward the manipulation of symbols by proposing to "study the relation of language to intelligence." He criticized the (cybernetic) brain-model approach by saying, "It seems clear that the direct application of trial-and-error methods to the relation between sensory data and motor activity will not lead to any very complicated behavior." He thought trial-and-error methods should be applied at a "higher level of abstraction," for example, with an English-like "artificial language which a computer can be programmed to use on problems requiring conjecture and self-reference."[31]

The proposal did not sit well with Morison at the Rockefeller Foundation. He told McCarthy in late September that the "proposal is an unusual one and does not fall easily into our program [in biological and medical research] so I am afraid it will take us a little time before coming to a decision." Not having heard from Morison by early November, McCarthy asked Shannon to approach Weaver and find out about the status of the evaluation. In late November, Morison finally told McCarthy that the Rockefeller Foundation could not grant the full funding of $13,500 for the seminar but could support a smaller effort at $7,500. "I hope you won't feel that we are being overcautious," Morison wrote, "but the general feeling here is that this new field of mathematical models for thought, though very challenging for the long run, is still difficult to grasp very clearly. This suggests a modest gamble for exploring a new approach, but there is a great deal of hesitancy about risking any very substantial amount at this stage."[32]

What is remarkable about Morison's letter is not that the interdisciplinary-minded Rockefeller Foundation hesitated to support a new interdiscipline, but that Morison and Weaver had jointly funded a related project in a grand manner—Wiener's and Rosenblueth's research on mathematical biology, which led to the book *Cybernetics*. Apparently Morison and Weaver did not consider "mathematical models for thought" to be part of the cybernetics project, despite the fact that Wiener himself discussed chess-playing programs in *Cybernetics*, and that McCulloch, who chaired the Macy conferences on cybernetics, and Pitts had created an influential logical model of neural nets, which represented a large part of the McCarthy proposal. It is unclear if the popular interpretation of cybernetics as the science of robots worked for or against the grant proposal. Another contributing factor may have been that the proposal was made only to Morison's division, and not to Weaver's division as well, as had been the case for the cybernetics grant. Morison supported that grant primarily because of the potential of bringing mathematics to bear on experimental neurophysiology, the research area of his friend Arturo Rosenblueth.[33]

Whatever the reason for the Rockefeller Foundation to cut the Dartmouth funding in half, McCarthy and the other organizers acquiesced and said they could get by with $7,500.[34] In deciding whom to invite, McCarthy made what turned out to be a fateful trip to visit Herbert Simon and Allen Newell at Carnegie Institute of Technology in February 1956. By that date, Simon and Newell, working at the Systems Research Laboratory at RAND, had essentially completed the Logic Theorist, one of the earliest software programs in symbolic AI. Newell recalls that he was inspired to work in this direction after hearing Selfridge give a talk on a pattern-recognition program at RAND in 1954.[35]

During his visit with Simon, McCarthy showed him and Newell the Rockefeller proposal for the Dartmouth seminar. Simon wrote to Shannon after the visit, reintroducing himself by reminding Shannon of a RAND report Simon had sent him on a chess-playing program, and enclosing a copy of an article of his that bore on the research Shannon proposed to do that summer at Dartmouth. Then, Simon took the liberty of critiquing Shannon's proposed research. In light of the work he had been doing with Newell, Simon said Shannon's two separate projects—on using information theory to analyze computing with unreliable components, and modeling the brain's simplest environments—were actually one project. "Because of the rapid progress we have made [on the Logic Theorist and a list-processing language] in the last two months," Simon concluded, "we are inclined to reverse the strategy expressed in the last sentence of your proposal—we believe that we can start with some of the most advanced human activities—i.e., proving theorems—and work back to the 'simplest'—e.g., learning nonsense syllables." Newell planned to "spend part of the summer at Dartmouth, and I shall plan at least a week there at the beginning, if this is agreeable to the 'proposers.'"[36]

McCarthy recommended Simon and Newell to the other proposers, and Newell and Simon wrote up their plans for doing research at Dartmouth. They listed four areas of their current research (chess machines, mathematical machines, learning theory, and simple theories) and would work on whichever one seemed more promising at the time of the seminar.[37] Shannon was slow to respond to Simon's challenging letter. Two months later, in late April, he wrote to Simon, apologized for the delay, and admitted he had not done much research on learning machines in the past two years. Shannon was on the committee of a master's student, Trenchard More, at MIT, where Shannon had recently been appointed a visiting professor. More, Shannon related, "has been developing a program quite similar to that you speak of for proving theorems in the propositional calculus." As with the program of Newell and Simon, More's could be simulated by hand. Perhaps to mollify Simon and Newell, Shannon included an announcement of the call for papers for the Symposium on Information Theory, sponsored by the IRE Professional Group on Information Theory, to be held at MIT in September 1956, and suggested "that perhaps some of the work you or Newell have been doing would be suitable for presentation" in a special session on learning machines.[38]

The invitation pleased Simon and Newell, but the implication that a master's student of Shannon's was doing work comparable to theirs irked them enough for Newell to bring up the letter with some passion in an interview conducted thirty-five years later. "So we have a letter [from Shannon]. And in this letter there is some stuff about a guy by the name of Trench Moore [sic], Trench was a student of Shannon's, and in fact Shannon thought that what Trench was doing was infinitely further ahead than what we were doing. So there's this funny little exchange and counterexchange and so forth that relates to what Trench is doing and so forth."[39]

By May 1956, McCarthy had formalized the arrangements for the seminar, which was to be held from mid-June to mid-August at Dartmouth College. McCarthy told the Rockefeller Foundation that the participants for the full period would be himself, Minsky, MacKay, Ray Solomonoff at MIT, John Holland from Rochester's group at IBM, and Julian Bigelow, a founding member of the Macy conference on cybernetics and von Neumann's chief engineer on the computer project at Princeton. Those who planned to attend for two weeks at the beginning of the summer at Dartmouth and two weeks at the end were Shannon, Rochester, and Selfridge. Newell and Simon would attend for two weeks at the beginning. To help compensate for the budget cut made by the Rockefeller Foundation, IBM and Lincoln Lab would pay the salaries of their employees while they attended the seminar. McCarthy hoped the conference would focus on his area of research, telling Morison, "At present it looks as though we shall concentrate on a problem of devising a way of programming a calculator [computer] to form concepts and to form generalizations."[40]

The conference did not work out as planned, neither in regard to who actually attended, nor to the group's focusing on one research project. Several of the participants listed by McCarthy did not attend. MacKay, who was probably added at the request of the Rockefeller Foundation, and whose travel from Britain the foundation was funding from another account, cancelled at the last moment. He decided to stay with his wife, who was having their first child.[41] That removed an outspoken advocate of the cybernetic, combined analog and digital approach to brain modeling from the seminar. Holland at IBM decided not to attend because of previous commitments, a decision he later regretted. Shannon did not make it either and probably sent his student Trench More in his place, most likely to continue the conversation with Simon and Newell. Bigelow, whom Shannon had met at the Macy conferences on cybernetics, does not appear in accounts of the Dartmouth seminar. New attendees included Herb Gerlertner and Arthur Samuel from IBM, and Selfridge. (The ten official participants were thus Gerlertner, McCarthy, Minsky, More, Newell, Samuel, Selfridge, Simon, Solomonoff, and Rochester.)[42] Bernard Widrow, who received his Ph.D. from MIT that summer, and a colleague heard about the conference, drove from Cambridge to Dartmouth on their own initiative, and were welcomed by the group. Widrow, who developed adaptive neural nets in the 1960s, recalls being "fascinated by the things I heard [at Dartmouth] about artificial intelligence. I knew that I was going to dedicate the rest of my life to that subject."[43]

One reason there is uncertainty about who attended the event is that McCarthy did not prepare a report on it for the Rockefeller Foundation. He submitted only a paper written by Solomonoff. In a letter promising to write a report, McCarthy told the foundation that at the session on automata at the IRE Symposium on Information Theory, to be held in September, "all of the speakers [Simon, Newell, Rochester, Minsky, Solomonoff, and McCarthy] will be participants in our summer project (some for only a short time)." Because they missed a deadline, only the papers by Simon and Newell (on the Logic Theorist) and by Rochester and his group (on brain modeling) would appear in the announcement for the symposium and in its proceedings.[44]

Simon recalls some "tough negotiations" about the session. McCarthy wanted to chair the session and summarize the research done at Dartmouth. "And we [Simon and Newell] allowed as how that wasn't going to happen and so poor Walter Rosenblith [a colleague of Wiener's at MIT] . . . who was supposed to chair the session, walked around with us around the M.I.T. campus, we strolled down the Memorial Boulevard and so on, negotiating this. . . . And finally it was agreed that John McCarthy would get up an give a general speech about what went on and then Al would present our work in particular." Simon and Newell were satisfied with the arrangements but felt the symposium's audience was not ready to understand their symbolic approach to artificial intelligence. McCarthy later admitted that Newell and Simon "felt, perhaps quite correctly, that the situation was anomalous."[45]

In later years, the 1956 information theory symposium at MIT came to be seen as the founding event for the new field of cognitive science because of papers given there by linguist Noam Chomsky and psychologist George Miller, in addition to that by Simon and Newell (see chapter 4). Miller also said the event led to the decline in the usage of the term *cybernetics* for this general area of research.[46] Shannon had stopped doing research on automata at this time, and he gave a paper at the 1956 symposium on coding for a noisy channel, which became a classic in the booming field of information theory.[47]

The Rockefeller Foundation was unsure at the time about the success of the Dartmouth conference. It probably liked the fiscal responsibility shown by the organizers and Dartmouth College, which returned about one-fourth of the grant as unspent funds.[48] But the foundation was dissatisfied with the lack of reporting about the seminar. When Selfridge spoke to Morison about the possibility of the foundation's providing travel money for U.S. researchers to attend a conference on artificial intelligence at the National Physical Laboratory in Britain, which was planned for fall 1957, with Selfridge and Minsky as the U.S. organizers, Morison had misgivings. As recorded in his diary, he told Selfridge, "there has been a little feeling here that we were not entirely well informed as to what went on at the conference held in Hanover [New Hampshire] last summer." Yet Morison came away from the interview with an "excellent impression of S. [Selfridge], and there is no doubt about the qualifications of most of the other people who would be invited to the conference." The list included Minsky and McCarthy, as well as such cybernetics researchers as Pitts and Lettvin from McCulloch's lab at MIT, and Ashby and Uttley in Britain. Shannon and MacKay were on the advisory committee. Morison now acknowledged that the "field is an important one and [is] developing very rapidly." But the foundation apparently did not fund the conference.[49]

Held in Teddington, England, in late 1958, the conference on "Mechanisation of Thought Processes" marked a milestone in the history of artificial intelligence.[50] Organized by Uttley in Britain and Selfridge in the United States, the conference gathered together proponents of the main approaches to machine intelligence of the day. Papers were given by Minsky, McCarthy, and Selfridge on symbolic AI; by McCulloch, Uttley, and Frank Rosenblatt of Cornell University on neural nets; and by Ashby, MacKay, and British cybernetician Gordon Pask on non-neural-net self-organizing systems. Minsky alluded to his conversion from neural nets to the symbolic method by noting that his proposed heuristic computer program for solving problems in plane geometry was based on notes he had circulated at the Dartmouth conference.[51]

In 1968, when Minsky credited cybernetics with spawning three branches of machine intelligence, he reduced the meaning of the term *cybernetics*, which had referred to all approaches to machine intelligence before the Dartmouth conference, to brain modeling. He effectively limited the label *cybernetics* to the work conducted on self-organizing systems and neural nets at the Ratio Club in Britain and the Biological Computer Laboratory at the University of Illinois.

### Bionics: From Living Prototypes to Human Augmentation

Not only was the Biological Computer Laboratory (BCL) a center for neural nets, it pioneered in the related area of bionics, a Cold War interdiscipline (a fusion of biology and electronics) that designed machines on the model of humans and other living organisms. In 1965, a popular-science writer praised bionics as an "offshoot science" of cybernetics, one that had a "more apt and readily understood name." Bionics later became known as the "engineering term for working on the idea of cyborgs," and, as an adjective, for a cyborg figure, a fusion of human and machine, as in the Bionic Man.[52] Yet the original purpose of the field was to imitate organic systems in the design of complex electronic systems for the military, to borrow ideas from "living prototypes" to make those electronic systems more reliable, not to create cyborgs.

The beginnings of bionics owe a debt to the research of Warren McCulloch and Walter Pitts on neural nets, Norbert Wiener on cybernetics, and to the BCL. Established in 1958 by Heinz von Foerster, the chief editor of the proceedings of the Macy conferences on cybernetics, the BCL was a small interdisciplinary lab. It was not as well known as MIT's Research Laboratory of Electronics, which McCulloch had joined in 1951. Von Foerster used the term *biological computer* to mean a computer that mimicked the information-processing functions of biological organisms, such as the pattern recognition performed by a frog's eye, the subject of a well-known paper by McCulloch's group at MIT.[53] The BCL and other centers of bionics, such as those at Cornell University, AT&T's Bell Telephone Laboratories, General Electric, and the Radio Corporation of America, sought to build computers from artificial neural nets, not from biological elements.[54]

Von Foerster established the laboratory as a center for cybernetics on the basis of a grant from the Office of Naval Research; it received most of its funding from the air force and some from the National Science Foundation. W. Ross Ashby was funded by the laboratory after he moved from Britain to join the University of Illinois in 1961. McCulloch consulted on the project, and his staff at MIT sent electronic neurons (to make neural nets) to Illinois.[55] Military agencies funded the Illinois lab and other projects in bionics because they thought biological organisms—which had adapted robustly to their environments through evolution—could provide clues on how to solve the reliability problems endemic to the complex electronic systems, such as guided missiles and avionics, used to fight the Cold War.[56]

Major Jack Steele of the air force's Aerospace Medical Division recalls coining the term *bionics* from Greek roots in the late 1950s to mean "using principles derived from living systems in the solution of design problems."[57] Seven hundred scientists and engineers from several disciplines in the Cold War military-industrial-academic complex attended the first Bionics Symposium, held at the Wright-Patterson Air Force Base in Ohio in 1960. Steele and John Keto, chief scientist at the Wright Air Development Division, organized the symposium, McCulloch chaired a technical session, and von Foerster wrote the preface to the conference proceedings. The air force sponsored three more symposia in the 1960s, which popularized bionics as a new area flush with military funding, reported to be $100 million in 1963.[58]

In his keynote address to the first symposium, Keto said bionics aimed to "cross-couple the know-how we have achieved, or are achieving, concerning live prototypes toward the solution of engineering problems." In an encyclopedia article on bionics, von Foerster defined it more extensively as "a new engineering science that in general applies organizational principles of living organisms to the solution of engineering problems. In particular, it considers living organisms as prototypes in dealing with the theory, circuitry, and technology of information-processing electronic components, systems of such components, and compounds of such systems."[59] Thus, the founders of bionics did not view the purpose of this merger of biology and electronics—whose symbol was a mathematical integration sign holding a scalpel at one end and a soldering iron at the other—to be the production of cyborgs, as implied by later interpretations of the word "bionics."[60] The present-day cyborgian meaning of bionics, the technological enhancement of humans to give them superhuman capabilities, dates to the popular television show, *The Six Million Dollar Man* in the mid-1970s.[61]

The contrast between the scientific and popular meanings of *bionics* is evident in one of the experimental projects von Foerster's lab completed in the early 1960s. The Numa-Rete, built in 1961, used a twenty-by-twenty array of photocells connected to a network of artificial neurons to detect edges of two-dimensional convex objects placed over the cells. By summing the differences in the number of edges detected and dividing, Numa-Rete could "count" the number of objects in its field of vision. Von Foerster's lab built the device from elements that resembled biological organisms—electronic neurons—rather than programming a digital computer to simulate perception, the competing method of symbolic artificial intelligence.[62]

Most of the participants at the air force bionics symposia in the 1960s did this type of research. They focused on the theory of neural nets and self-organizing systems, experiments on pattern and speech recognition in animals and machines, and artificial intelligence. Their efforts to create cyborgs (see the next section) existed on the margins of the conferences, in efforts to design prosthetic devices and human augmentations, often to operate weapons systems. At the first bionics symposium in 1960, Keto noted the military promise of bionics, then listed several "humanitarian" uses, similar to those proposed by Wiener in the 1950s: "Prosthetic devices to assist the crippled; aids to the blind to permit them to perform in a more normal fashion; means for restoring man's capabilities that deteriorate with age or due to disease—hearing, seeing and others."[63]

A few instances of this type of research, termed "medical bionics" by the air force, were presented at the symposia. At the third symposium in 1963, researchers at the Stanford Research Institute described a way to present spatial images by tactile means to assist jet pilots dealing with information overload. At the Spacelab company in California, researchers developed a myoelectric servo control system that would enable a pilot to "move his arm to certain positions in a space capsule under heavy g loads." The system operated much like a Russian artificial arm (see below).[64] At the fourth symposium in 1966, a researcher at the Philco Corporation described a joint project with Temple University, funded by the U.S. Vocational Rehabilitation Administration, that used the new technology of integrated circuits to provide pattern recognition of electromyographic (EMG) signals to control a powered prosthetic arm.[65]

More research was conducted on prosthetics and human augmentation than that presented at the bionics symposia. The Stanford Research Institute was working on updated versions of Wiener's projects: a tactile hearing aid and a photocell device for navigation by the blind. The navy funded a so-called amplified man, who would myoelectrically control "powerful mechanical arms and legs, not with levers and switches, but with thoughts."[66] The army funded "giant walking machines" for soldiers, what contractor General Electric called "cybernetic anthropomorphic machines."[67] In the mid-1960s, the National Aeronautics and Space Administration (NASA) and the Atomic Energy Commission funded a study of such "teleoperators," defining *teleoperator* as a "*general purpose, dexterous, cybernetic machine.*" The study noted that GE called the field "*mechanism cybernetics.*"[68] Nonmilitary projects included pacemakers, baropacers to regulate blood pressure, and an artificial arm developed at Case Institute of Technology, which was funded by the U.S. Department of Health, Education, and Welfare.[69]

In 1968, science writer Daniel Halacy briefly described these fusions of humans and machines (these cyborgs) in his popular book, *Bionics*. He defined *bionics* as "the science of machines and systems that work in the manner of living things."[70] The definition was broad enough to include the interpretation of the founders of the field—McCulloch, von Foerster, and Steele—as well as the new view of bionics as human augmentation, the science and engineering of cyborgs.

## Cyborgs: Fusing Humans and Machines

Although humans have created cyborgs for generations, the term itself has a specific Cold War pedigree.[71] In 1960, the same year the word *bionics* was coined at the height of the space race, Manfred Clynes, chief research scientist at the Rockland State psychiatric hospital in New York, introduced the term *cyborg* in a paper presented at a military conference on space medicine, which he coauthored with Nathan S. Kline, director of research at Rockland and a specialist in therapeutic drugs. "For the artificially extended homeostatic control system functioning unconsciously, one of us (Manfred Clynes) has coined the term Cyborg. The Cyborg deliberately incorporates exogenous components extending the self-regulatory control function of the organism in order to adapt it to new environments." Clynes and Kline created the cyborg technique as a means to alter the bodies of astronauts so they could survive the harsh environment of outer space, an alternative to providing an Earth-like environment for space travel.[72]

Clynes and Kline introduced the term as an abbreviation for "cybernetic organism." They used *cybernetic* in the sense defined by Wiener, as an adjective denoting the "entire field of control and communication theory, whether in the machine or in the animal."[73] At first encounter, *cybernetic organism* seems like a misnomer, because all organisms are cybernetic in that they interact with the world through information and feedback control, the key concepts in cybernetics. The usage by Clynes and Kline becomes clearer when we consider the laboratory mouse that they implanted with an osmotic pump. Drugs are injected into the mouse at a biological rate controlled by feedback. The researcher monitors and sets the rate of the pump. The mouse with the implanted pump is thus a *cybernetically extended organism*—an organism extended by means of cybernetic technology—what they called a cyborg.

Rather than being the central concern of cybernetics, as today's field of cyborg studies contends,[74] cyborgs were a minor research area in the cybernetics moment, usually classified under the heading of "medical cybernetics," in the United States and Britain in the 1950s and 1960s. Most cybernetics research focused on the analogy between humans and machines—the main research method of cybernetics—not the fusion of humans and machines—the domain of cyborgs. Most researchers created models of human behavior and humanlike machines, instead of enhancing human capabilities through cyborg engineering.

In addition to bionics, we can find cyborgs in three areas of cybernetics: prosthetics, bioastronautics, and technology policy. Although later writers and film directors often blurred the boundaries between robots and cyborgs—in the first *Terminator* movie, for example, where the Cyberdyne System Model 101, identified as a cyborg in the movie, is a "barely organic" cyborg, "merely a human skin over a complete robot"[75]—I will not discuss early robots in cybernetics in this chapter. The exemplars discussed in earlier chapters—Wiener's moth/bedbug, W. Grey Walter's tortoises, W. Ross Ashby's homeostat, and Claude Shannon's electromechanical mouse Theseus—did not have an organic component and, consequently, would not fall under the (rather broad) scholarly usage of the term *cyborg*. As noted by Katherine Hayles, these early robots were "cybernetic mechanisms," not cyborgs.[76]

Wiener's work on prosthetics was an early area in which a prominent cybernetician combined humans and machines into integrated information systems, what would later be called cyborgs. The hearing glove, described in chapter 3, is a good example of what Hayles calls technical cyborgs; she mentions the device as a prosthesis, but not explicitly as an example of a cyborg. Information is extracted from sound waves in a disembodied form so it can travel across the boundary between the machine (the electrical filters) and the organism (the human hand). In fact, Wiener described the glove's operation in terms of "amount of information," a key concept he and Shannon had independently developed in information theory. Hayles identifies the theory as the site for the scientific disembodiment of information, a prelude to creating electronic cyborgs.[77] Wiener called the hearing aid an "artificial external cortex." This is the type of comment that inspired media theorist Marshall McLuhan, who admired Wiener, a decade later to talk about telecommunications as the artificial nervous systems that humans wear outside of their bodies.[78]

At the same time he was working on the hearing glove, Wiener started thinking about another way to create cyborgs: artificial homeostasis, the external cybernetic control of a homeostatic physiological function in animals. In 1951, he described the recent invention of a "mechanical anesthetist" that automatically regulated the administration of anesthesia to an animal or human based on feedback from an electroencephalogram (EEG). Wiener called it an "artificial chain of homeostasis combining elements in the body and elements outside," and noted that the principle could be applied in other areas, like medicating the heart. He later predicted (correctly) that this form of "artificial homeostasis" would be used to treat diabetics with insulin.[79]

Toward the end of his life, Wiener worked a great deal in these areas, placing prosthetics and artificial homeostasis in the area of "medical cybernetics" and the analysis of brain waves, for example, in the area of "neurocybernetics."[80] In 1965, Ronald Rothchild, a master's student in mechanical engineering at MIT, designed and built an artificial arm controlled by amplified electric potentials from the amputated muscle, electromyographic (EMG) signals. The resulting "Boston Arm" was inspired by Wiener's ideas on the subject in the early 1960s.[81] In 1963, Wiener proposed the idea of implanting a syringe into diabetes patients to give them automatic injections of insulin based on feedback monitoring, which Wiener again referred to as "artificial homeostasis." He may have discussed this type of cyborg in conversations he had with a Lockheed scientist on applying cybernetics to space flight earlier in 1963, or with Manfred Clynes at a control-systems conference in Russia in the summer of 1960, when Clynes was in the midst of his research on cyborgs and bioastronautics.[82] What better way to describe the material cyborg, in fact, than an organism with artificial homeostasis?

The debt of Manfred Clynes and Nathan Kline, the creators of the cyborg technique, to cybernetics is clear. In May 1960, shortly before they delivered their paper to a symposium on the psychophysiological aspects of space flight, held at the air force's School of Aviation Medicine in Texas, a reporter asked Kline how they came up with the cyborg concept. "We were asked to present a paper on drugs for space flight, . . . and this naturally led to a question of how they would be administered. This would have to be done automatically, of course, and this led us to applications of cybernetics to the problem. From this we established a whole new approach based on adapting the man to the environment rather than keeping him in a sort of environment to which he was naturally adapted."[83]

Clynes and Kline proposed that humans could endure the rigors of long space flights, to Mars for example, by becoming cybernetically extended organisms. Like the cyborg mouse, humans would be unconsciously injected with drugs to control their physiological functions—a form of artificial homeostasis—so they could explore the vastness of space without cumbersome space suits and other life-support equipment. Artificial organs would further reduce their physiological needs. Ironically, Clynes and Kline thought that becoming a cyborg in this manner would thus *free* humans from their machines, from all the equipment needed to create an Earth-like environment in space. In a recent interview, Clynes said he did not think that joining humans to machines in this manner would change the nature of being human,[84] the concern of science fiction writers, social scientists, and humanists since the 1960s.

The partner most familiar with cybernetics was Clynes. After receiving a bachelor's degree in physics from the University of Melbourne in 1945, Clynes, an accomplished pianist, took courses on physiological acoustics and psychomotor coordination at the Julliard School of music, where he obtained an M.S. in 1949, then studied the psychology of music on a Fulbright fellowship. He joined Rockland State Hospital in 1956, as chief research scientist in charge of the Dynamic Simulation Laboratory. At Rockland, Clynes specialized in applying computer techniques and feedback theory to understanding homeostatic physiological functions, a field that was becoming known as "biocybernetics." Soon after joining Rockland, he met Warren McCulloch, who had worked at the hospital in the 1930s. A leading figure in cybernetics following the Macy conferences, McCulloch was impressed with Clynes's research; he gave his grant application to the National Science Foundation the highest rating and supported his application for senior membership in the Institute of Radio Engineers. McCulloch was also impressed with how Kline had put the Rockland hospital on the research map after the war. By 1961, Clynes had published almost a dozen papers on the application of control-system theory to physiology, and Kline was well known for his work on psychiatric drugs.[85]

Theirs was a fruitful collaboration for creating the radical idea of the cyborg technique for space medicine, of implanting cybernetic devices into astronauts so they could endure long space flights and explore planets. Clynes and Kline called the optimistic enterprise "participant evolution," and predicted that this human-controlled endeavor would drastically reduce the time it would take natural evolution to adapt humans to the environments of outer space. For them, "The challenge of space travel to mankind is not only to his technological prowess, it is also a spiritual challenge to take an active part in his own biological evolution."[86]

The term *cyborg* and representations of the space cyborg quickly entered popular culture. A few days before the air force symposium at which Clynes and Kline introduced the term in May 1960, the *New York Times* published a layperson's definition of the cyborg in an article about their paper, based on a press release and interviews with the authors. "A cyborg is essentially a man-machine system in which the control mechanisms of the human portion are modified externally by drugs or regulatory devices so that the being can live in an environment different from the normal one."[87] In July, an artist illustrated the futuristic vision of Clynes and Kline for a photo essay in *Life* magazine, nearly a year before the Soviets launched the first human into space (fig. 7). In the drawing, two cyborg astronauts, part-human, part-machine, explore the moon's surface in skin-tight space suits. Their lips sealed, but their eyes open (probably to give them a more human appearance), the cyborgs "breathe" by artificial lungs and communicate through radios activated by voice nerves. An array of tubes on their belts infuse chemicals to control homeostatically their blood pressure, pulse, body temperature, and radiation tolerance.[88]

The illustration seems to come straight out of a science fiction novel. Indeed, it was more futuristic than most contemporary science fiction in the United States, which had depicted cyborglike entities mainly as disembodied brains from the 1930s to the 1950s. One novel published in 1948, *Scanners Live in Vain,* by Cordwainer Smith, did portray entities similar to the cyborgs of Kline and Clynes. In the novel, future humans elect to have their bodies altered as cyborg "scanners" in order to travel in space. The sensory inputs to their brains are bypassed, so they do not feel pain, and are sent instead to a chest "brainbox," which the cyborgs continuously monitor (scan) for their physiological conditions while exploring outer space. Cyborgs are not depicted as specially fitted space explorers again in American science fiction until the mid-1960s.[89]

Although clearly futuristic, the *Life* illustration accurately depicts the technical proposals made by Clynes and Kline in their 1960 symposium paper. The cyborg concept was too drastic, however, for one reader of *Life.*

FIGURE 7. This futuristic drawing of astronauts exploring the moon without spacesuits, published in *Life* magazine, accurately portrays the "cyborg" concept (fusing humans with machines) that Manfred Clynes and Nathan Kline announced in a scientific paper in 1960. From *Life*, July 11, 1960, artwork by Fred Freeman.

A self-identified "technologist" wrote to the editor that he "was profoundly shocked by the inhuman proposal . . . for the manufacture of 'Cyborgs,' artificially de-humanized, mechanized monsters." The editor reassured him and other readers that "Cyborgs would be humans with some organs only temporarily altered or replaced by mechanical devices. On returning to earth the devices would be removed and normal body functions restored."[90]

As radical as these ideas seemed at the time (and perhaps even today), the space-medicine community took them seriously. A trade journal account of the 1960 air force symposium said that most of the participants recommended surrounding astronauts with as much of an Earth-like environment as possible such as breathable air and artificial gravity. But a "minority report filed by several of the experts questioned whether it might not be wiser to change man, making him more adaptable to space conditions as they are." A psychologist suggested using hypnosis; a professor of surgery recommended hypothermia. "The most imaginative alteration in man" was the cyborg concept proposed by Clynes and Kline."[91] Another writer described how the "Cyborg, a man-machine system," would help solve the vexing problem of protecting astronauts from the radiation in outer space. "A servo-mechanism would signal an increase in radiation count, and trigger the administration of anti-radiation drugs."[92]

NASA took notice and funded research on the cyborg technique. The United Aircraft Corporation in Connecticut presented a proposal to the life sciences unit at NASA's Ames Research Center in April 1962.[93] That August the newly formed Division of Biotechnology and Human Research, a branch of a reorganized Office of Advanced Research and Technology at NASA headquarters, signed an eight-month contract with United Aircraft's bioastronautics unit to conduct a study of cyborgs in space.[94] Heading a group of seven researchers, including medical doctors, physiologists, and engineers, director Robert Driscoll issued an interim report in January 1963 and a final report, titled "Engineering Man for Space: The Cyborg Study," that May. The lengthy document presented the results of Phase I of the contract, a feasibility study of five aspects of the cyborg concept: artificial organs, hypothermia, drugs, sensory deprivation, and cardiovascular models. Although the study referenced the symposium paper by Kline and Clynes only once, it explicitly stated their concept of the cyborg technique and its broad implications. "Circumventing the slow process of natural selection by integrating man with machine makes possible the special man with increased functional capabilities. This is the Cyborg, the cybernetically controlled man who functions servomechanistically to cope with environments he does not fully comprehend."[95]

The optimistic goal was introduced in the section on artificial organs (lungs, heart, and kidney). The section concluded, however, that the extensive equipment required to support artificial organs at that time would not permit them to be used in space flights in the near future. "The real significance of research into artificial organs lies in their use as experimental analogs for substitution into test conditions for evaluation *without* risking human life." The report was more optimistic about hypothermia, predicting that the bulky equipment required to support it could be reduced in size and that the process could be automated for space travel within five to fifteen years. More research was needed on drugs to induce hypothermia, as well as using drugs to control the psychophysiology of astronauts. The Cyborg Study also argued that sensory deprivation was an important factor to consider because of the recent experience of astronauts orbiting the earth.[96]

Moving from literature surveys and theoretical speculations to experimental research, UAC built electrical and mechanical models of the human cardiovascular system, which it verified through experiments on animals. The goal was to understand how human physiology fared in simulated space environments, to establish a medical basis that could be used to create the type of cyborgs advocated by Clynes and Kline. In this regard, they referred to Clynes's recent research on the biocybernetics of cardiovascular systems.[97]

For Phase II of the Cyborg Study, which began in May 1963,[98] United Aircraft dropped the areas directly related to building cyborgs (artificial organs, hypothermia, and drugs), proposed to continue their work on space medicine (the biocybernetic modeling of cardiovascular and other systems), and offered to design systems that addressed pressing needs of the space program (ways to overcome sensory deprivation and the observed loss of calcium during space flights). Although the life-support systems contradicted the "cyborg technique" of Clynes and Kline by providing Earth-like environments, the report restated the goals of that technique, albeit in a qualified manner. "Out of the CYBORG program we will be able to understand considerably more about man, his systems and his subsystems. Methods for augmenting and extending his limitations, which will be compatible with the state of the art and the applicability of man in a space mission[,] will be derived from CYBORG in an effort to obtain the maximum integration of man into a man-machine complex."[99]

While the eventual goal of this integration may have been the radically augmented and extended cyborg of Clynes and Kline, the researchers emphasized their plans to conduct long-term research in "Biocybernetics," of creating models to study human physiology in simulated space environments. Modeling resonated well with Wiener's definition of cybernetics. This emphasis is evident in the report's concluding lines. "A significant number of experiments will be performed on animals and man throughout this program to verify the modelling concepts which have evolved from the CYBORG theory," an allusion to Clynes's biocybernetics papers, not to the cyborg paper of Kline and Clynes. "In this way CYBORG will have accomplished its mission by providing a better understanding of the biological design of man and relating the impact of this understanding to compatible hardware systems."[100]

Apparently neither the long-term scientific goal nor the short-term design proposals were enough to continue the Cyborg Study.[101] United Aircraft does not seem to have issued a report on Phase II. It is not clear why the project was discontinued,[102] but NASA was dissatisfied with it. In August 1963, three months into Phase II, United Aircraft submitted a three-page progress report to Frank Voris, director of the Human Research section of NASA's Office of Advanced Research Technologies, detailing three research projects: in "biocybernetics" (intensive experiments on blood-pressure information sent to a dog's brain); mineral metabolism (early stage of human experiments on loss of calcium under immobilization); and sensory deprivation (proposed human experiments on psychological effects).[103]

The report did not satisfy. Two weeks later, Voris asked three researchers in space medicine—at a private laboratory, the Lockheed Company, and Brooks Air Force Base—to review the Cyborg Study. Voris acknowledged that changes in NASA management during the course of the project had resulted in a change of direction for it, then laid out his concerns. "Presently there exists in the minds of some of us a question of whether the company has produced results commensurate with the monies spent. Also, there is some doubt as to the capability of the company to successfully pursue further work under this contract." He asked for their "expert opinions as to whether the NASA should continue to support this effort," and, if so, to what extent.[104]

Despite these problems, Warren McCulloch, who was an adviser to Voris's division, asked NASA for a copy of the cyborg report in December 1964. McCulloch's involvement is not surprising because of his ubiquitous presence in the field of cybernetics in the United States. More specifically, he was a member of the Biocybernetics Committee of the Aerospace Medical Association, which Eugene Konecci, director of NASA's Biotechnology and Human Research Division, chaired. The fact that Konecci, a proponent of biocybernetics, resigned from NASA in 1964 may help to explain the demise of the Cyborg Study.[105] NASA went on to build many cyborgs—notably, by attaching life-support equipment to astronauts, the core of its "manned" space program—but not the type of cyborg proposed by Clynes and Kline.

A striking instance of cyborg imagery exists in the personal correspondence between cyberneticians. In April 1969, Walter Pitts wrote to his former collaborator on the theory of neural nets, Warren McCulloch, who was in the hospital recovering from a heart attack at the age of seventy. "I understand you had a light coronary . . . that you are attached to many sensors connected to panels and alarms continuously monitored by a nurse, & cannot in consequence turn over in bed. No doubt this is cybernetical. But it all makes me most abominably sad."[106] Walter thought he and Warren could perhaps one day draw up their wheelchairs and chat about old times. I interpret Pitts to mean that being cyborglike in this manner was of scientific interest, that it was "cybernetical" and therefore worthy of study. But the human aspect was sad. Pitts died a month later of complications from liver disease, followed by McCulloch in September.[107]

That brings us to the topic of cyborgs as a concern in technology policy, which I'll discuss by comparing two books published in the mid-1960s. In *God and Golem* (1964), Wiener briefly discussed prostheses under the category of the coordination of humans and machines. He thought that a Russian-built artificial arm, which operated from the amputee's EMG signals "really makes use of cybernetical ideas." He praised it as an example of the "construction of systems of a mixed nature, involving both human and mechanical parts." Although Wiener had warned the public for more than a decade about the possible adverse consequences of cybernetics, especially those caused by automatic factories and military applications, and although he had mentioned potential dangers in human augmentation in 1950, he did not warn readers of *God and Golem* about the dangers of a "new engineering of prostheses."[108]

The thrust of a more sensational, journalistic book, Daniel Halacy's *Cyborg—Evolution of the Superman* (1965) was to educate readers about the promises and dangers of the evolution of humans into cyborgs and the cyborg into a "superman." Recognizing that humans have linked themselves with machines for centuries in a cyborglike manner, Halacy worried about a speed-up in this process in the present. "For better or for worse we are committed to what Clynes and Kline have termed 'participant evolution.' Man himself is now an important factor in his own development." Scientists and engineers had lately turned science fiction into fact by creating artificial arms, pacemakers, and remote-controlled drones. Although Halacy imagined a bleak future in which cyborgs warred against humans—the theme of the *Terminator* movie twenty years later—he was not worried about the fate of humans. Because we cannot stop participant evolution, he reasoned, it was best to guide it in humane ways.[109]

The science fiction writer Arthur C. Clarke was even more optimistic in 1962: "I suppose one could call a man in an iron lung a Cyborg, but the concept has far wider implications than this. One day we may be able to enter into temporary unions with any sufficiently sophisticated machines, thus being able not merely to control but to *become* a spaceship or a submarine or a TV network. This could give far more than purely intellectual satisfaction; the thrill that can be obtained from driving a racing car or flying an airplane may be only a pale ghost of the excitement our great-grandchildren may know, when the individual human consciousness is free to roam at will from machine to machine, through all the reaches of sea and sky and space. . . . If this eventually happens—and I have given good reasons for thinking that it must—we have nothing to regret, and certainly nothing to fear."[110] These hopes and fears were amplified and reworked in the 1970s by other science fiction writers, futurists, and academics building a new field called Science, Technology, and Society.[111]

Modeling humans as machines in the social sciences and biology, creating humanlike machines in AI and bionics, and fusing humans and machines as cyborgs mark the height of the cybernetics moment in the United States. From the 1950s to the mid-1960s, these modeling and design practices went far to enact cybernetics as the universal science its proponents claimed it to be. By adopting cybernetic models in a wide variety of fields—psychology, linguistics, management science, political science, anthropology, bionics, and cyborg engineering—researchers demonstrated that cybernetics and information theory could indeed model organisms at all levels, from the cell to society. They helped create the "Machine in Man's Image," as one magazine titled an early review of Wiener's book *Cybernetics*.[112]

# CHAPTER SEVEN

# Cybernetics in Crisis

THE PROGRESS THAT CYBERNETICS MADE toward becoming a universal science in the United States by modeling humans as machines in the 1950s had an unintended consequence: the proliferation of meanings of cybernetics. Biophysicist Walter Rosenblith, one of Norbert Wiener's colleagues at MIT's Research Laboratory of Electronics, captured the tenor of those meanings in a memorial tribute to Wiener published in 1965. "To Wiener and many of his colleagues," Rosenblith wrote, "communication was clearly the cement of the nervous system, of society, of any completely organized structure. There was less than unanimity among them and the scientific community at large as to whether cybernetics was a unifying science, a common basis for thought, a convenient common language for functionally related problems, a set of analogies, or a program. How cybernetics will ultimately be viewed depends upon the scientific fruit it will bear."[1]

Ironically, that ambiguity infuriated many scientists and signaled the end of the cybernetics moment in the United States, as the movement to create a universal science fell apart in the 1960s. In *The Study of Information* (1983), a collection of essays that surveyed the information sciences, information theorist Peter Elias at MIT contrasted the history of cybernetics in the United States with that in the Soviet Union. In America, Elias claimed, "there is no present designation that has widespread professional acceptance and denotes approximately the particular mix of topics and techniques denoted by cybernetics here and by information theory in England in the early 1950s." The biologists and social scientists "who might have called themselves cyberneticists in the early 1950s had mostly returned by the 1960s to their original disciplines" and integrated cybernetic concepts into those fields. "In the Soviet Union, however, cybernetics became an important label just as its use was fading in the West."[2]

Rather than narrate the success story told by most historians of cybernetics, or the rise-and-fall story told by Elias and his contemporaries, I examine the mixed fate of cybernetics in the United States during the 1960s and 1970s. In that period of crisis, American cyberneticians lamented the discrediting of their field as a unifying discipline and its fragmentation into separate fields, attempted to revive its fortunes by establishing new professional societies for it, and reinvented cybernetics as a science of social systems and as a relativist epistemology. The cybernetics moment came to an end in the United States, not with the death of cybernetics, but with its decline, revival, and reinvention.

## Cybernetics Discredited

As we have seen, cybernetics, the new science with the mysterious name, attracted criticism from the very beginning. By the late 1950s, it had become a discredited label among information theorists in England. At Imperial College, Denis Gabor remarked in a 1958 book review that the "term [*cybernetics*] is strongly disliked and carefully avoided in British and American 'quantitative' circles," that is, in the physical sciences, engineering, and mathematics.[3] The following year, Donald MacKay at King's College, London, wrote to Heinz von Foerster at the University of Illinois that he (MacKay) was dismayed that von Foerster, chief editor of the Macy conference proceedings on cybernetics, had lent his name to ARTORGA (the Artificial Organism Research Group), which MacKay considered to be disreputable. "It's for just this kind of reason," wrote MacKay, "that folk like Gabor, [Albert] Uttley [a researcher in automata for the British military], [Colin] Cherry & I are chary of using the word 'Cybernetics' nowadays, and I do hope that the work of someone of your calibre won't lose some of the attention it deserves by this new connection." MacKay was probably incensed over the fact that ARTORGA, which began as an investment club of British cyberneticians, measured its success in becoming a collective "artificial organism" by the group's ability to make money! In a memorial tribute to Wiener, published in 1965, MacKay said, "Initial overselling by some of the devotees whom Wiener had such a gift of inspiring did something to discredit the cybernetic label, and much appearing under that label today does him less than justice."[4] Looking back on a field he had helped to create, British cybernetician W. Grey Walter noted in 1969 that a "peculiar gap between theory and practice is a feature of cybernetics, and may account for the disrepute which has accumulated around the term. So often has a cybernetic analysis merely confirmed or described a familiar phenomenon in biology or engineering—so rarely has a cybernetic theorem predicted a novel effect or explained a mysterious one."[5]

Cybernetics also had its difficulties in the United States. It found a home at only two American laboratories at this time—MIT's Research Laboratory of Electronics, and von Foerster's Biological Computer Laboratory at the University of Illinois—both of which were heavily funded by the military. While the RLE established cybernetics research groups in communications (supporting the work of Wiener and his protégé Yuk-Wing Lee), biophysics (headed by Rosenblith), and neurophysiology (headed by Warren McCulloch), the BCL, as we have seen, focused on bionics and artificial neural nets.

The decline in the status of cybernetics was noted by many American scientists. "As for myself," John McCarthy recalled, "one of the reasons for inventing the term 'artificial intelligence' [during the planning of the Dartmouth Conference on artificial intelligence (AI) in 1956], was to escape association with 'cybernetics.' Its association with analog feedback seemed misguided, and I wished to avoid having either to accept . . . Wiener as a guru or having to argue with him."[6] As noted earlier, physicist John Pierce at Bell Labs drew sharp boundaries in the early 1960s to separate cybernetics from information theory. Few scientists, he said, would lend their names to the glamor field of cybernetics because of the shallowness of its universal claims. Although political scientist Charles Dechert at Purdue University praised cybernetics for its ability to analyze social groups, he noted in 1966, "In the United States, scientists and engineers working in the theory and applications of self-regulation tend to avoid the term cybernetics" in favor of terms that referred to specific fields like bionics.[7] A year later, Jerome Rothstein, a physicist who corresponded regularly with Wiener about information theory, privately wrote to a colleague, "I think I have an appreciation of what cybernetics is likely to accomplish, but also of the kinds of expectations doomed to disappoint. I think there has been too much loose talk and sloppy thinking, almost if pronunciation of the word 'cybernetics' was a magic incantation."[8] These attitudes annoyed Wiener. A colleague recalled that Wiener "repeatedly complained [to him] that Americans did not like to use the word 'cybernetics,' whereas Europeans, East or West, adopted the word more open-mindedly."[9]

Even Warren McCulloch, who chaired the Macy conferences on cybernetics, had become leery of the term. In a 1964 speech to the inaugural dinner of the American Society for Cybernetics (ASC), McCulloch said, "I have never used the word 'cybernetics' until this year, unnecessarily. I have never used it, and I doubt that any of you have either, seriously, until recently. I have never used it—and for one good reason. I'm afraid of a word like 'cybernetics' because of the tendency of people to use it without ever bothering to understand the nature of the ideas inherent [in it], the effort required to achieve forward progress, and the team play necessary between the mathematician, the logician, the electrical engineer, the biologist, the physicist, the chemist, and so on."[10] He thought the ASC would help discipline the proper usage of the word. Surprisingly, McCulloch had used the term sparingly in print before 1964.[11] His remarks capture the many problems associated with the word *cybernetics* at this time.

Two other charter members of the Macy conferences were disappointed in the field they had helped to create. In a keynote address to the ASC's first annual symposium in 1967, Margaret Mead, who coedited the proceedings of the Macy conferences, said that she and other social scientists at the meetings "were impressed by the potential usefulness of a [cybernetic] language sufficiently sophisticated to be used to solve complex human problems, and sufficiently abstract to make it possible to cross disciplinary boundaries. We thought we would go on to real interdisciplinary research, using this language as a medium. Instead, the whole thing fragmented." Mead alluded to the failure of a project in the early 1960s that proposed cybernetics as the basis for cross-disciplinary communication between social scientists in the United States and Russia. Mead attributed the failure to fears of a "cybernetics gap" with the Soviets (see the next section).[12]

Also, in 1968, Julian Bigelow, coauthor with Wiener of a founding article in cybernetics, complained to an interviewer about the decline in the field's scientific status in the United States. Bigelow thought the ASC was a "philosophical split-off group, who don't really know what is going on" in research. "I think the thing which makes me not want to follow [cybernetics today], not want to call myself a cybernetician, is that it has moved away from research. There is nobody doing research in cybernetics in this country." No one was doing the type of interdisciplinary research discussed at the Macy conferences on cybernetics. "The good people don't publish their stuff as cybernetics," Bigelow complained. Instead, U.S. researchers worked in such separate areas as physiology, engineering, instrumentation, information theory, and computers.[13]

Bigelow gave what became a common explanation of the decline in cybernetics as a universal science in the United States: its fragmentation into separate disciplines, most of which also depended on military funding.[14] The result was often to drop the label "cybernetics." This occurred in the early 1960s, for example, with the rise of symbolic AI at the expense of research on neural networks, and with the creation of the interdiscipline of bionics, both of which grew out of cybernetics. In *The Study of Information*, biomedical engineer Murray Eden—former editor in chief of *Information and Control*, which published work in several areas of interest to a broad interpretation of cybernetics—interpreted the fragmentation of cybernetics teleologically: "The notions of cybernetics have permeated many disciplines—computer science, information theory, control theory, pattern recognition, neurophysiology, psychophysics, perceptual psychology, robotics, and the like. Having been integrated into them, cybernetics has performed the function for which it was proposed" by Wiener.[15] The editors of *The Study of Information* endorsed an alternative view, which Eden and other authors in the volume had proposed to explain the "virtual disappearance of cybernetics from the Western academic establishment. According to that interpretation, cybernetics is a part of general systems theory and has been completely absorbed by this much broader scientific research program."[16]

Contemporaries often offered two other explanations for the decline of cybernetics: its association with fantastic claims made in science fiction and its adoption by so-called fringe groups. The science fiction character of cybernetics was singled out as a factor in 1964 by philosopher of science Yehoshua Bar-Hillel at Hebrew University, who had worked in the Wiener-inspired communications group at MIT in the 1950s and had attended a Macy conference on cybernetics. Bar-Hillel thought the "popularity of 'cybernetics' declined rather quickly in the States, probably due to its having been usurped there by overt or covert science fiction."[17] Although many scientists and engineers were fans of science fiction, including Wiener himself, this association was a long-standing problem. As we have seen, in the 1950s, newspapers and magazines reported that cybernetics—interpreted as the science of robots—aimed to mechanize all human endeavor, and controversial groups such as Dianetics and Technocracy publicly aligned themselves with cybernetics.

In 1968, British cybernetician John F. Young at the University of Aston, in Birmingham, warned, "There is a danger that cybernetics will become generally regarded as an up-to-date form of Black Magic, as a sort of twentieth century phrenology. If this unfortunately happens, then one whole field of research, at the border of many disciplines, will be robbed of the support it needs." A couple of years later British cybernetician Michael Apter lamented, "Cybernetics also seemed to attract a lunatic fringe among scientists, particularly those with a penchant for the obscure and a facility for creating neologisms."[18] In his 1983 review of cybernetics, Murray Eden thought Young's prophecy had come true. For evidence, he named some recent books with cybernetics in their titles, including video artist and media activist Paul Ryan's *The Cybernetics of the Sacred* (1974), which dealt with the "cybernetic extension of ourselves through videotape" as a liberating experience.[19]

We can better understand the problematic nature of these associations by drawing on the rhetorical strategy of "legitimacy exchange." In this strategy, as explained by Geof Bowker, an "isolated scientific worker making an outlandish claim could gain rhetorical legitimacy by pointing to support from another field—which in turn referenced the first worker's field to support its claims. The language of cybernetics provided a site where this exchange could occur." As Bowker argues, legitimacy exchange worked well to claim that cybernetics was a universal discipline in the United States in the 1950s and to enable several fields to adopt its concepts. Fred Turner shows that legitimacy exchange also helped spread cybernetics to the counterculture when material on cybernetics was juxtaposed with that on heterogeneous subjects in the pages of Stewart Brand's publications, the *Whole Earth Catalog* and *CoEvolution Quarterly*, in the 1960s and 1970s.[20] But from the point of view of many scientists, engineers, and philosophers, such as MacKay, Eden, and Bar-Hillel, this rhetorical strategy proved disastrous to the scientific status of cybernetics when exchanges were initiated by such groups as Dianetics, ARTORGA, science fiction writers, and video artists, even though respected cyberneticians such as W. Ross Ashby and Heinz von Foerster supported some of these endeavors.[21]

Another problematic association was the embrace of cybernetics by the Soviet Union in the late 1950s, a dramatic reversal of its reputation there as a "reactionary pseudoscience" earlier in the decade. Journal articles and the translation of the first volume of *Cybernetics at the Service of Communism*, edited by Aksel' Berg, chair of the Soviet Council on Cybernetics, in 1962, publicized the high-status of cybernetics in the Soviet Union.[22] Many U.S. scientists and engineers, such as Dechert, Bigelow, Eden, and Elias, contrasted the "fall" of cybernetics in the United States with its "rise" in the Soviet Union. Usually, they juxtaposed the two phenomena, without drawing a causal link between them.[23] Computer scientist M. E. Maron at the RAND Corporation made an indirect link in the late 1960s by stating that the "vagueness about the meaning of 'cybernetics' has been accompanied by the fact" that the Russians interpreted the field broadly. As late as 1997, Peter Elias ridiculed the Soviet enthusiasm for cybernetics by saying that in a 1960s speech praising cybernetics, premier Nikita Khrushchev "sounds almost like an American salesman for a new management fad—cybernetics as, say, reengineering."[24]

Although none of these commentators explicitly identified the Soviet adoption of cybernetics as the reason why it was discredited in the United States, the announcement that cybernetics was the organizing principle of science in the Soviet Union, what Slava Gerovitch has called "cyberspeak," was bound to create suspicions about cybernetics in Cold War America, where loyalty oaths and security-clearance investigations were still common in science and engineering.[25] It probably did not help matters that Norbert Wiener was feted by Russian scientists as the founder of cybernetics during a month-long visit to the Soviet Union in 1960. Gerovitch argues that Wiener's refusal to accept military funding after the atomic bombing of Japan and U.S. fears of a "cybernetics gap" (see the next section) "tinged this field with the red of communism, and set hurdles for federal funding of cybernetics research."[26]

## Revival: The CIA and the American Society for Cybernetics

To revive the reputation of cybernetics, save it from disintegrating, and nurture the field as an umbrella discipline (its meaning in Europe)[27] organizers in the United States created a professional society in 1964 that is still active today: the American Society for Cybernetics. The impetus to create the ASC did not come from the scientific community, but from a covert government patron of science, the Central Intelligence Agency (CIA).

Despite the problem of gaining access to classified documents, scholars have shown that the CIA secretly supported science to combat communism through several avenues. In the 1950s and 1960s, it funded research at universities, such as social science at MIT's Center for International Studies; channeled money through dummy foundations, and often through them to the legitimate Ford, Rockefeller, and Carnegie Foundations, to fund the arts, humanities, and social sciences in the United States and Europe; funded physiological and social science research to develop mind-control drugs in the MKUltra project; and cooperated with professional societies, such as the American Anthropological Association.[28] In-house, the CIA established the Office of Scientific Intelligence in 1948, mainly to collect intelligence on Soviet science. The office, which hired its own scientific staff and let consultant contracts to university researchers, was placed under the CIA's Directorate of Science and Technology in the early 1960s.[29]

The CIA had links to several cybernetics researchers in the United States before helping to establish the American Society for Cybernetics. Jerry Wiesner, the MIT engineer and administrator who had worked with Wiener on communications projects, was a founding member of the Boston Scientific Advisory Panel, established in 1950 to assess Soviet weapons development for the CIA.[30] Frank Fremont-Smith, the Macy Foundation officer who organized its cybernetics conferences, channeled the CIA's funding of research on LSD in the MKULTRA project through the foundation in the early 1950s. He also organized Macy conferences on LSD research, which were attended by CIA technical staff.[31] Warren McCulloch, it will be recalled, was a fervent anticommunist and cold warrior; he did research on biological warfare and helped to establish the military-funded field of bionics and the NASA-supported field of biocybernetics. He also had ties to the CIA.[32]

The CIA had a specific national security goal in mind when it organized the American Society for Cybernetics: to counter the threat posed by a cybernetics gap that supposedly existed between the United States and the Soviet Union. The chief proponent of this view in the CIA was John J. Ford, a Russian expert in the Office of Scientific Intelligence who had been studying Soviet cybernetics since the late 1950s. During a meeting with high officials in President Kennedy's administration in October 1962, a meeting interrupted by news of the detection of Soviet missiles in Cuba, Ford briefed the gathering on what he saw as the "serious threat to the United States and Western Society posed by increasing Soviet commitment to a fundamentally cybernetic strategy in the construction of communism."[33]

Once the Cuban missile crisis eased in November, President Kennedy asked Wiesner, then his science adviser on leave from MIT, to convene a Cybernetics Panel under the President's Science Advisory Committee (PSAC) to investigate the CIA's assessment of Soviet cybernetics. Convening panels of scientists and engineers to write a report was the typical way that PSAC, founded in 1957 in the wake of the Sputnik crisis, formulated its advice to the President.[34] Chaired by Walter Rosenblith, who had worked at MIT with both Wiesner and Wiener,[35] the Cybernetics Panel held its initial meeting at MIT in the spring of 1963. Its membership reflected the interdisciplinary character of cybernetics, coming from such fields as information theory, statistics, computer science, brain research, and cognitive psychology. The other participants in the study, as was the custom in PSAC's two-tiered approach to panels, were appointed for their area expertise.[36] Panel members offered to contact outside experts to write reports in the above fields, as well as in industrial control, operations research, bionics, linguistics, economics, hospital controls, human-machine communications, and information retrieval. The list reflects the broad conception of cybernetics prevalent in the Soviet Union. In keeping with PSAC's practice of working with the CIA, the "Panel requested the staff to check with other government divisions in the CIA particularly, as they relate to communications and computers."[37]

Ford wrote a lengthy report in May 1963 that drew on CIA research to document the concerns he had raised to the Kennedy administration. He was worried that the Soviets had adopted cybernetics as a comprehensive political and scientific program to provide "optimal control of industrial, military, educational, medical, economic, and all the other complex processes essential in the functioning of social systems." He warned that it was this comprehensive "'system' and not the computers or pattern recognition devices which threatens 'to bury' the U.S." The Soviet program was all the more troubling because of the decrepit state of cybernetics in the United States. "Cybernetics, conceived in the West, is no longer a very influential part of the United States intellectual fabric," Ford stated, "and in its absence no other 'concept organizer' for national problem-solving has been formulated which could unify efforts of scientists working in the general area of communication and control, but in compartmentalized fields."[38]

In 1964, however, PSAC's cybernetics panel concluded that there was no cybernetics gap. "Soviet research in nearly all scientific areas reviewed lags the United States by a few years," mainly because of their lack of digital computers. While noting that the Soviets interpreted cybernetics "even more broadly than Wiener" and used it as a motivating slogan, the report did not criticize cybernetics as a communist ideology. Instead, the panel thought the "Soviet program will strengthen Soviet science and technology and make the Soviet economy more rational. It will also bring Soviet thinking in many fields closer to that of the West." The panel recommended overtly monitoring Soviet cybernetics in order to safeguard U.S. interests.[39]

While the Soviets treated cybernetics as a universal discipline,[40] Wiesner's advisory panel did not. These experts viewed it, as did most cybernetics researchers in the United States and Britain, from the point of view of their own research areas, as a loosely connected group of separate disciplines, in which "Western" science, that is, a non-Stalinist style of science, was evident. In the panel's eyes, cybernetics had fragmented in the United States; its revival in the Soviet Union as a universal philosophy was not something to be feared, just to be monitored.

Ford completed his CIA report at the same time. According to Flo Conway and Jim Siegelman, biographers of Norbert Wiener who obtained Ford's CIA reports through the Freedom of Information Act, "Ford acknowledged that, in most practical applications, the Soviets were still well behind the United States, but he was quick to warn against the dangers of Western complacency in the contest he saw as an escalating cybernetics race between the superpowers with global implications."[41] The Chief of Life

Sciences at the CIA thanked Rosenblith for the panel's work in the fall of 1964, saying it would provide the basis to establish a cybernetics program in his area at the agency.[42]

Publicly, Ford took steps to revive cybernetics in the scientific community. In 1964, under the cover of a teaching appointment at American University in Washington, DC, he gave a talk on Soviet cybernetics at a conference on the social implications of cybernetics, organized by political scientist Charles Dechert. That year he helped organize a local Ad Hoc Group on the Larger Cybernetics Problem, in Washington, and also the American Society for Cybernetics to reinvigorate cybernetics in the United States.[43] Ford was the main force behind the ASC in the 1960s, serving as an incorporator, membership chair, and executive director.[44] It was an inauspicious time to embark on the project, because revelations of the CIA's covert funding of academic research and public intellectuals made newspaper headlines in the mid-1960s.[45]

Finding national leadership for the ASC was difficult. Wiener had died in the spring of 1964 and, in any event, he most likely would have opposed CIA sponsorship because he strongly opposed classified research and the military funding of science, so much so that he had been investigated for this by the FBI.[46] Ford and his colleagues began on a high note by inducing ten cybernetics veterans to be Honorary Founders of the Society, including Bigelow, McCulloch, von Foerster, and Fremont-Smith. Fremont-Smith invited Bigelow and McCulloch to give speeches at the ASC's inaugural dinner held in the fall of 1964.[47]

The organizers assembled an impressive gathering for the dinner, including representatives from federal scientific agencies, local universities, and other professional societies. Rosenblith, as chair of the Cybernetics Panel, represented PSAC.[48] Before the dinner, McCulloch had told the ASC's acting president, "I am heartily in favor of the creation of the American Society for Cybernetics, the more so as it is a fitting tribute to Norbert Wiener, whose [recent] death we, who loved him, all deplored."[49]

But McCulloch changed his mind about the ASC after the dinner. He wrote to Ford and two other organizers, "It's no secret that I have been and still am thoroughly disturbed by the way in which the American Society for Cybernetics got off to what I believe is a false start." At the inaugural dinner, he was "particularly upset by a lawyer [whom he did not identify], who was principally concerned with matters of money, land for a building and general control of property. When I found he was in the nucleus of the group, I almost sent in my resignation." A subsequent board meeting further annoyed McCulloch. "This is not the way to create such a society of scientists."

McCulloch agreed to "organize at least half a dozen small groups of cyber-neticians who can give the society a respectable membership. But I will not, on mature consideration, lift my little finger in its behalf—rather my thumb against it—until it cleans house. I know well that the three of you have the courage of your convictions and the good of science at heart, and I trust that you will tackle this problem so that it no longer looks like another prostitu-tion of cybernetics, Wiener's Word, to some get-great-on-the-hay-scales = some money-making device = some operator's paradise." McCulloch thought that Bigelow, a permanent member of the prestigious Institute for Advanced Study at Princeton, having been elected to that post in 1950 while designing the IAS computer, "is quite right in not wanting to be associated with it as it now appears. But he is the only proper first president."[50]

A few days later, Bigelow explained his reservations, which mainly dealt with issues of interdisciplinarity. He worried that "in the U.S.A. Cybernetics does not clearly define an area of disciplinary activity, though it does (some-how or other) *'span'* many disciplines, such as control and regulation theory and systems, computation, applied mathematics, information theory, etc." Where would the ASC draw its membership if cybernetics was not a well-defined discipline? He did not want to open up membership to just anyone interested in the field, "because I believe that the rank-and-file membership for the first few years would [then] become heavily laden with cranks, dil-ettantes, and people who are not scientific in their approach, are misin-formed, and unrealistic."[51] Bigelow and McCulloch were thus apprehensive that the ASC would encourage the unsavory tendencies of the field that cyberneticians had been lamenting since the late 1950s.

Ford called a planning conference of honorary founders, directors, and charter members in the fall of 1965 to save the ASC (and also his efforts to counter the threat of Soviet cybernetics).[52] At the meeting, chaired by Mc-Culloch with Ford as rapporteur, the group agreed that "we are entering a cybernetic era," which required cyberneticians, cybernetic institutes, a cy-bernetics journal, and an American Society for Cybernetics to direct it. But the group could not agree on whether cybernetics was a metascience or not, and it appointed a committee to carry out Bigelow's idea to canvass possible members.[53]

Bigelow, however, did not act, and in the spring of 1966 a frustrated Ford begged McCulloch to reconsider. "Game as you've always been for trying to save lost causes, you may not have given the Society up for dead as yet. If this is the case, would you be willing to make one final resuscitative effort to save it by becoming the first president of the American Society for Cyber-netics?"[54] McCulloch, the old cold warrior, accepted. Von Foerster, who was on the board of directors, was delighted. He pledged his support and that of his "cybernetic friends," and suggested ways to strengthen scientific representation in the ASC.[55] Lawrence Fogel, president of a defense engineering consulting firm in California, agreed to serve as vice president. Fogel shared Ford's interest in the "relative status of East versus West" in cybernetics.[56]

During the next two years, McCulloch and Fogel reorganized the ASC and worked to turn it into what they considered to be a respectable scientific society.[57] One effort was to call on colleagues and former students to join the ASC and set-up local chapters.[58] In the fall of 1967, McCulloch and several other Macy veterans participated in the ASC's first annual symposium, a conference of two dozen invited speakers and other attendees. McCulloch chaired the meeting; Margaret Mead gave a keynote address; and von Foerster edited the proceedings. Bigelow apparently did not attend. If he had, he might have expressed a better opinion of the ASC when he was interviewed in 1968.[59]

The ASC changed leadership in 1969 when McCulloch died and John Ford moved from the CIA into a policy position at the White House.[60] Fogel presided over an emergent, but still struggling, American Society for Cybernetics. It had about four hundred members, two chapters (one in Chicago and one in Washington), had held four annual symposia in the late 1960s with financial support from the National Science Foundation and the National Institutes of Health, and it had awarded medals for papers presented at the symposia, one of which honored Wiener. Several members of the ASC had strong ties with scientific and engineering agencies in the federal government, such as NASA and the National Academy of Sciences, which they called on to send high-level administrators to the symposia.[61] In 1971, the ASC started the *Journal of Cybernetics*, edited by Fogel, which began touting cybernetics as a systems science that could solve the urgent social problems of the time.

### Reinvention: Cybernetics as a Science of Social Systems

In advertising the ability of cybernetics to solve social problems, the American Society for Cybernetics participated in a reinvention of cybernetics that flew in the face of Norbert Wiener's well-known skepticism about applying the mathematics of cybernetics to the social sciences, which he maintained until his death. Undaunted, another new professional society joined the ASC's efforts in this regard: the Institute of Electrical and Electronics Engineers (IEEE) group on Systems Science and Cybernetics (SSC). Established in 1965 by combining the IEEE's Systems Science Committee and its Cybernetics Committee, the SSC combined two strands in engineering: systems science *and* cybernetics. W. D. Rowe, a systems engineer at an electronics defense contractor and the first chair of the SSC, drew some disciplinary boundaries by distinguishing between these overlapping fields in the first issue of the group's *Transactions* in 1965. "Systems Science participants [in the SSC] approach problems from an optimization point of view, i.e., the system is described analytically by a set of cause and effect relationships whose parameters can be varied to optimize a particular measure of effectiveness. Cybernetics participants approach the same problems in terms of models (real or postulated) of natural systems, systems whose variables are not readily describable in analytic terms." By "natural systems," Rowe meant those involving humans: "biological, behavioral, sociological, political, legal, and economic systems." System scientists in the SSC tended to deal with physical systems, such as those in communications, transportation, industry, and the military, often using computer simulations. Rowe thought that the new high-level computer languages enabled cybernetics to make more rapid progress than in the past, especially in AI and bionics.[62]

Systems science, as defined by Rowe, appears to have had the upper hand over cybernetics in the SSC. W. Ross Ashby, who had joined von Foerster's laboratory at the University of Illinois, was an associate editor of the SSC's *Transactions* and published an early paper in that journal on "The Cybernetic Viewpoint."[63] But only three cyberneticians were on the fifteen-member Administrative Committee in 1967. The majority of the group's leadership came from systems science, including several engineers at defense contractors.[64]

Both factions in the SSC dealt with the analysis of social systems at this time. Editorials in the late 1960s increasingly mentioned social issues, and the *Transactions* published a special issue in 1970 on urban and public systems. Containing articles on public safety, environmental pollution, and traffic control, the issue featured a lead article by Jay Forrester at MIT's Sloan Business School. The founder of system dynamics, Forrester reported on the extension of that method, which drew on cybernetics, from industry to urban planning.[65]

The American Society for Cybernetics was even more involved in turning cybernetics into a science for analyzing social systems. The *Journal of Cybernetics* enlisted several influential scientists and engineers to write editorials during its early years. Several wrote about the ability of cybernetics to solve the vexing social problems of the 1970s. Dropping the criticism of cybernetics he had made a decade ago, Denis Gabor now praised the expansion of cybernetics into the social sciences. "With this extension," Gabor wrote, "we are now justified in considering cybernetics as deserving first priority among all the hard sciences. It may have come just in time to harden the regrettably soft social sciences and to save our free industrial society from the twin dangers of drifting into anarchy by its instabilities, or stiffening into a totalitarian system." Gabor thought that a "scientific theory of social systems" such as cybernetics was a "*necessary* condition" to achieve economic and political stability in these turbulent times.[66] Heinz von Foerster, who had been president of a European anthropological foundation in the 1960s, made Gabor's dramatic plea into a responsibility of cyberneticians. "My suggestion is that we apply the *competencies* gained in the hard sciences—and not the method of reduction—to the solution of the hard problems in the soft sciences. . . . I submit that it is precisely *Cybernetics* that interfaces hard competence with the hard problems in the soft sciences." Cybernetics fit this bill because it "has ultimately come to stand for the science of *regulation* in the most general sense."[67]

The new view in the ASC did not originate with Gabor and von Foerster. It arose, instead, in the context of the rise of systems thinking. Howard Brick notes that systems theory "was a watchword of the 1960s." The "broad application of the concept was itself cause for admiration, hope or horror" in academia, government, and the arts in the United States.[68] Although Brick said systems thinking originated with Wiener's cybernetics and biologist Ludwig von Bertalanffy's "general systems theory," the movement had wider roots.

In engineering and the sciences a protean "systems approach" grew out of the need to develop and manage large weapons projects during World War II and the early Cold War. Systems engineering firms applied their management methods to social problems in the late 1950s, when cybernetics began to be identified with systems modeling.[69] Some key players in the broad systems movement were researchers at the RAND Corporation who transferred "systems analysis," which they created to evaluate weapons systems, to President Kennedy's Department of Defense in the early 1960s and then to President Johnson's Great Society programs; economist Kenneth Boulding at the University of Michigan, who extended von Bertalanffy's general systems theory to economics and peace studies; and Jay Forrester, who used his method of system dynamics to model industrial, urban, and world systems on the digital computer.[70] In the mid-1950s, von Bertalanffy and Boulding helped establish the Society for General Systems Research to promote these and other approaches to systems theory.[71]

Cybernetics systems modeling was prevalent in urban planning. Jennifer Light notes that by the early 1960s, journals and conferences in this area "made frequent reference to systems analysis, cybernetics, operations research, and computers," owing in part to RAND, the consulting firm Thompson Ramo Woolridge, and other military contractors moving into the area of "civil systems" when defense spending declined at this time. Urban rioting during the long hot summers in the years from 1965 to 1968 increased the urgency to bring systems thinking, often going under the label of "cybernetics," to bear on an urban crisis now reframed as a "national security crisis."[72]

This interpretation of cybernetics was evident at the ASC's second annual symposium in 1968, which had a section devoted to urban systems,[73] and in the *Journal of Cybernetics*. In addition to the editorials by Gabor and von Foerster, the journal published similar ones by the national science policy elite in the early 1970s. Edward David, Jr., science adviser to President Nixon, even expanded the social role of cybernetics. Because of advances in computers, David reasoned that "we are on the verge of being able to attack the really large problems which are associated with today's world—weather prediction, economic modeling, environmental simulations, and many others." He hoped the "establishment of the Journal will coincide with a new era of success in bringing cybernetic thinking into the mainstream of the nation." It was quite a change of heart for David, who had published a satire on the shallow interdisciplinarity of bionics—a branch of cybernetics—a decade earlier. He acknowledged inspiration for that article from John Pierce, his colleague at Bell Labs and fellow satirist of cybernetics.[74] William McElroy, a biologist and director of the National Science Foundation (NSF), sounded a note of caution in his editorial in 1971. McElroy thought cybernetics could help solve such pressing social problems as overpopulation, pollution, urban decay, and resource depletion, but noted "some profoundly pessimistic projections from the cybernetics community," some of whom, such as Stafford Beer in England, had described the intractable problems of modeling these complex, nonlinear feedback systems. Yet McElroy agreed with David's optimism that an "extraordinary multidisciplinary effort" would create better models, and thought the best way to do this was "through accelerated interaction between the biological and cybernetic disciplines."[75] McElroy's successor at the NSF, H. Guyford Stever, also thought the modeling problems could be solved: "This is one reason why cybernetics, as a science and a way of thinking, has taken on a new and perhaps central role in our lives."[76]

The editorials, in addition to similar ones by the chief scientist at IBM who had inspected computing facilities in the Soviet Union during the cybernetics scare in 1964, and by the dean of engineering at Harvard,[77] mark a public relations coup for the *Journal of Cybernetics*. They indicate that some of the most prominent science policymakers of the day had come to consider cybernetics to be central to bringing science to bear on the vast social problems of the 1970s. It was a far cry from hearing the loud laments of the cyberneticians and the satire from its critics a decade earlier, which still echoed in some quarters.

Despite the common research interests of the American Society for Cybernetics and the IEEE group on Systems Science and Cybernetics, the two organizations had little contact. The Washington chapters of both groups held at least one joint meeting, in 1969, and the national societies cosponsored an International Conference on Cybernetics and Society in 1972, in which the IEEE took the leading role.[78] But the ubiquitous von Foerster and Ross Ashby are the only common links between the two societies that I have been able to identify. The two organizations were essentially independent societies existing on the margins of science and engineering, whose publications consisted of highly mathematical models of automata and of biological and social systems. The public relations coup in science policy did not ensure a thriving ASC. A sociological study concluded in 1972 that cybernetics had a "lack of success in institutional terms" in the United States and Europe, as compared with major sciences like physics, chemistry, and biology.[79] In contrast, several postwar interdisciplines, such as operations research (OR) and materials science, did obtain substantial institutional support in American universities. These fields focused on specific problem areas, a characteristic of interdisciplines that helped them gain academic and research support in this period.[80]

The relationship between cybernetics and other systems sciences was a matter of dispute in the 1960s. In his 1962 presidential address to the Operations Research Society, Merrill Flood, a former researcher at RAND, thought the boundaries were blurred between cybernetics, OR, and other systems approaches. "Among companion developments [to OR] are the very active fields of human engineering, econometrics, cybernetics, information theory, automata theory, management science, systems engineering, statistical decision theory, and various other scientific approaches to the understanding, design, and management of complex man-machine systems. Indeed, there is little hope of distinguishing each of these fields from the others, or from OR, even though each effort has its own characteristic flavor and special group of adherents and supporters."[81] Flood thought a "systems science" was emerging from this intellectual ferment.

Von Bertalanffy did as well, but he and other founders of general systems theory (GST) drew an inclusive boundary around it and claimed that cybernetics was part of their field. In the 1960s, von Bertalanffy, then at the University of Alberta, in Canada, classified cybernetics as one of the postwar fields created to analyze the "organized complexity" of large-scale, human-made systems, a problem Warren Weaver had identified in the late 1940s. Von Bertalanffy critiqued cybernetics for dealing with closed systems, as opposed to his open-systems approach, in which the organism was open to transfers of matter and energy from the environment. To him, cybernetics was a special case of general systems theory, applicable to closed, self-regulatory systems, such as guided missiles and thermostatically regulated furnaces.[82] Boulding, who had talked with Wiener in the 1950s about cybernetics, listed cybernetic systems as third in his hierarchy of the nine levels of systems considered by GST.[83]

Jay Forrester was more accommodating. In the late 1960s and early 1970s, he said his modeling technique, system dynamics, was a "clarification and codification of many ideas that run through cybernetics, servomechanisms theory, psychology and economics," and that it "belongs to the same general subject area as feedback systems, servomechanisms theory, and cybernetics."[84] The link to cybernetics was clear as Forester, a control-system engineer and digital computer pioneer who moved from MIT's Servo-mechanisms Laboratory to its Sloan Business School, scaled up his models of production control in manufacturing companies (1960) to model the growth and decay of cities (1969), and then the world's population and natural resources (1971), a model the Club of Rome used in its (in)famous report *Limits to Growth* (1972). Forrester went beyond previous cybernetic researchers in the social sciences to model highly complex, nonlinear systems having multiple feedback-information loops, whose differential equations were intractable without the aid of the digital computer.[85]

The relationship between cybernetics and the social sciences was rocky in the 1960s. On the positive side, we have seen that several prominent social scientists used cybernetics to model individual human behavior and behavior in groups. Sociologist Talcott Parsons at Harvard, for example, relied heavily on cybernetic control to model the interaction between the various subsystems of his influential structural-functional theory of social action. Recognizing this affinity with their field, the American Society for Cybernetics invited Parsons to give a paper at its first annual symposium.[86]

On the other hand, several sociologists criticized Parsons. In promoting general systems theory as a means to supplant sociological theory in 1967, sociologist Walter Buckley, then at the University of California, Santa Barbara, discounted Parsons's theory as a conservative, stability-seeking, closed systems approach, which he thought Parsons had not developed much from when he derived it in the 1930s and the early 1950s. Buckley, a proponent of cybernetic modeling, thought von Bertalanffy's open system approach overcame those objections by drawing on cybernetics and other "modern" theories to account for change and growth. As noted by sociologist Robert Lilienfeld at the City College of New York, Buckley ignored Parsons's later adoption of cybernetic principles. Harshly critical of what he viewed as the hubris and ideological, technocratic character of all systems theory, including cybernetics, Lilienfeld classed Parsons and Buckley together to illustrate the problems in applying systems theory to the social sciences.[87]

Yet, cybernetic concepts found a secure home in the social sciences in the United States in the 1960s, as it had in biology. In addition to sociology, cybernetic ideas were adopted in political science, cognitive psychology, and structural linguistics.

### Reinvention: Second-Order Cybernetics

In the 1970s, both systems theory in the social sciences and the American Society for Cybernetics fell on hard times. Structuralism, including systems theory, was sharply criticized. In sociology, for example, Parsons's theory was challenged by phenomenology, symbolic interactionism, and ethnomethodology.[88] The ASC cosponsored some meetings with other groups in the 1970s, but it did not hold an annual conference in the second half of the decade. The Society for General Systems Research served as the ASC's temporary home in this period; indeed, two veterans of cybernetics, Margaret Mead and Heinz von Foerster, were president of the parent society in the 1970s.[89]

One result of this turmoil was to provide intellectual and institutional support for a major reinvention of cybernetics by a splinter group in the ASC. Instigated by von Foerster and Chilean biologist Humberto Maturana, the reinvention resulted in a relativist epistemology known as "second-order cybernetics." As with most innovations in early cybernetics, the impetus came from physical and biological scientists, not from social scientists or humanists.

All accounts agree that the starting point for second-order cybernetics was the research done by Maturana and his colleagues on the patterned phenomenon of a frog's vision in Warren McCulloch's laboratory at MIT in the late 1950s. Maturana and von Foerster then developed their parallel, related approaches, aided by many conversations with each other. They talked at Maturana's lab in Chile, at conferences and at von Foerster's Biological Computer Laboratory in Illinois, and on the ill-fated project to create a cybernetics system to manage the Chilean economy in the 1970s.

While working in Chile on color vision, Maturana began studying cognition from the point of view of the observer. He developed a theory of the biology of cognition based on the "circular organization" of a biological system, which he published as a BCL report in 1970. In Chile, Maturana and his student Francisco Varela developed the theory further by creating a language of "autopoiesis" in an essay published in Spanish in 1973. It characterized the self-creating, self-referencing, autonomous systems central to their theory of living systems and their observer-centered epistemology.[90]

During this period von Foerster also began to incorporate the observer into cybernetics while analyzing and building self-organizing systems (mainly artificial neural networks) with funding from the military. He coined the term *second-order cybernetics* in a BCL report of 1974 and described the enterprise at an ASC conference held that year. "I submit that the cybernetics of observed systems we may consider to be first-order cybernetics; while second-order cybernetics is the cybernetics of observing systems." He related the new approach to the idea that cybernetics could solve current social problems, which he had expressed in his editorial in the *Journal of Cybernetics* noted earlier. But now he said that "social cybernetics must be a second-order cybernetics—a *cybernetics of cybernetics*—in order that the observer who enters the system shall be allowed to stipulate his own purpose" and assume responsibility for social and technical decisions.[91] Varela later noted that von Foerster created a "framework for the understanding of cognition. This framework is not so much a fully completed edifice, but rather a clearly shaped space, where the major building lines are established and its access clearly indicated."[92]

In 1980, the work that Maturana and Varela had done to erect and fill-in that edifice became more widely known when they published *Autopoiesis and Cognition: The Realization of the Living*. Based on their earlier writings, the book presented a biological theory of the organization of living systems and discussed the broad implications of its relativist epistemology. Maturana and Varela maintained that all living systems are autopoietic, producing themselves as organizationally closed, autonomous, self-referencing entities, by which they interact as observers in a structural coupling with the environment. The environment triggers an observational reaction (as it did when the laboratory frog noticed darting objects, but not slow-moving objects, in its field of vision). They argued that "perception should not be viewed as a grasping of external reality, but rather as the specification of one."[93]

Second-order cybernetics and autopoiesis are a far cry from the scientific realism held by all early cyberneticians, including a younger von Foerster. It resonates, instead, with the associations between cybernetics and the counterculture, which Murray Eden lamented in 1983. Foremost in that regard is Gregory Bateson's *Steps to an Ecology of Mind* (1971). As noted earlier, Bateson extended cybernetics to create a theory of mind "immanent in the total interconnected social system and planetary ecology," connected though pathways completing a circuit between the individual's mind and the "larger Mind" of God. Bateson's epistemology of cybernetics inspired a large following in the 1970s, including the media utopianism depicted in Paul Ryan's *Cybernetics of the Sacred*, one of the book titles that incensed Eden.[94] As shown by Fred Turner, Bateson became a guru to Stewart Brand and other members of the "cybernetic counterculture," who transformed the military-funded cybernetics into an alternative philosophy with which to fashion a new and more ecologically sustainable society.[95]

Brand's *CoEvolution Quarterly*, modeled on his *Whole Earth Catalog*, covered the new trends in cybernetics, featuring articles by Bateson, its guiding light, and also by the newcomer Varela.[96] In 1975, von Foerster referred readers to an article by Maturana and Varela on autopoiesis and said that a published dialogue between Bateson and Brand came under the second generation of cybernetics, the "cybernetics of observing systems," that is, second-order cybernetics.[97] Several avant-garde musicians, writers, and artists in the United States also appropriated cybernetics as a liberating and guiding force in their work. In addition to Paul Ryan in video art, these included John Cage in experimental music, William Burroughs in Beat poetry, and acid-guru Timothy Leary in "flicker," a system of strobe lights that produced visual effects.[98]

These new cultural meanings of cybernetics spilled over into the sciences and engineering and led to more instances of "legitimacy exchange" that damaged the scientific reputation of cybernetics. Stuart Umpleby, a social scientist at Georgetown University who did his Ph.D. under von Foerster, recalls that many scientists and engineers who heard von Foerster speak at conferences in the 1970s did not like his challenge to scientific realism and thought he was a charlatan.[99] The cybernetics unit in the IEEE, which changed its name to the Society on Systems, Man, and Cybernetics in 1972, acquired a "flaky image" among engineers. C. Richard Johnson, a professor of electrical engineering at Cornell University, recalled that as a graduate student in control engineering in the late 1970s, his friends would say the name of the new society with a hippie inflection as the Society on Systems, *Man*, and Cybernetics.[100]

In the late 1970s, Umpleby helped revive the American Society for Cybernetics, which had ceased holding an annual meeting, by bringing former BCL researchers into the group. They transformed the ASC into an organization to promote second-order cybernetics. When asked recently in an interview, "What is left from the BCL?" which closed when military funding dried up and von Foerster retired in 1973, Umpleby replied, "The American Society for Cybernetics." Umpleby noted that von Foerster and the BCL crowd excluded anyone "who wasn't a second-order cybernetician." The boundary work "was brutal, intolerant," especially toward logical positivists. Umpleby preferred a more eclectic approach and thought the method of von Foerster and Maturana was closed and ungenerative.[101] Yet the second-order approach was fruitful, as several social scientists in the United States and Europe adopted the observer-centered epistemologies of von Foerster and Maturana in the 1980s.[102]

## The Perils of Legitimacy Exchange and Patronage

But the dream that cybernetics would become a universal discipline had collapsed by the 1970s with the crisis in cybernetics and the lack of long-term academic support for it in the United States. Cybernetics survived as a specialty in three professional societies: as second-order cybernetics in the American Society for Cybernetics; as part of general systems theory in the Society for General Systems Research, and as the field dealing with biological and social systems in the IEEE's Society on Systems, Man, and Cybernetics.[103] What explains this remarkable turn of events?

The perils of legitimacy exchange with so-called fringe groups help explain why cybernetics was perceived by physical scientists and engineers to be in crisis. While liberating for artists, musicians, and the counterculture, these exchanges damaged the scientific reputation of cybernetics in the United States.

Patronage patterns illuminate other aspects of the checkered career of cybernetics in the Cold War. Hunter Crowther-Heyck argues that two successive patronage systems, not a unitary system, supported the postwar behavioral and social sciences in the United States. In the first system, which thrived from 1945 to the mid-1960s, military agencies and foundations, as the principal patrons, promoted interdisciplinary research centers over disciplinary departments, and favored a general systems approach. In the second patronage system, which became dominant around 1970, civilian federal agencies such as the National Science Foundation were the principal patrons. They favored disciplinary research that focused on such techniques as computer modeling. This shift, often accompanied by the devolution of research centers back into disciplinary departments, contributed to concerns among social scientists about the fragmentation of their field after 1970.

Crowther-Heyck claims that these patronage patterns also characterize many postwar physical and biological sciences.[104]

The early history of cybernetics fits the first patronage system well. The interpretation of cybernetics as a universal theory of feedback-information flows in adaptive systems was used by social scientists to shape the mathematical, behavioral, and systems orientation of this patronage system. Crowther-Heyck notes that the orientation was partially shaped by "the incredible array of work in sociology, psychology, and anthropology that was influenced by cybernetics."[105]

The early funding of cybernetics also followed that of the first patronage system. Military agencies such as the Joint Services Electronics Program, the Office of Naval Research, and the air force funded research on communications, biophysics, neural nets, and bionics at the two main interdisciplinary centers for early cybernetics in the United States: McCulloch's lab at MIT and von Foerster's lab at Illinois. More widely, NASA funded research in biocybernetics, and the CIA tried to revive cybernetics as a universal discipline. The Rockefeller and Macy Foundations provided crucial support for the development of cybernetics as an interdisciplinary synthesis in the 1940s and 1950s.

In regard to the second patronage system, funding for cybernetics declined when the military, CIA, and foundational support for it dropped dramatically in the 1970s.[106] But there were significant differences between the case of cybernetics and that of the social sciences. The loss of this patronage was nearly fatal for the field. McCulloch's group virtually disbanded after he died in 1969, the BCL closed when military funding declined in 1973, and the ASC was inactive in the 1970s when John Ford moved from the CIA to the White House. The NSF, one of the new patrons of the second system, did not step forward. Although the NSF had funded early ASC symposia, and its leadership had praised cybernetics in the early 1970s, it did not fund large programs under the label of "cybernetics." Instead, the NSF joined the military in funding the disciplines into which cybernetics had fragmented— such as neuroscience, information theory, and control engineering.

The uncertain identity of cybernetics—whether it was an interdiscipline devoted to solving specific problems or a universal discipline that could solve problems in all fields—undoubtedly exacerbated the funding problem as well. One reason was that interdisciplines have traditionally had more difficulty than disciplines in securing institutional support in the United States.[107]

In effect, the struggles to revive and redefine cybernetics in the 1960s and 1970s created two independent yet overlapping subfields: first-order and second-order cybernetics.[108] In regard to patronage and scientific standing, they exist today as diminished interdisciplinary endeavors that have found institutional homes in the United States on the margins of biology, engineering, and the social sciences. First-order cybernetics resides in the IEEE's Society on Systems, Man, and Cybernetics and the Society for General Systems Research; second-order cybernetics in the American Society for Cybernetics. They remind us of the disunity, rather than the unity, of cybernetics in Cold War America.

# CHAPTER EIGHT

# Inventing an Information Age

IN 1986, Princeton sociologist James Beniger published a list of seventy-five "modern societal transformations" that had been identified since 1950. Ranging from the popular to the obscure, the list included the "Computer Revolution" (1962), the "Knowledge Economy" (1962), the "Global Village" (1964), the "New Industrial State" (1967), the "Technectronic Era" (1970), "Compunications" (1971), "Postindustrial Society" (1971), the "Information Revolution" (1974), the "Telematic Society" (1978), the "Computer Age" (1979), the "Microelectronics Revolution" (1980), the "Third Wave" (1980), the "Information Society" (1981), the "Information Age" (1982), and the "Computer State" (1983). Beniger argued that the variety of labels referred to different aspects of a lengthy "control revolution" in the United States. In this transformation, which extended from the late nineteenth century to the 1980s, new technologies and organizations were created to deal with the "crisis of control" caused by modern industrialization. What postwar pundits were trying to capture with their frenzy of labels was simply the latest stage of the Control Revolution, which had been accelerated by microelectronics.[1]

Beniger's list is a time capsule, a reminder that it was not clear in the mid-1980s what label, if any, would emerge to mark the new era. In fact, my survey on Google Books of volumes published in American English shows that the phrases *postindustrial society* and *computer age* were more popular in the 1970s than *information age*, *information society*, and *information revolution*, which came to dominate how we talk about the present era.[2] *Cybernetic age* was slightly more popular than *information age* from the mid-1960s to the mid-1970s. But cybernetics did not make its way onto Beniger's list, even though he used cybernetics to construct his theory of the control revolution.

How do we explain the rise of a popular information discourse in this period and its triumph over a cybernetics discourse? While cyberneticians struggled to survive the crisis in cybernetics in the 1960s and 1970s, enthusiastic academics, policymakers, and business leaders in the United States confidently predicted, theorized, and created the language of an information age. Expanding the purview of the informational metaphor beyond the biological and social sciences, these groups invented an appealing discourse— a techno-revolutionary narrative—which claimed that computers, microelectronics, and satellite communications were creating a new society, a new world based on the processing of information rather than energy. A handful of social scientists and humanists responded with a counternarrative that criticized the concept of an information age, while others spoke in terms of the cybernetic age. But they could not overcome the dominant information discourse in Cold War America.[3]

In the vocabulary of this future-oriented rhetoric at the height of its popularity during the dot-com boom of the 1990s, the *information revolution*, based on *information science* and *information technology*, was transforming industrial society into the *information economy* and the *information society*, the hallmarks of the *information age*. The new way of speaking had become so widespread by the 1980s that British Prime Minister Margaret Thatcher proclaimed 1982 as "Information Technology Year" ("IT-82") to reinvigorate Britain's electronics industry and keep it from falling behind the United States and Japan, the first "information societies." To celebrate IT-82, the British government issued an IT stamp celebrating submarine telegraphy as Britain's pedigree in information technology and commissioned an IT play and an IT ballet.[4] In the United States, the *New York Times* had covered the new narrative emanating from government, business, and academia since the 1960s, using such phrases as *postindustrial society, information society*, and *information revolution*. This rhetoric flourished in the *Times* in the 1980s, when the personal computer was commercialized, and it expanded dramatically in the 1990s, when the privatization of the Internet led to the dot-com boom. Between 1980 and 2000, the use of *information technology* and *information age* in the *Times* grew exponentially at the expense of such terms as *computer age*, whose usage remained fairly constant. My survey of books published in American English confirms these trends.[5] By the end of the millennium, *information* served not only as a keyword in science and engineering, but as a popular, all-purpose label to mark the new technological era, what Fred Turner has called "digital utopianism."[6]

Changes in the meaning of the word *information* are central to understanding this discourse. Its traditional meanings—the "action of informing" and "knowledge communicated concerning some particular fact, subject, or event"—expanded dramatically with the development of electronic computers, cybernetics, and information theory in the United States after World War II. By 1970, discourse communities in academia, government, and business had given a variety of new meanings to *information*, which they usually then attempted to appropriate for their own discipline.[7] For physical scientists, communications engineers, and many social scientists, such as George Miller and Roman Jakobson, *information* was a mathematically defined, nonsemantic quantity related to entropy—what literary critic N. Katherine Hayles has called the disembodiment of information.[8] In the new fields of information science and management science, *information* was defined semantically as the middle term between "data" and "knowledge" in a hierarchy of cognition.[9]

For all of these groups, as well as those in business and policymaking, *information* was what was transmitted, stored, and processed by computers, communications systems, living things, and society.[10] Additionally, all groups defined *information* as being ubiquitous, even though these definitions differed both semantically (e.g., as a form of knowledge or a meaningless string of bits) and materially (e.g., as a commodity or a disembodied pattern transferred across the boundaries between biological and nonbiological entities).[11] By claiming the word *information*, these groups attempted to show that they had the expertise to determine its scientific definition, to reinterpret their own field in terms of information flow, to devise artifacts and systems to solve the "information crisis," and to understand (and thus perhaps control) the essential commodity and technological basis of the future.[12] By adding the adjective *information* to such older keywords as *science, technology, revolution, society,* and *age,* these groups created powerfully resonant tropes that enabled *information* to become the keyword of our time.[13]

This chapter examines the techno-revolutionary narratives and counternarratives of the information age by focusing in the United States from about 1960 to 1990 on the discourses surrounding two of its most popular terms: *information technology* and *information society.* While *information technology* is now ubiquitous and is, by far, the most common information trope, having moved from academia, business, and policy circles to everyday speech, *information society* remains closer to the realm of the social sciences, reflecting its origins as a social theory. Debates surrounding these terms reveal contestations and counternarratives that have been forgotten in the rush to present a mirror-smooth interpretation of the inevitability of the information age. This pervasive technological-determinism—the idea that technology drives history in a prescribed manner[14]—denies the existence of past controversies and alternative ways of conceiving and talking about the present era.

## Information Technology

The genealogy of *information technology* is more contested than one might assume for what has become a matter-of-fact phrase.[15] In the late 1950s, when the enthusiasm for cybernetics and information theory had not yet waned, discourse communities in management and business introduced the phrase *information technology* and gave it two distinct meanings. Management scientists coined the term to refer to a form of knowledge—a set of mathematical techniques that utilized the computer to assist, or even replace, mid-level management. Business groups began to use *information technology* to refer to artifacts and systems. The first usage drew on a common nineteenth-century meaning of the keyword *technology*; the second usage drew on a meaning of *technology* that came into vogue after World War II.[16] By the early 1970s, *information technology* came to be seen as a social force that derived its power from information coursing through computer and communications systems.

The meaning of *information technology* as a management technique led a healthy life during the 1960s, although it largely disappeared in subsequent years. In late 1958, management professors Harold Leavitt of the Carnegie Institute of Technology (now Carnegie-Mellon University) and Thomas Whisler of the University of Chicago predicted in a much-cited article in *Harvard Business Review* that management would be revolutionized during the 1980s by the computer and sophisticated mathematics. This "new technology," they said, referring to data processing, mathematical methods for decision making such as operations research, and artificial intelligence, "does not yet have a single established name. We shall call it *information technology*." Tracing the origins of the proposed field to information theory, cybernetics, and game theory, Leavitt and Whisler expected companies to adopt such a (then) esoteric approach for "its implicit promise to allow the top [management] to control the middle [management] just as Taylorism allowed the middle to control the bottom."[17]

In 1960, as editors of a work on the computer's effects on management, Whisler and his colleague George Shultz, then professor of industrial relations at the University of Chicago, defined *information technology* more specifically than had Leavitt and Whisler in 1958. For Whisler and Schultz, the term included three areas: "(1) the use of mathematical and statistical methods, with or without the aid of electronic computers; (2) the use of computers for mass integrated data processing; and (3) the direct application of computers to decision-making through simulation techniques." They also coined the term *information technologists* to refer to those who applied the techniques of the new discipline, rather than to those who created computer hardware and software.[18]

Management specialists were excited about the proposed discipline of information technology popularized by Leavitt, Whisler, and Shultz. Herbert Simon at the Carnegie Institute of Technology, a colleague of Leavitt's at the university's Graduate School of Industrial Administration, used the term in this manner. So did less-eminent figures such as John Burlingame, a consultant in operations research at General Electric.[19] In 1970, management consultant Edward Tomeski published *The Computer Revolution: The Executive and the New Information Technology*, crediting Simon, Norbert Wiener, Claude Shannon, John von Neumann, and others with creating the elements of an approach to management that was based on information theory, communications, game theory, and feedback systems. The present meaning of *information technology* as artifacts and systems should not blind us to Tomeski's use of the term in the title of his book. For Tomeski, it referred to the computer-based management discipline advocated by Whisler: "*Information technology, as viewed by this book* ... consists of the disciplines of planning, systems design, systems analysis, operations research, and computer programming." The primary product that "information technologists" supplied to "administrators is information—information to facilitate and sharpen the administrator's functions."[20]

Experts in the field of public administration also wrote about the new discipline of *information technology* and its distinctive terminology in the 1960s. Comments were not always favorable. New York urban planner William Levine observed that "information technology, primarily mathematical, seems, at least in the short run, to demand action counter to what human-relations research requires" in regard to participative management. Ida Hoos, an urban planner at the University of California at Berkeley who was skeptical about the current fad of applying aerospace systems analysis to governmental functions, nevertheless used the new terminology to describe that effort.[21] On the other hand, proponents of the technique waxed enthusiastic about its promise. William Gore, professor of government at Indiana University, argued that "though it is of little help in selecting the relevant facts, information technology holds the promise of a very much broader basis of fact in decision-making."[22]

Discourse communities outside management science and public administration adopted this usage of the new phrase in the early 1960s. In 1964, Gilbert Burck, a business writer who thought Leavitt and Whisler's predictions were being fulfilled, proclaimed in *Fortune* magazine, "As the power plant of the new so-called information technology, the computer is steadily raising high management's power to make accurate decisions."[23] In a study of work and leisure published in 1961, University of Michigan political scientist Harold Wilensky referred to the new field—which he identified as "computers, mathematical programming and operations research"—as one that was routinizing administrative jobs.[24]

Despite this attention, the meaning of *information technology* as a management discipline seems to have all but disappeared by 1970, Tomeski's *The Computer Revolution* being perhaps its last vestige. Large numbers of business and government entities adopted ever more powerful data processing systems, but not the full range of mathematical decision-making techniques that Tomeski thought were revolutionizing management.

The change in the meaning of *information technology* is revealed by another management book published in 1970, Thomas Whisler's *Information Technology and Organizational Change*. One of the originators of the management meaning of the term, Whisler now changed direction and defined it as an industrial art—and the artifacts and systems produced by that art: "Information technology is defined here as the technology of sensing, coding, transmitting, translating, and transforming information. More specifically we are interested in the newest elements of technology—the computer and the program written for it, data transmission networks, and sensing and translating devices such as optical scanners." He did not refer to the previous uses of the term that he, Leavitt, and Shultz had promoted in the late 1950s and the early 1960s. Instead, he emphasized the computer element of his former interpretation; and, to place *information technology* in historical perspective, drew on the growing use of the keyword *technology* to refer to the products of an industrial art: "Information technology is as new as the computer and as old as the signal drum and the abacus."[25]

In elevating *information technology* to the category of an industrial art that spanned human history, Whisler added his voice to a discourse that had arisen in the business community in the 1960s. This usage is evident in the writings of John Diebold. United States director of the International Association of Cybernetics, founder of a management consulting firm, and admirer (and occasional critic) of Norbert Wiener, Diebold wrote articles and popular books on the coming age of automation.[26] Despite his profession, he rarely used the term *information technology* to denote a field of management science.[27] For him, it was a discipline, an industrial art, "built upon the twin foundations of theory and of physical advances in electronics, optics, and other related sciences." A 1967 advertisement for the Diebold Group's biweekly newsletter called it a publication for the manager "who is vitally concerned with the business significance of development in computer and information technology." Diebold predicted in 1969 that "Information technology is leading us to the construction of machines that exhibit most of what we have previously meant by 'intelligence'—machines that can truly be said to learn and machines that not only respond intelligently to speech commands but also speak."[28]

Other business writers spoke during the 1960s of *information technology* as an industrial art. In his best-selling book, *The Age of Discontinuity* (1969), Peter Drucker highlighted the recent growth of an "information industry" based on the computer, which, he admitted, was creating some technological unemployment. "But at the same time," Drucker reassured his readers, "the information technology also creates a great many more highly skilled and demanding jobs."[29] That same year, Carl Heyel marveled at the expansion of the new technology. "It is clear that the new information technology—largely either computer based, computer related, or computer influenced—today reaches into all kinds and sizes of enterprises, in every conceivable industry and specialized activity, and with a multiplicity of options from which a user can choose whatever suits his particular needs."[30]

The situation was more complex in management science because of the competing meaning of *information technology* as a management discipline. Thomas Haigh observes that in the 1960s, "systems men," who advocated the reform of management based on the computer, called their field "Management Information Systems" (MIS) instead of adopting Leavitt and Whisler's definition of *information technology*.[31] Indeed, management publications favored the term *information system* until the 1980s, when the use of *information technology* finally became common. Like Diebold and Drucker, however, many management specialists employed *information technology* to mean an industrial art. For example, Robert Head, who admired and criticized MIS, said in 1967: "In seeking to chart a course of action, management men sometimes become understandably confused about just what their systems people are trying to do in the field of information technology."[32]

The discourse of experts in public administration followed a similar trend. The state of the subject during the late 1960s is seen in *Information Technology in a Democracy* (1971), a work edited by political scientist Alan Westin, which grew out of a research project Westin began in 1967 for Harvard's interdisciplinary Program on Technology and Society. In this book,

Westin published descriptions and critiques of the application of information technology by municipalities as well as by the U.S. State Department and the Department of Defense. Although it included an article by Harold Wilensky that used the management-discipline meaning of *information technology*, overall, the book indicated that the primary meaning of the term as used in public administration was that of an industrial art.[33]

In the 1960s, authors in the social sciences and humanities also began to speak of *information technology* in this way, especially when predicting the future. Ulric Neisser, a founder of cognitive psychology who then worked at MIT's time-sharing lab, Project MAC, foresaw a time when students would use widely available computer terminals in the home "to do their homework in every field—from history and Latin to information technology." More skeptically, political scientist Robert Pranger observed: "Disturbed by the potentialities of programmed information technologies, certain moralists, including the founder of cybernetics himself [i.e., Wiener], anxiously envisage mankind's prospects under ultra-rationalized, over-organized regimes."[34] Samuel Miles, a technical writer, defined the term broadly as "the art and science of creating and processing information," in order to identify his field with the more glamorous one of *information technology*.[35] Historians climbed aboard the bandwagon by embracing cliometrics. In 1967, a review of the application of computers to historical research predicted, "Modern information technology, and the automated data archives that this technology facilitates, will allow historians to use evidence of this sort more extensively and effectively, and in a more sophisticated and productive manner."[36]

Practitioners of the new field of "information science"—an amalgam of European documentation, specialized library science, and computer science, and thus a discipline in which a heavy usage of *information technology* could be expected—began to call themselves "information scientists" after the American Documentation Institute changed its name to the American Society for Information Science in 1968. The paradox created by practitioners' continued preference for the phrases *information systems* and *information retrieval* may be explained by their adherence to the ideal of "pure science," in which basic science is viewed as the fount of all new technology. Despite a few exceptions, information scientists thus did little to promote *information technology* as a keyword in the 1960s.[37]

The meaning of *information technology* shifted with the advent of microprocessors, cable television, VCRs, fax machines, computer networks, and increased satellite communications during the 1970s and 1980s.[38] As technologies that processed information became more visible to professionals and the public, the management-science meaning of the term died out, as we have seen, leaving behind two related, dominant meanings: a specific industrial art and the artifacts and systems produced by that art.

The latter meaning was often accompanied by the idea that because *technology* was a social force, *information technology* must be one as well. John Diebold stated this relationship as early as 1962. "This new technology will produce profound change in all human activity wherever information, its communication, and its uses occur. . . . Technology is truly an explosive agent of social change." Building on Wiener's concept of a second industrial revolution based on computerized automation, Diebold thought that the current "technological revolution will run even deeper" than the first one. Like others who deployed the rhetoric of technological determinism in business, he advocated adopting the new technology in order to survive its predicted, sweeping social changes.[39]

While this shift in meaning of *information technology* to artifacts and systems in the 1970s occurred among all of the discourse communities examined in this chapter, the shift was most striking when policy analysts talked about the future. An early example is a 1968 report on the proposed development of the third world written by Lewis Bohn for the Hudson Institute. He explained that "by 'information technology' . . . we mean: TV, radio, motion pictures, teletype, telephone, sound recording equipment, facsimile systems, computers, information storage and retrieval systems, data links, teaching machines, radar, sonar, communication satellites of various kinds, and the like. We would not even exclude new means of printing books or periodicals. We refer not only to existing 'conventional' systems and techniques, but to more advanced technology such as holography, lasers, and light pipes."[40]

This usage was common during the early 1970s when analysts called for political action at the national level. At Stanford University, Edwin Parker included a section titled "Information Technology Policy" in a summary paper on information and society that drew on sociologist Daniel Bell's idea of a postindustrial society. Identifying such "key" information technologies as "cable television, communication satellites, computers, and a cluster of video technologies (tapes, cassettes, cartridges, disks)," Parker recommend that the federal government increase productivity by improving the "infrastructure of information technology." He and a colleague pushed the idea of an "information utility" that would allow the public to access information from home via cable TV. In a third paper published in the early 1970s, Parker drew on the then new field of technology assessment. "The assessment of information technology may be particularly significant because of the potentially far-reaching effects of changes in access to information on redistribution of political and economic power. . . . The sooner we study each new information technology, the greater the chance of being able to use the research results to influence policy in a meaningful way."[41]

Political scientist Nicholas Henry focused on the growing concerns about copyright issues. In the mid-1970s, he advocated changing copyright laws to catch up with such "information technologies" as "cable television, photocopying, and computer-based information storage and retrieval systems." Commenting on the passage of the new copyright act, he remarked in 1977 that with "the advent of new information technologies in the twentieth century, however, both the term and the concept of publication have lost their significance."[42] Like Whisler in 1970, Henry and Bohn placed *information systems* under the more general, and seemingly more powerful, rubric of *information technology*.

Another legal analyst worried about privacy, one of the "enormous long-range social implications [of the] new information technology."[43] Indicative of this trend was the establishment in 1973 of Harvard's Program on Information Technologies and Public Policy. Computer scientist and linguist Anthony Oettinger, who had criticized computers in education in an earlier project funded by Harvard's Program on Technology and Society, directed the new program.[44]

Diebold's techno-revolutionary language often accompanied the artifactual meaning of *information technology*. Parker spoke of a "revolution in information technology" in 1974.[45] That same year, Australian economist Donald Lamberton edited a special issue of the *Annals of the American Academy of Political and Social Science* titled "The Information Revolution," noting that in the 1960s, attention was "focussed on the new information technologies—for example, computers and satellites—which seemed to symbolize the movement of society into a new industrial revolution: the information revolution." He also quoted Karl Marx's technologically determinist aphorism about the windmill producing the feudal lord and the steam engine, the industrial capitalist: "Today, we are attempting to analyze the beginnings of the information revolution. . . . We seek to know what kind of society is being created by the computer, the satellite, television and a host of other devices to which we refer collectively as modern information technology."[46]

Policy analysts who used *information technology* to mean an industrial art also spoke in revolutionary terms. Cornell political scientist Ted Lowi began a 1975 paper by quoting from Vonnegut's *Player Piano*, which references Wiener's claim of an emerging "second industrial revolution." "There can no longer be any question that the industrial nations of the world are producing another technological revolution of historic importance," Lowi continued, adding that the "trigger is the revolution in information technology."[47] In an article on war and peace in the information age, systems analyst Anatol Rapoport at the University of Toronto spoke of a "second Industrial Revolution—as Wiener once called the new development now called the information revolution." "The possible positive contributions of the information revolution in alleviating other forms of human conflict," Rapoport said, "depend upon the uses to which information technology will be put."[48]

The practice of combining the revolutionary meaning of *information* with the technologically determinist, artifactual meaning of *technology* in the compound term *information technology* intensified during the 1980s when the phrase came into widespread use. The number of English-language books with *information technology* in their titles increased dramatically at this time, as did references to the term in the *New York Times*.[49] One trend in the latter period was the identification of *information technology* with the rapid growth of microelectronics.[50]

With this history in mind, we can understand how Margaret Thatcher drew on a discourse that had been building up during the 1970s when she proclaimed 1982 as "IT-82"—"Information Technology Year"—in Great Britain. At the same time, she contributed to this narrative by popularizing the acronym *IT*, which was quickly picked up by academics. British operations researcher Alec Lee used it extensively in 1983: his article comparing the electronics industries in Japan and the United Kingdom was confidently titled "The Age of Information Technology."[51]

Similarly, journalist Tom Forester called his 1985 anthology of articles *The Information Technology Revolution*, thus indicating the growing popularity of the social-force meaning of *information technology*. Forester drew on an older meaning of the term in the introduction when he said that "Information technology in its strictest sense is the new science of collecting, storing, processing, and transmitting information." But in the book itself, he applied the term to artifacts and systems.[52] Several of the articles in the book used the term in this manner—most often as *information technologies*—when discussing education, business applications, development of the third world, and impacts of intelligent machines on the workforce, sexual roles, and lifestyles.[53]

The debates over the meaning of *information technology* from the late 1950s to the 1970s indicate what was at stake for professionals in this discourse. Management scientists drew on cybernetics and quantitative concepts of information to craft a new discipline, based on related mathematical techniques and the digital computer, that they called *information technology*. Business groups, social scientists, and policy analysts took a parallel discursive path, combining the new meanings of *technology* and *information* to create a different meaning of the term *information technology*: an industrial art that produces artifacts and systems having the power of a social force. In using this phrase in lieu of, or in addition to, such alternatives as *information systems* and *information retrieval systems*, professional groups elevated *information technology* to a general historical category that included these rivals under it as the latest examples of *information technology*.

The fact that *technology* and *information* had independently gained the meaning of a social force strengthened this aspect of the new phrase. By the early 1980s, *information technology* was widely viewed as unstoppable. It was also seen as something that had to be cultivated if the United States and Europe were to survive the economic threat of Japan; had to be regulated if the rights of privacy and free speech were to be protected; and had to be mastered if a major overhaul of corporate management was to be prevented. Condensing *information technology* to *IT* during the 1980s made the keyword even more abstract, to the point where it became even more autonomous in everyday speech. In many cases, this techno-revolutionary narrative signaled that technology was out of control. In making this claim, business consultants, policy analysts, and social scientists could then propose their own pet projects—whether management systems, public policies, or social theories—to solve such perennial problems as the "information crisis."[54]

## Information Society

The phrase *Information society* has a much different genealogy from that of *information technology*. While management scientists, business entrepreneurs, and policymakers introduced the latter term and revised its meaning from that of knowledge to a social force, social scientists debated whether or not a new society was being created by the new technology and, if so, what to call it. *Information society* was not their first choice. They proposed competing social theories with their own labels to describe the predicted social transformations that would occur with the proliferation of computers and communications.

The origins of the phrase *information society* were also disputed. James Beniger claimed in 1986 that "one major result of the Control Revolution had been the emergence of the so-called Information Society," a concept he dated to economist Fritz Machlup's research on the "knowledge economy" in the early 1960s.[55] In contrast, Dutch futurist Michael Marien claimed in

1983 that the "term 'information society' has only been used for about a decade, superseding the less specific label 'service society,' and the even more ambiguous 'post-industrial society.' Information Society was apparently first used in Japan in the late 1960s. . . . Forerunners to 'information society' include the terms 'age of cybernation' (used widely in various forms during the 1960s), 'electronic age' and 'age of information' (both proposed by Marshall McLuhan in 1964), 'knowledge society,' described by Peter Drucker in 1969, and the ungainly 'technetronic society,' suggested by Zbigniew Brzezinski in 1970."[56]

Beniger correctly attributed the economic interpretation of an information society to Machlup, but Marien correctly said that the term itself was introduced into English in the 1970s,[57] most likely from Japan. Yet Marien and Beniger do an injustice to the complex history of the information-society concept by referring to "forerunners" and subsuming the diversity of its discourse under the rubric of the control revolution. This section examines a more contested genealogy by considering talk of a second-industrial revolution, cybernation, and postindustrial society in the 1950s and 1960s, and the discourse about an information society and an information economy in the 1970s.

Norbert Wiener started a debate about the consequences of a second industrial revolution in *Cybernetics* (1948) and *The Human Use of Human Beings* (1950). Although he implied that the transformation of information was at the heart of this revolution—an idea made explicit by a science writer in 1949—newspapers and magazines focused on Wiener's warnings that the automatic factory would cause vast technological unemployment. They ignored the information-processing aspect of automation. The term *second industrial revolution* thus became a synonym for computer-controlled automation. By 1960, a vice president at Michigan Bell Telephone Company could tell a national conference on banking, "It is one of the cliches of our literature that automation is accounting for the phenomenon known as the second Industrial Revolution."[58] Cliché though it was, talk of a second industrial revolution was popular in circles related to technology and the law, the social sciences and humanities, the spread of computers, and other areas during the 1960s.[59]

Increasingly, however, social theorists adopted the term *cybernation* in place of *second industrial revolution*. Drawing on Wiener's work, social psychologist Donald Michael coined *cybernation* in a widely read report published by the Center for the Study of Democratic Institutions in 1962. He coined the term "in order to eliminate the awkwardness of repeating the words 'automation' and 'computers' each time we wish to refer to both at the same time, and in order to avoid the semantic difficulties involved in using the one term or the other to mean both ends of the continuum." Under the label *cybernation*, Michael referred to automation in general, computerized automation, and any use of the computer to replace labor, a broad usage that shaped postwar debates about automation and technological unemployment in the United States.[60]

Media guru Marshall McLuhan quickly picked up the new vocabulary, but he did not share the economic concerns animating Michael and other social scientists. McLuhan drew on cybernetics and information theory in *Understanding Media* (1964) to analyze the cultural effects of information flow in modern media in terms of his famous idea that the "medium is the message" (that a medium's ability to extend human senses, not the content of the medium itself, conditions human life). "We live today in the Age of Information and of Communication," McLuhan wrote, "because electric media instantly and constantly create a total field of interacting events in which all men participate." In the new age, information circulates instantly in the human-made "nervous information system" of electrical and electronic communications (by means of telephones, radios, televisions, and satellites) that humans now wear outside their bodies, creating a Global Village of tribalism and sensory numbness. McLuhan did not ignore economics. He claimed that as cybernation "takes hold, it becomes obvious that *information* is the crucial commodity, and that solid products are merely incidental to information movement."[61] For McLuhan, however, the key transformations were cultural, not economic. In a 1966 article, "Cybernation and Culture," he referred to cybernation as the latest information environment. "Cybernation seems to be taking us out of the visual world of classified data back into the tribal world of integral patterns and corporate awareness."[62]

The New Left linked cybernation with ideas about a postindustrial society. In 1964, the "Ad Hoc Committee on the Triple Revolution," consisting of such notable activists as Erich Fromm, Todd Gitlin, Michael Harrington, Gunnar Myrdal, and Robert Heilbroner, issued a manifesto declaring that "three separate and mutually reinforcing revolutions are taking place," the "Cybernation Revolution," the "Weaponry Revolution" in nuclear arms, and the "Human Rights Revolution." Focusing on cybernation, the manifesto proposed making a guaranteed income a right in order to distribute fairly the wealth produced by cybernation, which broke the connection between jobs and income. Although the authors used the term *cybernated society*, rather than *postindustrial society*, they alluded to a tenet of the latter theory by saying that the new era's "principles of organization are as different from those of the industrial era as those of the industrial era were different from the agricultural."[63]

The manifesto of the Triple Revolution is a key document supporting Howard Brick's argument that the "idea of postindustrial society was a joint project, or shared discourse, of liberals [such as sociologists David Riesman and Daniel Bell] and the New Left in the early and middle 1960s." The shared discourse lasted about a decade, from 1958, when Riesman introduced the phrase *postindustrial society*, to 1967, when Bell developed a mature theory of it and the New Left had become disillusioned with the concept. In this period—before Bell popularized the phrase *postindustrial society* in the early 1970s—proponents did not base the concept solely on economics and technology, but on the principles of fraternity and the abundance generated by cybernation, from which "new forms of community emerged as counterweights to market based forms of organization."[64]

A decade in the making, Bell's tome, *The Coming of Post-Industrial Society: A Venture in Social Forecasting* (1973), became a key text in framing the later discourse of an "information society." Bell introduced his concept of postindustrial society at a 1962 conference on technology and social change, chaired by Heilbroner, a signer of the Triple Revolution manifesto, and further developed it in the mid-1960s as head of the Commission on the Year 2000, sponsored by the American Academy of Arts and Sciences.[65] In his earliest publications on the topic, Bell disagreed with the New Left's insistence on politically controlling cybernation; he posited theoretical knowledge instead of technology as the lynchpin of the postindustrial society.[66]

*The Coming of Post-Industrial Society* presents a comprehensive theory of postindustrialism. By dating the birth of postindustrial society to the first decade after World War II and predicting that it would be realized in the first decade of the new millennium,[67] the book is, in many ways, Bell's contribution to the Commission on the Year 2000. Employing the methodology of a macrosociology based on axial principles and structures, Bell divides society into three parts: the polity; culture; and the social structure, which comprised the economy, technology, and the occupational system. Each part is ruled by an axial principle: participation for the polity; self-fulfillment and self-enhancement for culture; and economizing for the social structure. The concept of postindustrial society is a social forecast, one that "deals primarily with changes *in the social structure*, the way in which the economy is being transformed and the occupational system reworked, and with the new relations between theory and empiricism, particularly science and technology." Rejecting economic and technological determinism, Bell does not claim that changes in the social structure determined changes in politics or culture. "Rather, the changes in social structure pose *questions* for the rest of society."[68]

Five dimensions make up Bell's postindustrial society: the economic sector (the change from a goods-producing to a service economy); the occupation distribution (the preeminence of the professional and technical class); the axial principle (the centrality of theoretical knowledge as the source of innovation and of policy formulation for the society); the future orientation (the control of technology and technology assessment); and decision making (the creation of a new "intellectual technology").[69]

Where does "information" fit into this elaborate scheme? Theoretical knowledge and intellectual technology seem more important than the capacity of computers and communications systems to process and transmit bits of information. One clue can be found in the middle of the book, where Bell identifies the technology of postindustrial society as "Information." In contrast, the technology of preindustrial society is "Raw materials," and that of industrial society is "Energy."[70] The evolution from energy to information invokes Wiener's concept of the "second industrial revolution," but Bell translates the idea into the language of game theory. The design of a "post-industrial society is a 'game between persons' in which an 'intellectual technology,' *based on information*, rises alongside of machine technology."[71] In the postindustrial service economy (the game between persons), "what counts is not raw muscle power, or energy, but information. The central person is the professional. . . . Information becomes a central resource, and within organizations a source of power."[72]

In this scheme, Bell relates information to intellectual technology in a qualitative manner rather than with a quantitative measure such as the amount of data processed by computer and communications systems. Bell defines "intellectual technology" as the "substitution of algorithms (problem-solving rules) for intuitive judgments" in the decision-making dimension of postindustrial society, and contrasts it with "social technology" (e.g., the organization of work) and "machine technology" (e.g., artifacts or systems). Often embodied in computer systems, intellectual technology arose, according to Bell, from applying information theory, cybernetics, decision theory, and game theory to solve the "major intellectual and sociological problems of the post-industrial society." Bell identifies those problems with the "organized complexity" of human-made systems, which Warren Weaver had identified in the late 1940s. Cybernetics also helped give birth to postindustrial society. "If the atom bomb proved the power of pure physics," Bell wrote, "the combination of the computer and cybernetics has opened the way to a new 'social physics'—a set of techniques, through control and communications theory, to construct a *tableau entière* for the arrangement of decisions and choices."[73]

Although Bell did not attend the Macy conferences on cybernetics, he had interacted with researchers in cybernetics and information theory in the 1960s, as chair of the Commission on the Year 2000. Lawrence Frank, an organizer of the Macy conferences, suggested to the American Academy of Arts and Sciences that it establish the commission. Frank served with Bell on the planning group for the commission and participated in its sessions. He wrote a paper for the commission, in which he argued that Ludwig von Bertalanffy's general systems theory and Wiener's cybernetics should inform the political theory needed for the year 2000.[74] The commission also included Macy veteran Margaret Mead; Karl Deutsch and George Miller, who had applied cybernetics and information theory to political theory and psychology; and information theorist John Pierce, who had criticized cybernetics. Although they did not mention cybernetics in their published papers or during the proceedings of the commission, Pierce did discuss Shannon's theory of information in his paper.[75]

Although the Commission on the Year 2000 probably helped to convince Bell that cybernetics was an element of the theoretical knowledge forming the postindustrial society, he criticized its application to the social sciences. As chair of the commission, Bell dutifully mentioned "cybernetics models" as one of the methodologies advocated by a commission member (Lawrence Frank) to create social forecasts. Yet he downplayed the approach: "There are few such large-scale social models in existence, though Soviet economists and mathematicians are now drafting such cybernetic models for the Soviet economy."[76] He was more critical of cybernetics in *The Coming of Post-Industrial Society*, where he faulted sociologist Amitai Etzioni for analyzing postindustrialism in terms of consciousness and cybernetics: "Cybernetic models were necessarily closed and mechanistic, while the conscious control of humans and nature implied an open system."[77]

In the book, Bell briefly compared his conceptual scheme to other theories, usually to dismiss them. He criticized his former colleague Zbigniew Brzezinski's idea of the "technetronic society" for its technological determinism, which elevated technology over theoretical knowledge as the cause of social change. He criticized the depiction of postindustrial society by Herman Kahn and Anthony Wiener of the Hudson Institute, which they had presented to the Commission on the Year 2000, for dealing almost exclusively with economics. He thought the Triple Revolution pinned its hopes on unrealistic predictions of the bounty of goods cybernation could produce.[78] Bell took more pains to distinguish his work from the pioneering efforts of economist Fritz Machlup to measure the production and distribution of knowledge. Machlup estimated that it was 29 percent of U.S. gross national product (GNP) in 1958. While Machlup defined knowledge broadly to include intellectual and everyday forms of unwanted knowledge, Bell narrowed his definition to "that which is objectively known, an *intellectual property*." Under this definition, "Any meaningful figure about the 'knowledge society' would be much smaller" than Machlup's estimate.[79]

The question of what to name the new era was important to Bell. He had been asked "why I have called this speculative concept the 'post-industrial' society, rather than the knowledge society, or the information society, or the professional society, all of which are somewhat apt in describing salient aspects of what is emerging." He admitted that the postindustrial society "is a knowledge society in a double sense." Knowledge was the main source of innovation, and the percentage of GNP devoted to its production and distribution was large.[80]

Bell did not cite the book that popularized the phrase *knowledge society*, Peter Drucker's *The Age of Discontinuity* (1969). Drucker argued that "knowledge has become the central resource of modern society." He called his era the "age of discontinuity" because, up to that time, the economy had developed in a continuous manner from late-nineteenth-century industries that relied on experience rather than knowledge. In the book, Drucker liberally used such terms as *knowledge society, knowledge worker, knowledge economy, knowledge industry* (which he attributed to Machlup), and computer-based *information industry*. Unlike Bell, Drucker did not propose a social theory or an extensive social forecast, nor did he talk about a postindustrial society, even though he cited the paper that Kahn and Wiener had written for Bell's Commission on the Year 2000. Drucker focused instead on the role of knowledge as a "factor of production" in the present and the creation of new industries (rather than an expanded service economy) in the future.[81]

Bell clarified his decision to use the term *postindustrial* in the 1976 foreword to *The Coming of Post-Industrial Society*: "I rejected the temptation to label these emergent features as the 'service society' or the 'information society' or the 'knowledge society,' even though all these elements are present, since such terms are only partial, or they seek to catch a fashionable wind and twist it for modish purposes."[82]

As Bell most likely knew, the fashionable wind blew from Japan. In the 1960s, Japanese journalists, social critics, and futurists began speaking about *joho shakai* (information society) and *johoka* (informationalization) and put the terms into wide circulation. The "information society fad" in Japan was intensified by the concept of a postindustrial society, after Japanese futurists translated articles presented at the Commission on the Year 2000 in 1965 and 1966, and after a joint United States–Japan Symposium, "Perspectives on Postindustrial Society," was held in Tokyo in 1968. Bell participated in the symposium by telephone from the United States. Politicians in the National Diet of Japan began talking about the *johoka shakai* (informationized society) in 1969.[83]

This discourse gained more currency when Yoneji Masuda, chief investigator of the Committee on Computerization at the nonprofit Japanese Computer Development Institute, submitted an information society development plan to the government in 1972. Stating matter of factly that advanced countries were "steadily shifting from the industrialized society to the information society," the report recommended that the Japanese government fund an expensive ($68 billion) two-stage plan that would guide the transformation of Japan to an information society. The goal was to reach an intermediate target, the "establishment of computer mind," by 1985, which meant computerizing medicine and education, updating management information systems, creating a national information network, and so forth. The report stressed the necessity of moving from industrialization to informationalization because of the limits on world resources predicted by the Club of Rome report, *Limits to Growth*, which made an impact in resource-strapped Japan.[84]

Translated into English as *The Plan for Information Society: A National Goal toward Year 2000* (1972), the report helped popularize the term *information society* in the United States. In 1973, economist Edwin Parker at Stanford, a key figure in the *information technology* discourse, was one of the first policy analysts to cite the report.[85] The phrase *information society* (but not necessarily the Japanese concept of *johoka shakai*) spread quickly. The American Society for Information Science issued a call for papers for a session on the "Information Society" to be held at its annual meeting in 1973.[86] The American Association for the Advancement of Science announced that a session, "America: The First Information Society," sponsored by the U.S. Department of Commerce, would be held at its annual meeting in 1976. The AAAS speakers included Parker; economist Marc Porat, who was studying the "information economy" for the Commerce Department; Harvard computer scientist Anthony Oettinger; and MIT information theorist Robert Fano.[87] In December 1977, the School of Communications at the University of Washington in Seattle organized a forum attended by fifty American and Japanese scholars (including the ubiquitous Parker) on "Information Societies: Comparing the Japanese and American Experiences."

According to a scholar of the information-society discourse in Japan, the achievement of this conference "was the formal and systematic introduction of the Japanese Shakai Approach" to the United States.[88]

Humanists also adopted the new vocabulary. In 1979, artist and sociologist John McHale, at the University of Houston, predicted in a McLuhanite manner "that the cultural configurations of 'an information society,' more dependent on visual communications, will be even more different from that which has preceded it—just as the print-oriented society differed from societies dependent upon the oral communication and transmission of cultural experiences."[89]

Many analysts in the 1970s related the new concept to Bell's postindustrial society. The editor of the *Journal of Marketing* imagined an evolution from the postindustrial society to the information society; a researcher at the Congressional Office of Technology Assessment linked the two societies together; and a library scientist said they were equivalent.[90] More analytically, Parker distinguished between quantitative and qualitative aspects of the concepts: "These statistics [from Machlup and other economists] provide a simple economic description of the dominant trend in the society: a shift from an industrial society to an information society. Such facts underlie the various qualitative discussions of what is described as the 'postindustrial society' or the 'knowledge society.' "[91]

Parker identified a turning point in the information-society discourse of the 1970s, one that hinged on the difference between the quantitative analysis of Machlup and the (mostly) qualitative analysis of Bell. The quantitative approach got a boost in the 1970s when the governments of Japan and the United States began measuring informational activity in earnest. In 1971, the Japanese government started assessing the country's progress toward becoming an information society by conducting an annual National Information Flow Survey. The survey calculated an "index of informationalization" based on four categories: volume of information, diffusion of information, level of information-processing capability, and a coefficient of information. The Ministry of Posts and Telecommunications' first white paper on communications, published in 1974, set Japan's informationalization index at 100 in 1970, which compared with 108 for the United Kingdom, 101 for West Germany, 96 for France, and 199 for the United States.[92] The calculation was part of a more comprehensive method begun in Japan in the late 1960s, the "information societies approach" discussed above. By the late 1970s, it included information measurement, impact on social change, study of the best use of information channels, and policy implications.[93]

The U.S. government funded an econometric analysis to establish a national information policy. Conducted for the Commerce Department's Office of Telecommunications by Marc Porat, who did his Ph.D. under Parker at Stanford, the exhaustive nine-volume study, *The Information Economy* (1977), concluded that 46 percent of the country's GNP in 1967 could be attributed to informational activities. Rather than measuring information flow, as was done in Japan, Porat used finer categories than Machlup's—to whom he acknowledged an intellectual debt—to create two new categories with which to measure informational activity. The primary information sector included "*those firms which supply the bundle of information goods and services exchanged in a market context*," such as computers, telecommunications equipment, finance, insurance, and education. The secondary information sector included "*all the information services produced for internal consumption by government and noninformation firms*," such as those involved with research and development, marketing, and advertising. Porat calculated that the primary information sector accounted for 25 percent of GNP in 1967, the secondary information sector 21 percent, adding up to the 46 percent figure mentioned earlier. "Information workers" made up nearly one-half of the workforce in 1970 and earned a little over one-half of all labor income in 1967. The figures prompted Porat to conclude that the "U.S. has now emerged as an information-based economy." Later in the report, he more cautiously said, "We are just on the edge of becoming an information economy."[94] Presumably, Porat would not classify the entire U.S. economy as an "information economy" until the combined information sectors accounted for more than one-half of GNP. John Richardson, Porat's supervisor on the project, was bolder: "Without such definition and measurement, I cannot imagine how we can formulate sound policy for an information society."[95]

Porat climbed on the "information society" bandwagon the next year. In a 1978 article titled "Global Implications of the Information Society," Porat summarized his research and concluded, "On the strength of the GNP and labor data, the U.S. can now be called 'an information society.'" Admitting that "there are many phrases for the same phenomenon," such as Brzezinski's *technetronic age* and Bell's *postindustrial society*, Porat effectively erased the differences between his quantitative approach and Bell's qualitative approach, even though he had expressed an intellectual debt to Bell in the 1977 volume.[96]

The idea of an information society was reified in academic and policy circles with the launching of the journal *Information Society* in 1981. Its editor, Joseph Becker, was president of an international consulting firm, had headed the CIA's records systems in the 1960s, and was a past president of the America Society for Information Science. The journal's associate editors included such prominent figures in information science as J. C. R. Licklider, Rob Kling at the University of California at Irvine, and Edwin Parker at Stanford. The journal's first volume included articles by Parker on the research conducted by his former student Marc Porat on the information economy, and an extract from *The Coming Information Age* by Wilson Dizard, Jr., a veteran State Department communications-policy analyst at the U.S. Information Agency. Recognizing that *information age* was a trendy, though convenient, "cliché label," Dizard grounded the rubric in Porat's work and in Bell's theory of a postindustrial society.[97] In 1982, the Office of Technology Assessment declared that the "United States has become an information society dependent on the creative use of and communication of information for its economic and social well-being."[98]

The clamorous talk about the "information society" proved to be too much for Daniel Bell to ignore. In a long essay titled "The Social Framework of the Information Society," commissioned for a 1979 volume on the future of the "computer age" edited by computer scientists at MIT, Bell engaged in the information-society discourse. But he did so on his own terms. He retained his favorite label, the *postindustrial society*, and only used the phrase *information society* for the title of the article and two section headings. The essay summarized Bell's book-length argument about the postindustrial society and updated it to consider Porat's research on the information economy and rapid technological changes that combined computers with telecommunications—what his Harvard colleague Anthony Oettinger was calling *compunications*. Rather than cede agency to technology, Bell included it in the structural framework of the postindustrial society. "This revolution in the organization and processing of information and knowledge, in which the computer plays a central role, has as its context the development of what I have called the postindustrial society." That context included three aspects: the "change from a goods-producing to a service society," the "centrality of the codification of theoretical knowledge for innovation in technology," and the "creation of a new 'intellectual technology' as a key tool of systems and decision theory."[99]

In considering computer technology more fully than before, Bell modified how information fit into the framework of *The Coming of Post-Industrial Society*. "Intellectual technology" replaced "information" as the "technology" of postindustrial society, and "information" was moved to a new category, becoming the "transforming resource" for postindustrial society. In the modified scheme, the "crucial variables of the postindustrial society are information and knowledge." He retained his definition of knowledge but redefined *information* to mean "data processing in the broadest sense; the storage, retrieval, and processing of data becomes the essential resource for all economic and social exchange." Delving into information theory, which he had only mentioned in the book, Bell criticized Shannon's mathematical definition of information, while accepting the related view that information was a "pattern or design that rearranges data for instrumental purposes." Bell concluded that information was not a typical commodity; it was a "collective good" that remained with the producer when it was sold. In regard to econometrics, Bell thought Shannon's nonsemantic theory of information, though useful in communications engineering, "is not in general appropriate for economic analysis, for it gives no weight to the value of the information."[100] Despite making these concessions to the importance of information, Bell did not fully embrace the discourse of an information society.[101]

## Counternarratives

Many commentators criticized the concepts of a postindustrial society and an information society. The criticism grew to such an extent in the 1980s and the early 1990s that it formed a counternarrative to the dominant discourse that the United States, Japan, and Europe were entering an information age.

An early critic was MIT computer scientist Joseph Weizenbaum, who commented on Bell's essay presented at the 1979 symposium held at MIT. Weizenbaum satirized the revolutionary hyperbole of the new information discourse: "The widely shared belief in technological inevitability, especially as it applies to computers, is translated by scholars and the popular media alike into the announcement of still another computer revolution. (It will be remembered that the past two decades have, according to these same sources, already witnessed one or two such revolutions.) This time, the much-heralded revolution will transform society to its very core, and a new form of society will emerge: the information society." Advertising agencies oversold computers and computer networks as solutions to imaginary problems; Bell and other sociologists aided that effort by theorizing an information society. Furthermore, Bell's definition of knowledge was "circular and incomplete," and he was too sanguine about the ability to code knowledge into computer programs.[102] Bell defended his definition of knowledge, and said Weizenbaum had misinterpreted his theory; he was not a technological determinist. "In my work on postindustrial society," Bell declared, "I have reiterated the point that a change in the technoeconomic order (and that is the realm of information) does not determine changes in the political and cultural realms of society but poses questions to which society must respond."[103]

Humanists and social scientists elaborated on this critique. In 1984, philosopher of technology Langdon Winner at Rensselaer Polytechnic Institute argued that the enthusiastic talk about a computer revolution amounted to "mythinformation." While some critics tried to develop an adequate theory of the information society, Winner disabused readers of the very idea. Mythinformation was an "almost religious conviction that a widespread adoption of computers and communication systems along with easy access to electronic information will automatically produce a better world for human living." Under this regime, "technological determinism ceases to be a mere theory and becomes an ideal." The widespread beliefs that the "use of computers will cause hierarchies to crumble, inequalities to crumble, participation to flourish, and centralized power to dissolve simply do not withstand close scrutiny. The formula information = knowledge = power = democracy lacks any real substance." Mythinformation was a self-serving ideology that expressed the beliefs of the computer industry and its allies, "those that build, maintain, operate, improve, and market these systems." For Winner, the computer revolution was thus a conservative revolution.[104]

In *The Cult of Information* (1986) historian and social critic Theodore Roszak, best known for coining the term *counterculture* in the 1960s,[105] satirized the euphoria surrounding the word *information*.

> The word has received ambitious, global definitions that make it all good things to all people. Words that come to mean everything may finally mean nothing; yet their very emptiness may allow them to be filled with a mesmerizing glamour. The loose but exuberant talk we hear on all sides these days about 'the information economy,' 'the information society,' is coming to have exactly that function. These often-repeated catchphrases and clichés are the mumbo jumbo of a widespread public cult. Like all cults, this one also has the intention of enlisting mindless allegiance and acquiescence. People who have no clear idea what they mean by information or why they should want so much of it are nonetheless prepared to believe that we live in an Information Age, which makes every computer around us what the relics of the True Cross were in the Age of Faith: emblems of salvation.[106]

Roszak cited Machlup's recent comments that Shannon's choice of the word *information* for the central concept in his theory of communication was "infelicitous, misleading, and disserviceable." It marked the beginning of *information*'s career as an "all-purpose weasel word."[107]

The most extensive counternarrative appeared a decade later when British sociologist Frank Webster published *Theories of the Information Society*

(1995). Webster framed his critique around five categories of criteria that contemporaries used to justify the existence of the information society: technological, economic, occupational, spatial, and cultural. Webster identified futurists and journalists with the technological explanation; Machlup, Porat, and Bell with the economic and occupational criteria; sociologist Manuel Castells with the spatial analysis (the spread of computer and other information networks); and sociologist Jean Baudrillard with the cultural (information as the proliferation of linguistic signs). Webster criticized the technological argument for its technological determinism, the economical and occupational arguments for their extreme subjectivity in classifying informational activities, and all of the arguments for using arbitrary quantitative measures to judge when a society had passed from the industrial to the information stage. He thought Shannon's theory of information lent credence to quantitative arguments at the expense of qualitative ones, especially in regard to the technological and spatial criteria. That was ironic because the quantitative measures were marshaled to argue that "society must encounter profoundly meaningful change."[108]

Webster preferred qualitative explanations, such as those by sociologist Anthony Giddens and philosopher Jürgen Habermas who analyzed the role of information in contemporary society in a skeptical and nontechnologically determinist manner, and stressed continuity with the past and the political, economic, and social contexts of the informatization process.[109] He liked the qualitative parts of Bell's theory that emphasized the role of theoretical knowledge but sharply criticized it for being technologically determinist, historicist, teleological, and neoevolutionary. Referring to the existence of an expanding service economy before World War II, Webster emphatically stated, "*There is no novel, 'post-industrial' society: the growth of service occupations and associated developments highlights the continuities of the present with the past.*"[110]

Webster's critique found a large audience in the academic world. *Theories of the Information Society* became a popular college textbook, going through three revised editions by 2006. The editors of the *International Encyclopedia of the Social and Behavioral Sciences*, a classic reference work, invited Webster to write the entry on "Information Society" for its third edition in 2001. In addition to repeating his criticisms of theories of the information society, Webster denigrated the concept itself by saying it "appears to have limited value for social scientists."[111]

Although these criticisms may have had some influence in academia, the idea that the United States had become an "information society" did not disappear from public discourse. The number of articles using the phrase in the *New York Times*, for example, remained fairly constant from the 1980s to the 1990s. It rose in the mid-1990s when reporters quoted Newt Gingrich, Speaker of the House of Representatives, on his enthusiasm for the utopian rhetoric in Alvin Toffler's futuristic book *The Third Wave* (1980). Yet talk of the *information age* in the *Times* overwhelmed that of the *information society* in the 1990s. In the diffusion of the information discourse from academia, government, and business to the *Times*, the phrase *computer age* was replaced with the label *information age*, rather than *information society*. On the other hand, the *Times* picked up the talk of *information technology* from those circles without missing a beat. In fact, the *Times* amplified that usage by speaking of *information technology* as being both a new industry and an unquestioned social force. A similar trend is evident in my survey on Google Books.[112]

Part of the reason was the entrenched meaning of the keyword *technology* as a social force, whereas the keyword *society* had more diffuse meanings during this period. In the end, U.S. journalists followed a twentieth-century tradition and spoke of an "age" rather than a "society" marked by a dominant technology (e.g., the "automobile age" and the "space age," rather than the "automobile society" and the "space society"). They proclaimed that the rapid development of computers and communications, which put more and more information into circulation at work and at home, was creating the "information age."

The alternative discourse of cybernetics had many fewer proponents, despite the role of cybernetics in creating the information discourse. My search on Google Books reveals that the terms *cybernetic age* and *cybernetic society* peaked in the late 1960s and early 1970s, with the former polling higher than the latter throughout. As noted earlier, *cybernetic age* was even more popular than *information age* from 1965 to 1975. Its popularity was aided by the translation into English of science fiction writer Stanislaw Lem's *Cyberiad: Fables for the Cybernetic Age* in 1974, before being swamped by the information discourse that arose in the 1980s. The phrase *cybernetic society* got a boost from Michael Arbib, a computer scientist at the University of Massachusetts–Amherst, whose textbook *Computers and the Cybernetic Society* (1977) went through a second edition in 1984. A latter-day cybernetician who made the transition from the world of Warren McCulloch to that of computerized neural nets, Arbib balanced Wiener's original meaning of cybernetics, as control and communication in the animal and the machine, with the new meaning that stressed the information-processing aspect of the field.[113]

After the cybernetics moment ended in the United States in the 1970s, the cybernetics-inspired term that survived in academic and popular discourse, of course, was not *cybernetic age* nor *cybernetic society*, nor even *cybernetics* itself, but the adjective *cyber*, which science fiction writer William Gibson popularized when he coined the term *cyberspace* in his novel *Neuromancer* in 1984, a genre that became known as *cyberpunk*. *Cyber* became a favorite adjective to describe the world of information flowing in vast computer networks, a truncated residue of what remained of the rich discourse of cybernetics in the information age.

# CHAPTER NINE

# Two Cybernetic Frontiers

I N THE SPRING OF 1976, as the cybernetics moment was ending, Margaret Mead sat at the great oak dining table in the house of her former husband and research partner, Gregory Bateson, and his third wife Lois, waiting to be interviewed by Stewart Brand, the founder of the *Whole Earth Catalog*. A fan of Bateson's philosophy of cybernetics, Brand brought Mead and Bateson together so they could reminisce about the Macy conferences on cybernetics and reflect on the legacy of the meetings they had worked hard to realize (fig. 8).

The two cybernetic veterans had gone their separate ways after the group photograph was taken at the final Macy conference nearly a quarter of a century earlier. Divorced during that entire time and now in their seventies, they stood at the pinnacles of two contrasting careers. A fixture at the American Museum of Natural History in New York City, Mead was a world-famous anthropologist who wrote books and magazine articles on child-rearing, racial strife, and international relations during the Cold War. She came to the interview as the president of the prestigious American Association for the Advancement of Science. While Mead stayed at the museum for more than a half-century, Bateson led the life of a peripatetic scholar. He moved from university to university, research center to research center, investigating topics ranging from alcoholism to animal learning. A visiting lecturer at the University of California at Santa Cruz, where he drew crowds of admiring students, Bateson was a celebrity to anthropologists and psychologists, a cult figure to the ecological wing of the counterculture, and a hero to second-order cyberneticists. Brand could claim some of the credit for that renown because he promoted Bateson's cybernetic epistemology as a means to solve the ecological crisis of the 1970s. Brand introduced the interview, published

FIGURE 8. Gregory Bateson and Margaret Mead, being interviewed by Stewart Brand, 1976. On the thirtieth anniversary of the founding of the Macy conferences on cybernetics, Brand asked them to reminisce about the conferences and to speculate about why cybernetics had not become "public knowledge" in the United States. Used with permission from *CoEvolution Quarterly* 10 (Summer 1976).

in his journal *CoEvolution Quarterly*, by saying that Bateson's book, *Steps to an Ecology of Mind*, "wowed me out of my shoes."[1]

We have seen that Bateson enthusiastically embraced cybernetics, to the extent of creating a holistic theory of immanent mind, which resided in the flow of messages (information) through circular pathways between humans and their environment. Mead referred to cybernetics only a few times in her work. Yet during the interview, she and Bateson agreed on many things about the Macy conferences and the fate of cybernetics. They invoked some of the excitement social scientists had felt about cybernetics in the 1950s, when, for the first time, they had a theory of circular causality that could model the purposeful behavior of individuals and organizations. Brand was curious, however, why cybernetics, as a discipline, had never obtained in the United States the high scientific and political status that it enjoyed in the Soviet Union, and why "conceptual cybernetics" failed to become "public knowledge" in the United States. Bateson and Mead blamed this on the American love of machines and utilitarian thinking, which promoted the spread of general system theory instead of cybernetics. There were other reasons,

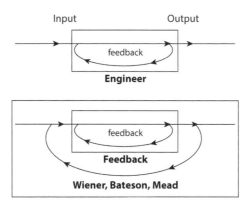

FIGURE 9. Gregory Bateson answered Stewart Brand's query about the decline of cybernetics by drawing two diagrams, revealing a split in the field. The top half represents first-order cybernetics, which found a home in the IEEE Society on Systems, Man, and Cybernetics; the bottom half represents second-order cybernetics, which found a home in the American Society for Cybernetics. Used with permission from *CoEvolution Quarterly* 10 (Summer 1976).

as well. Mead recalled that the tarring of cybernetics with the red brush of Soviet communism harmed her efforts to propose cybernetics as a common language to communicate with the Soviets. Bateson thought that cybernetics spread slowly in the behavioral sciences because "cyberneticians in the narrow sense of the word went off into [the] input-output" way of thinking, which was common among engineers and computer scientists.[2]

Bateson sketched two diagrams to explain his reasoning (fig. 9). The top diagram shows how engineers would draw a line around a system to create a black box that was separate from its environment, then analyze the feedback between the box's output and input. Their science was the science of the black box. In contrast, the bottom diagram showed that the "essence of Wiener's cybernetics was that the science is the science of the whole circuit. . . . What Wiener says is that you work on the whole picture and its properties . . . your ecosystem, your organism-plus environment, is to be considered a single circuit. . . . And you're not really concerned with an input-output, but with the events within the bigger circuit, and you are part of the bigger circuit. It's these lines around the box (which are just conceptual lines after all) which mark the difference between the engineers" and Wiener. Bateson and Mead identified with Wiener by placing their own subject position inside the larger box, because, like Wiener, they included the observer (themselves) in the system to be studied.[3]

In drawing these diagrams, Bateson distinguished between two styles of cybernetics. He contrasted the narrow input-output approach of engineers, computer scientists, and systems analysts with a broad ecosystems approach that included the observer. At this time, Heinz von Foerster and others were beginning to refer to these approaches as first-order and second-order cybernetics. In effect, Bateson enrolled Mead and Wiener (posthumously) into second-order cybernetics.[4]

This interview and other articles on Bateson in *CoEvolution Quarterly* left the impression that cybernetics had little to do with computers, despite the fact that Macy veteran John von Neumann had utilized cybernetic analogies to coinvent the ubiquitous stored-program design for digital computers, what (erroneously) became known as the "von Neumann architecture," and that another Macy veteran, Julian Bigelow, was chief engineer for von Neumann's Cold War computer project at Princeton.[5] In 1974, two years before the interview with Mead and Bateson, Brand discussed the relationship between cybernetics and computers in *II Cybernetic Frontiers*. In the booklet, Brand comments on an interview he conducted with Bateson in 1972 and an article he had written on computer laboratories, both of which are reprinted within it. In hindsight, we can see that the booklet sketches two diverging frontiers that computer scientists and cyberneticists explored from the mid-1970s to the present. The computer frontier led to the digital life we take for granted today as a "natural" development. Bateson's frontier led to an ongoing fascination with cybernetics in the humanities during the information age.

Brand introduced the two frontiers—which he called "machine cybernetics" and "organic cybernetics"—by saying that the "field of cybernetics is still busy finding out what it is" and was "jumping with fascinating activity." He acknowledged that his piece on computers was the "better article" and that its topic, making computers accessible, would have a direct bearing on people's lives. But he gave it second billing "behind the awkward piece on Gregory Bateson because what Bateson was getting at, I'm convinced, will *indirectly*, inform damn near everybody's lives." Computers were a "special case" of cybernetics, machine cybernetics; they did not deal with the broad philosophical issues tackled by Bateson in organic cybernetics.[6]

Brand's interview with Bateson had appeared in *Harper's* magazine in 1972 under the title "Both Sides of the Necessary Paradox." Though awkward and sketchy, it gives a flavor of Bateson's style of systems thinking, which Brand admired on this cybernetic frontier. In a wide-ranging conversation, edited by Brand, Bateson brought up a host of seemingly unrelated subjects: paralinguistics (his double-bind theory of schizophrenia), religion

(Taoism), anthropology (his fieldwork with Margaret Mead in New Guinea and Bali), and animal learning (research on how his concept of deutero-learning applied to porpoises). What held these diverse subjects together was Bateson's method of making analogies between organic and inorganic cybernetic systems, in which information flowed in circular-causal feedback paths.[7] The analogies formed the basis for a new epistemology of ecological, mindlike systems (those composed of humans, machines, and the environment).[8] Bateson did not explain (or even identify) these cybernetic concepts, nor discuss his cybernetic epistemology. Instead, he criticized "some very bad thinking inside cybernetics" by engineers who reduced its rich philosophy to an input-output method that pitted humans against the environment, a point he repeated in the joint interview with Mead.[9]

In *II Cybernetic Frontiers*, Brand admitted that the two frontiers of machine and organic cybernetics "are polar opposites in their relation to computing machines, [but] they are parallel in relation to Man's special pride, conscious purposefulness. Both subvert it."[10] On the organic side, Brand alluded to Bateson's view, stated in an earlier article, that computers could not be said to think. "The computer," Bateson said, "is only an arc of a larger circuit which always includes a man and an environment from which information is received and upon which efferent messages from the computer have effect. This total system, or ensemble, may legitimately be said to show mental characteristics."[11] In Bateson's cybernetics, computers were of interest as powerful machines inhabiting larger, mindlike circuits. In machine cybernetics, computers were powerful tools that created a new way of life.

Brand's article on machine cybernetics, published in *Rolling Stone* magazine in 1972 under the title "Spacewar: Fanatic Life and Symbolic Death among the Computer Bums," became a classic in computer journalism. It gives a good snapshot of the state of the art of research in human-computer interaction in three Bay Area labs. At Stanford's Artificial Intelligence Laboratory, Brand did not interview John McCarthy, the lab's director who helped establish the field of symbolic AI, or other computer scientists. Instead, he wrote about the "computer bums," the technical staff of "hackers" who wiled away the night playing Spacewar. In this early interactive computer game, run on the newly invented minicomputer with a cathode-ray-tube display, players fought battles with their spaceships as they flew across starry galaxies. He also discussed the main patron of the laboratory, the Department of Defense's Advanced Research Projects Agency (ARPA, or after 1972, DARPA), and ARPANET, the pioneering computer network that linked the computer research centers funded by ARPA. At the Xerox company's Palo Alto Research Center (Xerox PARC), which Brand satirically called "Shy Research Center" to protect its anonymity, he described Alan Kay's proposed Dynabook, an interactive, handheld sketchpad computer for kids. At the Stanford Research Institute (SRI), Brand mentioned Douglas Englebart's work on an online, multimedia text-management system, which he knew about because he had videotaped a demonstration of it in 1968. In addition, Brand identified two ways in which computers were becoming more accessible: through the programmable calculator and via community computer centers.[12]

All of these technologies have earned a prominent place in the history of computing. The government-supported ARPANET is the fabled forerunner of the Internet. The nerdy programmable calculator helped popularize the idea of personal computing. Englebart's now-famous demo utilized a computer interface consisting of mouse, chord keyset, keyboard, and windows. Kay's group at Xerox PARC transformed that complicated technology into today's graphical user interface of mouse, windows, and desktop icons for the prototype Alto computer. Kay's Dynabook was an inspiration for the Alto. After Steve Jobs visited Xerox PARC in 1979, Apple Computer adopted the Alto's graphical user interface and popularized it on the Macintosh in 1984. To an early-twenty-first-century reader, Kay's sketch of two kids playing Spacewar on their Dynabooks looks like two kids playing games on their iPad's.[13]

Brand made a couple of accurate predictions about computer networks. He thought newspapers would cease to exist in their present form if the Associated Press news-feed on the ARPANET was made available in the home, and that the sharing of digital music files on the ARPANET would do the same for record stores.[14] Yet Brand missed a major connection between cybernetics and computers. When he contrasted the two cybernetic frontiers, he did not mention that Engelbart drew on the cybernetic theory of coevolution to invent his graphical user interface, a research project he called the augmentation of human intellect.[15]

The Spacewar article sketched a frontier of machine cybernetics that was developed to its fullest extent by the U.S. government, computer manufacturers, hackers, and hobbyists. The editors of *Time* magazine were quick to recognize this cornucopia of innovation when they published a special section on "The Computer Society" in February 1978 (fig. 10). On the cover of the magazine appeared a drawing of a small group of men, women, and children, whose heads (their brains?) had been replaced by anthropomorphic information-technology devices. Human-like eyes and mouths appeared on a sheaf of punched IBM cards, a magnetic tape drive, a microelectronics chip, a digital watch, a computer terminal, a handheld electronic calculator,

FIGURE 10. The contrast between these covers of *Time* magazine illustrates the shift in public discourse on computers and daily life in the 1970s. The critical stance of the cybernated generation (*left*) gave way to the digital utopianism of the computer society, which prevailed at the end of the cybernetics moment (*right*). Used with permission from *Time*, April 2, 1965, and February 20, 1978.

and a universal-product-code strip. A robot from Star Wars (C-3PO), in its own body, gave the "bunny ears" sign over the "head" of the microchip. It was a happy family portrait of the computerization of society. The article focused on semiconductor technology, the "miracle chip" microprocessor at the heart of several devices on the cover, which *Time* correctly predicted would become ubiquitous in automobiles and in home appliances. A few years later, *Time* took the momentous step of naming the personal computer the "Machine of the Year" for 1982, in place of naming a human the "Man of the Year," as was its custom. A sculpture of a man sitting at a desk, looking at the screen of his computer graced the cover of the magazine. On the cover's inside flap is a companion sculpture of a woman sitting in a chair near her desktop computer. Coffee cup in hand, she looks away from the computer for the moment. The title announced, "The Computer Moves In."[16]

These stories contrast sharply with the way *Time* had treated computers more than a decade earlier, in April 1965, when it published a cover story on the social implications of computers called "The Cybernated Generation" (fig. 10). On the magazine's cover a multiarmed robot devours information from punch cards and spits it out via computer printouts to human underlings under the banner "The Computer in Society." The image resonated with Wiener's warnings that computer automation could cause social problems. In the story, the word *cybernetics* referred to the "science of computers" and served as a synonym for computerization.[17] In the 1978 cover story, the only allusion to cybernetics is in an interview with Stewart Brand, who said, "This is a story that goes back to the beginning of tool-using animals, back to the rocks the earliest man picked up in Africa. As soon as he started picking up rocks, his hands started changing, his brain started changing. Computers are simply a quantum jump in the same co-evolutionary process."[18] *Time* did not mention cybernetics when it named the computer the Machine of the Year in 1982.

The arc of Brand's career from the 1970s to the 1990s illustrates the mixed fate of cybernetics in the information age, even for a devotee of the subject. After publishing *II Cybernetic Frontiers*, he turned away from the computer frontier to promote Bateson's cybernetic epistemology for a decade.[19] In 1975, he arranged a lengthy interview between Bateson and California Governor Jerry Brown, held in Brown's office. A friend of the counterculture, Brown was sufficiently impressed to appoint Bateson as a regent of the University of California system.[20] The next year, Brand and Bateson organized a conference on the mind/body problem at the Zen Center in San Francisco. Chaired by Bateson, the conference brought together researchers on the frontiers of organic and machine cybernetics. Bateson, Francesco Varela, Heinz von Foerster, and Gordon Pask represented the second-order wing of cybernetics. Mary Catherine Bateson, the daughter of Margaret Mead and Gregory who had written a book on a European conference on cybernetics and ecology that her father organized in 1968, rounded out the researchers in organic cybernetics. From the computer side, Brand invited Alan Kay, whom he had interviewed at Xerox PARC for the Spacewar article, and Terry Winograd, a professor of computer science at Stanford.[21] Brand reported on his many other efforts to promote Bateson's cybernetics, including printing excerpts from Bateson's last book *Mind and Nature: A Necessary Unity* (1979), in the pages of *CoEvolution Quarterly*. When Bateson died in 1980, peacefully at the San Francisco Zen Center, the journal published the fond reminiscences of Bateson by one of his former students.[22]

Fred Turner has shown how a constellation of events in the early 1980s— Bateson's death, the collapse of the New Communalist movement, and the advent of personal computers—encouraged Brand to turn away from organic cybernetics and return to the machine cybernetics frontier. Ever the network entrepreneur, Brand cofounded, in the mid-1980s, the Hackers Conference and the WELL (the Whole-Earth 'Lectronic Link), a pioneering dial-up bulletin-board computer network. Used primarily by journalists and other professionals in the Bay Area, the WELL convinced Howard Rheingold to write about online "virtual communities." In 1985, Brand merged *CoEvolution Quarterly* with a new venture, *Software Review*, to form the *Whole Earth Review*. The new journal promoted personal computers and software as traditional "appropriate technologies," alongside the coevolutionary whole-systems philosophy.[23] In one of its first issues, the *Whole Earth Review* reported on debates at the Hackers Conference about whether or not information wanted to be free. Brand picked up the emerging discourse of an information society when he used the term *information economy* in the article's title and said that by "reorganizing the Information Age around the individual, via personal computers, the hackers may well have saved the American economy" from a recession.[24] When *Wired* magazine was founded in 1993, it had close ties with the Whole Earth network. Kevin Kelly, the computer editor for *Whole Earth Review*, became its executive editor. Brand and other journalists on the WELL wrote a couple of early articles for *Wired*, which became the apostle of digital utopianism, a vibrant form of information-age discourse.[25]

While extolling the bright future of computers, Brand did not forget his Bateson nor how to critique utopian claims about new technology. In his best-selling book, *The Media Lab: Inventing the Future at MIT* (1987), organic cybernetics found a niche in Brand's frontier of machine cybernetics. Commissioned by the Media Lab and based on Brand's three-month stint there researching electronic publishing, the book celebrates the vision of the lab's founder, Nicholas Negroponte, to merge old media (print and broadcasting) with the new media of personal computing. Brand's interviews with Negroponte, other leaders of the lab, and a dozen researchers and visitors paint a glowing picture of the enterprise. The book leaves the impression that Negroponte had replaced Bateson as Brand's mentor.[26]

Although hardly central to the book, cybernetics comes into the picture in a few places. In his interview with Jerry Wiesner—the former president of MIT and Negroponte's indispensable partner in raising corporate funds to build the Media Lab—Wiesner praised cybernetics as a precedent-setting research program in the 1950s. Yet Wiesner, perhaps drawing on his experience investigating for President Kennedy the alleged "cybernetics gap" with the Soviets, also noted that cyberneticists often made exaggerated claims about the analogy between the computer and the brain. He told Brand that Wiener's work on cybernetics had created a "golden age" of interdisciplinary research on communications at MIT's Research Laboratory of Electronics, when Wiesner promoted that work in the 1950s. But as director of the lab in the 1960s, Wiesner could not convince MIT to establish a Communications Science Center. Brand suggested that the Media Lab was the fulfillment of that dream. Wiesner disagreed, however, and emphasized Negroponte's idea of converging old and new media as the guiding force behind the Media Lab.[27] Brand brought up cybernetics a few more times in the book. He mentioned Wiener's argument that the centralized control of communications was unstable because it did not allow for feedback, and discussed a cybernetic analysis of inequalities among nations in the information system of global finance. He interviewed Warren McCulloch's old colleague Jerry Lettvin about their famous 1959 paper, "What the Frog's Eye Tells the Frog's Brain." But he did not relate the paper to McCulloch's cybernetics program at MIT, nor to second-order cybernetics.[28]

More surprisingly, Brand brought up Bateson without discussing Bateson's version of cybernetics, which he had done so much to promote a decade earlier. Instead, he praised Bateson's definition of information as "any difference which makes a difference" as a semantic alternative to the non-semantic definition in Shannon's theory of information. The remarkable aspect of Bateson's formulation, alluded to by Brand, is that it combines the linguistic elements of syntactics, semantics, and pragmatics in one pithy, easy-to-remember phrase. "The definition of information I kept hearing at the Media Lab," Brand noted, "was Bateson's highly subjective one. That's philosophically heart-warming, but it also turns out there's a powerful tool kit in the redefinition." Brand referred to the desire of Negroponte and Marvin Minsky—one of the founders of symbolic AI who was now a leading voice at the Media Lab—to use a semantic theory of information to improve data compression in TV and films.[29]

*The Media Lab* is emblematic of how Brand treated cybernetics when he returned to the computer frontier in the 1980s. He presented the science of cybernetics as a historical precedent, alluded to its usefulness in analyzing information systems, but failed to mention it when discussing two of his heroes in cybernetics, Warren McCulloch and Gregory Bateson. He may have been influenced by Minsky, who had relegated cybernetics to being a predecessor of AI in the 1960s. Bateson's pithy definition of information also fits well with the information-talk in *The Media Lab*. Brand reproduced a graph from Marc Porat's research on the information economy, and he titled the chapter on global finance "The World Information Economy."[30]

In contrast, Negroponte barely mentioned cybernetics in his best-selling book *Being Digital* (1995). Composed of monthly columns Negroponte wrote for *Wired* magazine, of which he was an early investor, *Being Digital* did not discuss Bateson's definition of information, which Brand had heard around the Media Lab. Negroponte only mentioned cybernetics when he said that his Architecture Machine Group at MIT, the predecessor of the Media Lab, had received funding from the Cybernetics Technology Office (CTO) in 1976 to research animated movies for the training of soldiers.[31] A branch of DARPA, the CTO's brief existence, from 1975 to 1981, has been overshadowed by the well-publicized success of DARPA's Information Processing Techniques Office (IPTO), which funded the design and construction of the ARPANET and research into computer graphics and artificial intelligence.[32]

The Cybernetics Technology Office was established in 1975 as a successor to ARPA's Behavioral Sciences Research Office at ARPA, which psychologist J. C. R. Licklider had created in 1962, the same year he founded IPTO. Social scientist Robert Young was CTO's first director. DARPA chose the cybernetics name in response to congressional criticism of the agency's support of the behavioral sciences and to indicate the funding of applications rather than basic research, a priority under DARPA's new leadership. The social scientists at CTO—Robert Young, his successor Stephen Andriole, and program officer Judith Ayres Daly—would have known about the movement to transform the meaning of cybernetics into a science that studied social systems.[33] In fact, CTO funded a good deal of research and development on computerized statistical systems for crisis management and decision making in foreign policy. One of the researchers, applied mathematician Paul Werbos, recalled that the system included a secret "worldwide conflict forecasting model for the Joint Chiefs of Staff." One tie to classical cybernetics in this program was that political scientist Karl Deutsch was a Ph.D. adviser to Werbos at Harvard.[34]

Another major research area at the Cybernetics Technology Office was in the new field of human-computer interaction (HCI). Program officer Craig Fields, who went on to become director of CTO and then of DARPA, funded several projects at the Media Lab, including the spatial management of information. Those projects enjoyed a good deal of success and are now seen as part of the storied history of HCI. In contrast, a more radical project, the Biocybernetics Program on Closed-Coupled Man/Machine Systems, was discontinued when the office was closed in 1981. The Biocybernetics Program funded a dozen projects at Harvard, MIT, SRI, UCLA, the University of Illinois, and other research centers to create psychophysiological links that would use brain signals and other means to interface humans with computers. The program dispersed more than two million dollars to researchers from 1975 to 1979. Because of funding cuts and administrative disruptions at DARPA, the Biocybernetics Program supported only one laboratory over the entire period of its existence, the Cognitive Psychophysiological Laboratory at the University of Illinois. A science fiction–like subvocal lab demonstration at the Stanford Research Institute attracted public attention in 1974, when *Time* magazine ran an article on it called the "Mind-Reading Computer."[35]

The contrast between the Cybernetics Technology Office and the Information Processing Techniques Office at DARPA illustrates the uncertain fate of cybernetics in the information age. While IPTO funded the information technologies in common use today—computer networks, computer graphical interfaces, and AI—the CTO funded many projects in the shadows—secret decision-making and forecasting systems for the Pentagon and a biocybernetics project that resonates more with the CIA than with the Media Lab. Even though the Aspen Movie Map project at the Media Lab was developed to train soldiers, it is celebrated today as contributing to the civil side of the information revolution, as a predecessor to Google Street Maps.[36] The breadth of the CTO programs symbolizes the breadth of cybernetics in the 1970s—computer modeling in the social sciences, biocybernetics, and more traditional HCI. Its dissolution in 1981, in the face of IPTO's success, symbolizes how difficult it was for cybernetics to survive in the 1980s, when the cybernetic moment had ended and talk of an information age was rampant.

The marginalization of cybernetics is evident in popular books that celebrated an information age in the 1980s and 1990s. Cybernetics and information theory, lauded in the 1960s as two new sciences ushering in the information revolution, do not make a significant impression in these books. Neither science is mentioned in information pundit Yoneji Masuda's *The Information Society as Post-industrial Society* (1981), journalist Tom Forester's edited volume, *The Information Technology Revolution* (1985), nor in sociologist Manuel Castells's three-volume set, *The Information Age: Economy, Society, and Culture* (1996–1998).[37] Cybernetics is mentioned several times in *The Computer Age* (1979), edited by two MIT computer scientists. Computer expert Michael Dertouzos, psychologist Licklider, and sociologist Daniel Bell identified cybernetics as an element of the computer age, while Marvin Minsky, once again, called it a "premature" precursor of AI.[38] Information theory is given its own chapter in *Silicon Dreams* (1989), written by Bell Labs vice president and electronics pundit Robert Lucky.

Although Lucky praised information theory as a high-level intellectual accomplishment that led to some practical applications in space communications and the coding of compact discs, he noted the exaggerated claims made for it in the 1950s and let readers judge its true value.[39] In the *Social Life of Information* (2000), John Seely Brown, the director of Xerox PARC, and his coauthor, humanist Paul Duguid, noted problems with Shannon's nonsemantic definition of information. They preferred Bateson's definition of information, but did not identify that formulation with cybernetics. Like Brand, they reduced Bateson's complex philosophy of cybernetics to a pithy information slogan.[40] It fit the discursive needs of an information age.

Ironically, the concepts of cybernetics and information theory were treated more fully by humanists and social scientists who criticized the rash talk about an information society. In her introduction to the edited volume *Myths of Information* (1980), Kathleen Woodward at the University of Wisconsin–Milwaukee even praised cybernetics as a method academics could use to combat the reigning ideologies of information. "Circulating through *The Myths of Information*," Woodward explained, "are two major propositions regarding cultural theory and action: we must think cybernetically, and we must trust in de-centered action in everyday life." Drawing on Bateson, Woodward said, "These two mandates are related, for cybernetic thought asserts that the part can never control the whole and that change in a system comes about from shifts in the structural relations, or constraints, of the system."[41] Theodore Roszak was more critical in 1986. Although he praised Wiener for being the conscience of cybernetics, for warning readers of its potential negative social consequences, he traced the origins of the "cult of information" to Wiener's 1950 book, *The Human Use of Human Beings*. The criticism was not new for Roszak. In his widely read *The Making of a Counter Culture* (1969), he listed cybernetics as one of the intellectual sources of the technocratic society against which the counterculture was fighting. Wiener's version of cybernetics, Roszak declared, "worked out one of the key propositions of technocratic managerialism: namely, that man and social life generally are so much communicating apparatus," and thus manipulable using the science of communication and control.[42]

Brand did not ignore these sorts of criticisms. The cover of the first issue of *Whole Earth Review*, published in January 1985, stated a provocative theme in large type, "Computers as Poison," below a favorite Whole Earth adage, "All panaceas become poison." Brand explained that the journal used the panacea-as-poison theme to attract loyal readers of the money-losing *CoEvolution Quarterly*, which had been merged into *Whole Earth Review*. "It seemed appropriate to combine the routine skepticism of *CQ* with the computer obsession of *Software Review* and let those perspectives fight it out for awhile," Brand concluded. The issue included an early version of Langdon Winner's critical article, "Mythinformation," and reprinted E. M. Forster's 1909 dystopian short story, "The Machine Stops," about a proto–information age gone amok, which Brand later discussed in *The Media Lab*. Yet in the software review part of the issue, Brand revealed that he had moved on to the machine cybernetics frontier: "Having presented harsh words about computers in general, we here reveal our true colors with a whole section of largely kind words about computers in particular."[43]

One social scientist employed cybernetics to explain the origins of the information society. In *The Control Revolution* (1986), James Beniger adopted the standard cybernetic view that information-feedback loops ensured the survival of all living systems, from the cell to the world economy, by negating the increase of entropy at the local level. Based on that premise, he charted the history of the control revolution from the nineteenth century to the microelectronics-based information society (see chapter 8). Cybernetics held a unique status in Beniger's account. Created by scientists and engineers who knew about the computers and communications systems invented in World War II, cybernetics, to Beniger, was a universal science that could explain the new society being built by those very innovations.[44]

Beniger knew his cybernetics. In the late 1970s, he identified his work as part a broad European movement in sociology called the "new cybernetics" or "sociocybernetics." The movement aimed to reform "classical cybernetics" by including the observer as part of the system being studied, and by considering system changes caused by its interaction with the environment. The work of English cybernetician Gordon Pask and German sociologist Niklas Luhmann along these lines later came under the rubric of "second-order cybernetics." Even though Beniger did not adhere to the radical epistemology of second-order cybernetics, his work fits into this broad reform movement because he proposed a theory of social control that synthesized purposive individual actions with autonomous systems.[45] He applied the general outlines of this theory of sociocybernetics in *The Control Revolution* to date the origins of the information society to the nineteenth century.

Second-order cybernetics in the United States did not address the convergence of computers and communications occurring on the frontier of machine cybernetics. Instead, it represents the exploration of the frontier of organic cybernetics. Second-order cybernetics has led a marginal disciplinary existence in the United States as a mainstay of the American Society for Cybernetics and of *Cybernetics and Human Knowing*, a journal sponsored by the ASC. It has been more widely adopted in Europe. In the third edition of the *International Encyclopedia of the Social and Behavioral Sciences*, published in 2001, the entry "Sociocybernetics" defines it as the "applications of first-order and especially second-order cybernetics to the social sciences, and the further development of the latter." The entry prominently cites the work of Luhmann and other Europeans.[46]

The encyclopedia also illustrates the decline in the interpretation of cybernetics as a universal discipline. While the second edition of this classic reference work, published in 1968, devoted a separate entry to "Cybernetics," which recounted the wide variety of research performed under that flexible and contested rubric, the third edition replaced that entry with one titled "Sociocybernetics."[47] The term *cybernetics* can be found in more than forty articles in this twenty-six-volume reference set. But usually it is cited as a historical precedent to current theories in the social sciences. The entry "Cyborg," for example, describes how the cyborg grew out of Cold War research in cybernetics, the proliferation of cyborgs in science fiction and fact, and how humanists have followed the lead of Donna Haraway to embrace the cyborg as an ironic myth for political action. Other entries note the importance of (first-order) cybernetics in the work of such eminent, deceased scholars as Karl Deutsch and Roman Jakobson, and in the history of such fields as cognitive science and family therapy.[48] The encyclopedia thus treated cybernetics as an outdated predecessor to current scholarship, even in the entry "Sociocybernetics."[49]

We have seen that many information theorists and computer scientists, such as John Pierce at Bell Labs and Peter Elias and Marvin Minsky at MIT, have treated cybernetics in a similar manner. A revival of sorts is under way, though, in the field of AI. Rodney Brooks, the head of the Computer Science and Artificial Intelligence Laboratory at MIT, has drawn inspiration from W. Grey Walter's robotic tortoises in the 1950s to create a new cybernetic form of AI. His lifelike robots explore and learn about their environment instead of relying on cognitive maps that are characteristic of symbolic AI.[50] In sociology, Andrew Pickering has utilized the history of British cybernetics to propose reforming modern science with a new vision: the "nonmodern performative ontology" of cybernetics, as exemplified by the work of Grey Walter, Ross Ashby, Stafford Beer, and Gordon Pask. Pickering does not aim to replace representationalist science with performative cybernetics. He offers an alternative vision that challenges the dominant way we understand the world.[51]

In the early twenty-first century, forty years after Stewart Brand published *II Cybernetic Frontiers*, it is obvious which frontier has blossomed. Machine cybernetics has filled our world with seemingly indispensable information gadgets. Organic cybernetics lives a marginal life as second-order cybernetics and sociocybernetics in the social sciences and humanities. First-order cybernetics has experienced a rebirth at MIT's Artificial Intelligence Laboratory, and information theory hums along as a subfield of electrical engineering, contributing the codes necessary to run a digital age. But the protean concepts of cybernetics and information theory that created the intellectual and cultural excitement of the cybernetics moment—the constructs of circular causality and negative entropy, the idea that society is held together by messages—have been flattened to a discourse of bits, Big Data, and the all-purpose adjective of *cyber* to label an underexamined information age.[52]

The sociotechnical transformation of old media into new media has been accompanied by a cultural transformation in how we talk about, and thus interpret, the major technology of our time. The cybernetics moment transformed the sciences into information sciences, inspired the invention of information technology, and provided a vocabulary to describe the new age. Ironically, the ending of the cybernetics moment—when cybernetics and information theory were no longer considered to be universal panaceas, separately or conjoined—permitted the rise of an impoverished information discourse about an information revolution, society, and age. In 1995, Negroponte summed up the essence of "being digital" as moving bits, not atoms, which a recent journalist has called the "motto of the Internet Age."[53] Although Negroponte was speaking metaphorically, the motto elides the enormous amount of telecommunications infrastructure, computer hardware, and "information labor" required to move bits of information so that it seems to be done without effort, cost, or environmental impact. Most Internet users, for example, are unaware that a Google search uses any energy at all.[54] The ideas that fill the writings of Wiener and Bateson—the unsuitability of information to be a commodity because of its equivalence to entropy, critiques of secrecy in scientific research and the unemployment caused by computerization, and that adopting a cybernetic epistemology would eliminate destructive attitudes toward the environment—are no longer part of a broad discourse on cybernetics and information science. The topics they address are debated as separate questions in separate disciplines, not in terms of what the information age should be.

Understanding how the cybernetics moment transformed the sciences, inspired the invention of technology, and gave our current age its name helps us understand the present by recovering the contestations of the past and the paths not taken. Perhaps, it will even help us make good decisions about the future of our age of information.

# ABBREVIATIONS

| | |
|---|---|
| *AHC* | *IEEE Annals of the History of Computing.* |
| APS | American Philosophical Society, Philadelphia. |
| ASC | American Society for Cybernetics. |
| AT&T | American Telephone and Telegraph Corporation. |
| ATTA | AT&T Corporate Archives, AT&T Archives and History Center, Warren, NJ. |
| BCL | Biological Computer Laboratory, University of Illinois, Urbana. |
| *BSTJ* | *Bell System Technical Journal.* |
| BTL | Bell Telephone Laboratories, AT&T. |
| CBIOH | Oral History Collection, Charles Babbage Institute, University of Minnesota, Minneapolis. Available online at http://www.cbi.umn.edu/oh/index.html. |
| CC | E. Colin Cherry Papers, Corporate Records Unit and College Archives, Imperial College, London. |
| *CCIT* | Colin Cherry, ed., *Information Theory: Papers Read at a Symposium on "Information Theory" . . . 1955* (London: Butterworths Scientific Publications, 1956). |
| *CESCP* | Claude E. Shannon, *Claude Elwood Shannon: Collected Papers*, ed. N. J. A. Sloane and Aaron D. Wyner (New York: IEEE Press, 1993). |
| CESLC | Claude E. Shannon Papers, Manuscript Division, Library of Congress, Washington, DC. |
| CESMIT | Claude E. Shannon Papers (MC386), MITA. |
| DG | Denis Gabor Papers, Corporate Records Unit and College Archives, Imperial College, London. |
| HAS | Herbert A. Simon Collection, University Archives, Carnegie-Mellon University. Available online at http://diva.library.cmu.edu/Simon. |

| | |
|---|---|
| HVF | Heinz von Foerster Papers, University of Illinois Archives, Urbana. |
| *HVFCyb* | Heinz von Foerster, Margaret Mead, and Hans Teuber, eds., *Cybernetics: Circular Causal and Feedback Mechanisms in Biology and Social Systems. Transactions* . . . , vols. 6–10 (New York: Macy Foundation, 1950–1955). |
| IEE | Institute of Electrical Engineers. |
| IEEE | Institute of Electrical and Electronics Engineers. |
| *IEEE Trans IT* | *IEEE Transactions on Information Theory.* |
| *IEEE Trans SMC* | *IEEE Transactions on Systems, Man, and Cybernetics.* |
| *IEEE Trans SSC* | *IEEE Transactions on Systems Science and Cybernetics.* |
| *IESB* | *International Encyclopedia of the Social and Behavioral Sciences*, ed. Neil J. Smelser and Paul B. Baltes, 3rd ed. (Amsterdam: Elsevier, 2001). |
| IRE | Institute of Radio Engineers. |
| *IRE Trans IT* | *IRE Transactions on Information Theory.* |
| JAC | Journalism and Communications, Institute of Communications Research, Record Series 13/5/1. University of Illinois Archives, Urbana. |
| JCRL | J.C.R. Licklider Papers (MC499), MITA. |
| JR | Jerome Rothstein Papers, Niels Bohr Library and Archives, American Institute of Physics, College Park, MD. |
| JVN | John von Neumann Papers, Manuscript Division, Library of Congress, Washington, DC. |
| JW | Jerome B. Wiesner Papers (MC420), MITA. |
| LB | Léon Brillouin Papers, Niels Bohr Library, American Institute of Physics, College Park, MD. |
| LNR | Louis N. Ridenour Papers, University of Illinois Archives, Urbana. |
| MIT | Massachusetts Institute of Technology, Cambridge. |
| MITA | Institute Archives and Special Collections, MIT Libraries. |
| MITMath | Department of Mathematics Records (AC123), MITA. |

| | |
|---|---|
| MM | Margaret Mead Papers, Manuscript Division, Library of Congress, Washington, DC. |
| NARCR | NASA Ames Research Center Records, Sunnyvale, CA. |
| NDRC | National Defense Research Committee. |
| NW | Norbert Wiener Papers (MC022), MITA. |
| *NWCW* | Norbert Wiener, *Collected Works with Commentaries*, ed. Pesi Masani, 4 vols. (Cambridge, MA: MIT Press, 1976–1986). |
| *NWCyb* | Norbert Wiener, *Cybernetics: Or Control and Communication in the Animal and the Machine* (Cambridge, MA: Technology Press; New York: John Wiley, 1948). |
| *NWHUHB* | Norbert Wiener, *The Human Use of Human Beings: Cybernetics and Society* (Boston: Houghton Mifflin, 1950). |
| *NWHUHB*, 2nd ed. | Norbert Wiener, *The Human Use of Human Beings: Cybernetics and Society*, 2nd. ed. rev. (Boston: Houghton Mifflin, 1954). |
| *NYT* | *New York Times.* |
| ONR | Office of Naval Research. |
| PE | Peter Elias Papers (MC606), MITA. |
| PGIT | IRE Professional Group on Information Theory. |
| RFA | Records of the Rockefeller Foundation, Rockefeller Archive Center, Sleepy Hollow, NY. |
| RF-Dartmouth | RFA, record group 1.0002, series 200, box 26, folder 219. |
| RF-MacKay | RFA, record group 1.2, series 401, box 19, folder 179. |
| RJ | Roman Jakobson Papers (MC72), MITA. |
| RLE | Research Laboratory of Electronics, MIT. |
| RMF | Robert Mario Fano Papers (MC413), MITA. |
| RVLH | Ralph Vinton Lyon Hartley Papers, Niels Bohr Library, American Institute of Physics, College Park, MD. |
| SH | Steve Heims Papers (MC361), MITA. |
| *Trans PGIT* | *Transactions of the IRE Professional Group on Information Theory.* |
| VB | Vannevar Bush Papers, Manuscript Division, Library of Congress, Washington, DC. |

WJ          Willis Jackson (Lord Jackson of Burnley) Papers, Corporate Records Unit and College Archives, Imperial College, London.

WJCT     Willis Jackson, ed., *Communication Theory: Papers Read at a Symposium on "Applications of Communication Theory"* ..., *1952* (New York: Academic Press, 1953).

WR          Walter Rosenblith Papers (MC055), MITA.

WSM        Warren S. McCulloch Papers, APS.

WSM—ASC    WSM, ser. II, American Society for Cybernetics (folder subtitle).

WSM—Macy   WSM, ser. II, Conference: Josiah Macy Foundation meeting (roman numeral for meeting number).

WSMCW      Warren S. McCulloch, *Collected Works of Warren S. McCulloch*, ed. Rook McCulloch, 4 vols. (Salinas, CA: Intersystems, 1989).

## Introduction

1. See, e.g., Stewart Brand, "Introduction to the Perennial Edition," in Margaret Mead, *New Lives for Old: Cultural Transformations—Manus, 1928–1953* (1956; rpt., New York: HarperCollins, 2001), xv–xvii.

2. My analysis of successes and failures in communicating, collaborating, drawing boundaries, and creating research centers across disciplines contributes to an emerging literature in the history of interdisciplinarity that speaks to policymaking in science. See Jerry A. Jacobs and Scott Frickel, "Interdisciplinarity: A Critical Assessment," *Annual Review of Sociology* 35 (2009): 43–65, on 54–61.

3. James Gleick, *The Information: A History, a Theory, a Flood* (New York: Pantheon, 2011), chaps. 8 and 13.

4. See, e.g., Flo Conway and Jim Siegelman, *Dark Hero of the Information Age: In Search of Norbert Wiener, the Father of Cybernetics* (New York: Basic Books, 2005); and M. Mitchell Waldrop, "Claude Shannon: Reluctant Father of the Digital Age," *Technology Review* (July/August 2001): 64–71.

5. Theodore Roszak, *The Cult of Information: The Folklore of Computers and the True Art of Thinking* (New York: Pantheon, 1986).

6. David E. Nye, *America as Second Creation: Technology and Narratives of New Beginnings* (Cambridge, MA: MIT Press, 2003), chap. 1.

7. See, e.g., David Pogue, "Steve Jobs Reshaped Industries," *NYT*, August 25, 2011. On the history of the terms *technology* and *information technology*, see Eric Schatzberg, "*Technik* Comes to America: Changing Meanings of *Technology* before 1930," *Technology and Culture* 47 (2006): 486–512; and Ronald R. Kline, "Cybernetics, Management Science, and Technology Policy: The Emergence of 'Information Technology' as a Keyword, 1948–1985," *Technology and Culture* 47 (2006): 513–535.

8. For example, Michael E. Hobart and Zachary S. Schiffman, in *Information Ages: Literacy, Numeracy, and the Computer Revolution* (Baltimore: Johns Hopkins University Press, 1998), question the idea that there has been one information age and chart the history of three of them: literacy (the invention of the alphabet), numeracy (the analytical mathematics of the nineteenth century), and computerization.

9. Most histories of cybernetics in the United States ignore the rise-and-fall of

the field, recounting instead its success as a postwar science. See, e.g., Steve J. Heims, *The Cybernetics Group* (Cambridge, MA: MIT Press, 1991); Peter Galison, "The Ontology of the Enemy: Norbert Wiener and the Cybernetic Vision," *Critical Inquiry* 21 (1994): 228–266; Geoffrey C. Bowker, "How to Be Universal: Some Cybernetic Strategies, 1943–1970," *Social Studies of Science* 23 (1993): 107–127; Galison, "The Americanization of Unity," *Daedalus* 127 (1998): 45–71; and Conway and Siegelman, *Dark Hero of the Information Age.*

10. Hans Wellisch, "From Information Science to Informatics: A Terminological Investigation," *Journal of Librarianship* 4 (1972): 157–187.

11. Georges R. Boulanger, "Prologue: What Is Cybernetics?," in *Survey of Cybernetics: A Tribute to Dr. Norbert Wiener,* ed. J. Rose (London: Iliffe, 1969): 3–9, on 3–4.

12. James Baldwin, *No Name in the Street* (1972), rpt. in Baldwin, *Collected Essays* (New York: Library of America, 1998), 349–476, on 384.

13. I use the term *disunity* to refer to the multiple interpretations and eventual fragmentation of cybernetics, rather than in the technical sense employed in the philosophy of science. Peter Galison, for example, argues that the disunity of physics—the intercalcation of theory, experiment, and observation, which is held together by pidgin and creole languages created in the "trading zone" between these subcultures—better explains the history of physics than Thomas Kuhn's paradigm shifts. See Galison, "Computer Simulations and the Trading Zone," in *The Disunity of Science: Boundaries, Contexts, and Power,* ed. Peter Galison and David J. Stump (Stanford, CA: Stanford University Press, 1996), 118–157; and Galison, *Image and Logic: A Material Culture of Microphysics* (Chicago: University of Chicago Press, 1997), chap. 9.

14. Andrew Pickering, *The Cybernetic Brain: Sketches of Another Future* (Chicago: University of Chicago Press, 2010), for Britain; Philipp Aumann, "The Distinctiveness of a Unifying Science: Cybernetics' Way to West Germany," *AHC,* 33, no. 4 (2011): 17–27; David A. Mindell, Jérôme Segal, and Slava Gerovitch, "From Communications Engineering to Communications Science: Cybernetics and Information Theory in the United States, France, and the Soviet Union," in *Science and Ideology: A Comparative History,* ed. Mark Walker (London: Routledge, 2003), 66–96; Gerovitch, *From Newspeak to Cyberspeak: A History of Soviet Cybernetics* (Cambridge, MA: MIT Press, 2002); and Eden Medina, *Cybernetic Revolutionaries: Technology and Politics in Allende's Chile* (Cambridge, MA: MIT Press, 2011).

15. On the conduit model, see Michael J. Reddy, "The Conduit Metaphor: A Case of Frame Conflict in Our Language about Language," in *Metaphor and Thought,* ed. Andrew Ortony (Cambridge: Cambridge University Press, 1979), 284–324.

16. On this form of mutual construction, see Paul N. Edwards, *Closed World: Computers and the Politics of Discourse in Cold War America* (Cambridge, MA: MIT Press, 1996), chap. 1.

## Chapter 1: War and Information Theory

1. Norbert Wiener to Arturo Rosenblueth, April 16, 1947, NW, box 5-77. On Pitts, see Neil R. Smalheiser, "Walter Pitts," *Perspectives in Biology and Medicine* 43 (2000): 217–226. Wiener told a former graduate student, Yuk-Wing Lee, that Pitts was the "most brilliant student I ever had," which must have annoyed Lee! See Wiener to Lee, October 22, 1945, NW, box 4-69.

2. Norbert Wiener to Walter Pitts, April 4, 1947 (quotation); Wiener to Pitts, April 10, 1947; Wiener to Arturo Rosenblueth, April 10, 1947; Pitts to Wiener, n.d., ca. April 10, 1947 (postcard); and Wiener to Pitts, April 16, 1947, all in NW, box 5-77. Flo Conway and Jim Siegelman, *Dark Hero of the Information Age: In Search of Norbert Wiener, the Father of Cybernetics* (New York: Basic Books, 2005), 203–204, interviewed Lettvin, who said he took the blame to relieve pressure on Pitts. On the homosocial environment surrounding Pitts, see Elizabeth A. Wilson, "'Would I Had Him with Me Always': Affects of Longing in Early Artificial Intelligence," *Isis* 100 (2009): 839–847. Although Wilson points out that such environments of close male groups often bordered on homosexuality, we do not have evidence for it in this group.

3. Norbert Wiener to Arturo Rosenblueth, April 16, 1947; and Wiener to Warren McCulloch, April 5, 1947, both in NW, box 5-77. He alluded to the lost manuscript episode in *NWCyb*, 24.

4. William Aspray, "The Scientific Conceptualization of Information: A Survey," *AHC* 7 (1985): 117–140; and Jérôme Segal, *Le zéro et le un: Histoire de la notion scientifique d'information au 20ᵉ siècle* (Paris: Éditions Syllepse, 2003).

5. Claude Shannon, "Part I. Foundations and Algebraic Structure of Secrecy Systems," attached to Ralph Bown to R. K. Honaman, February 26, 1947; Honaman to Bown, March 18, 1947, both in ATTA, FC 3137, vol. G; and Norbert Wiener to Walter Pitts, March 21, 1947, NW, box 5-76. On Wiener's hearing Shannon talk at Harvard, see Wiener to Francis Bello, October 13, 1953, NW, box 12-179. Shannon was scheduled to give two talks, one at Brown University on March 21, 1947, and another "a few days later" at Harvard (Bown to Honaman). Because Wiener wrote to Pitts on the day of the Brown talk, he may have heard the Brown talk instead of the Harvard talk that he recalled hearing in 1953.

6. P. S. Dwyer, "Report on the New York Meeting of the Institute," *Annals of Mathematical Statistics* 18 (1947): 468; and Norbert Wiener to Warren McCulloch, May 2, 1947, NW, box 5-78.

7. Claude E. Shannon, "A Mathematical Theory of Communication," *BSTJ* 27 (1948): 379–423, 623–656, on 393; and *NWCyb*, 76.

8. See, e.g., Robert M. Fano, "The Information Theory Point of View in Speech Communication," *Journal of the Acoustical Society of America* 22 (1950): 691–696, on 691; Henry Quastler, ed., *Essays on the Use of Information Theory in Biology* (Urbana: University of Illinois Press, 1953), 5; Colin Cherry, *On Human Communication: A Review, a Survey, and a Criticism* (Cambridge, MA: MIT Press, 1957), 176, 214, 219, 226; Fred Atteneave, *Applications of Information*

Theory to Psychology: A Summary of Basic Concepts, Methods, and Results (New York: Henry Holt, 1959), 8; Pesi R. Masani, Norbert Wiener, 1894–1964 (Basel: Birkhauser, 1990), 155; N. Katherine Hayles, How We Became Post-Human: Virtual Bodies in Cybernetics, Literature, and Informatics (Chicago: University of Chicago Press, 1999), 300n.2; and Lily E. Kay, Who Wrote the Book of Life? A History of the Genetic Code (Stanford, CA: Stanford University Press, 2000), 99. Steve J. Heims, The Cybernetics Group (Cambridge, MA: MIT Press, 1991), 97; and Paul N. Edwards, Closed World: Computers and the Politics of Discourse in Cold War America (Cambridge, MA: MIT Press, 1996), 200–201, credit both men equally but do not use the Wiener-Shannon or Shannon-Wiener attribution.

9. John von Neumann, "[Review of] Cybernetics . . . ," Physics Today 2 (May 1949): 33–34. On Szilard's work, see Segal, Le zéro et le un, 19–21.

10. See, e.g., Warren Weaver, "Recent Contributions to the Mathematical Theory of Communication," in Claude E. Shannon and Weaver, The Mathematical Theory of Communication (Urbana: University of Illinois Press, 1949), 93–117, on 95n; Denis Gabor, "A Summary of Communication Theory," in WJCT, 1–24; Robert M. Fano, Transmission of Information: A Statistical Theory of Communications (New York: MIT Press and John Wiley, 1961), 1–2; Aspray, "Scientific Conceptualization of Information"; and David A. Mindell, Between Human and Machine: Feedback, Control, and Computing before Cybernetics (Baltimore: Johns Hopkins University Press, 2002), 319–320. Weaver did cite von Neumann's reference to Szilard.

11. John R. Pierce, "The Early Days of Information Theory," IEEE Trans IT 19 (1973): 3–8; Sergio Verdú, "Guest Editorial," IEEE Trans IT 44 (1998): 2042–2043; Verdú, "Fifty Years of Shannon Theory," IEEE Trans IT 44 (1998): 2057–2076, on 2058n.5; and Bernard Geoghegan, "The Historiographic Conceptualization of Information," AHC 30 (2008): 69–81.

12. The term new science was commonly applied to cybernetics in the popular press; see, e.g., William L. Laurence "Cybernetics, a New Science, Seeks the Common Elements in Human and Mechanical Brains," NYT, December 19, 1948, E 9. For an early usage of this term for information theory, see Willis Jackson, "Information Theory," Nature 167 (1951): 20–23, on 20.

13. NWCyb, 19, 38.

14. Arturo Rosenblueth, Norbert Wiener, and Julian Bigelow, "Behavior, Purpose and Teleology," Philosophy of Science 10 (1943): 18–24.

15. NWCyb, 19; and Wiener, "Cybernetics," Scientific American, 179 (November 1948): 14–19, on 14. On the grant, see Rockefeller Foundation Executive Committee Minutes, January 16, 1947, RFA, RG 1.1, series 224, box 1-2; and Kay, Who Wrote the Book of Life?, 82–84, 349.

16. NWCyb, introduction, chaps. 1, 5–8.

17. Ibid., chaps. 2–4.

18. Ibid., 74–76. The formula given by Wiener was $\int p(x) \log p(x)\, dx$, where

p(x) is a probability density of a continuous function. It represents the weighted average of the logarithm of p(x).

19. Ibid., 78, 104.

20. Ibid., 183–184.

21. Ibid., 50, 155, 187.

22. Harry M. Davis, "Mathematical Machines," *Scientific American* 180 (April 1949): 29–38, on 29.

23. Shannon, "A Mathematical Theory of Communication." For a good exposition of the theory, see John R. Pierce, *Symbols, Signals, and Noise: The Nature and Process of Communication* (New York: Harper & Brothers, 1961), chaps. 3–8.

24. Claude Shannon to Norbert Wiener, October 13, 1948, his emphasis; and Wiener to Shannon, n.d., both in NW, box 6-85.

25. Shannon, "A Mathematical Theory of Communication," 379.

26. Ibid., 380–381, 392–396, quotation on 393. The equation for the discrete case was $H = -\sum p_i \log p_i$, where $p_i$ is the probability of the occurrence of a discrete symbol. It represents the weighted average of the logarithm of $p_i$. For the continuous case, Shannon gave the same equation as Wiener (see note 18 above), but with a negative sign, rather than a positive sign (628).

27. Ralph V. L. Hartley, "Transmission of Information," *BSTJ* 7 (1928): 535–563, on 539–540; Aspray, "Scientific Conceptualization of Information," 120–121; and Segal, *Le zéro et le un*, 53–59, 128–130. Hartley defined amount of information as the "logarithm of the number of possible symbol sequences" (540), i.e., $H = n \log s, = \log sn$.

28. *Bell Laboratory Record*, March 1946, 122. In a 1946 Bell Labs report, Hartley called his work "information theory," an early usage of this term. See Hartley, "Television from the Information Viewpoint," May 29, 1946, RVLH, series I, box 2-37.

29. Shannon, "A Mathematical Theory of Communication," 380, 407–410. All uncertainties are calculated using entropy equations.

30. Ibid., 393.

31. Ibid., 410–413. A simple coding example for a noiseless discrete channel, drawn from the paper (404–405), illustrates Shannon's approach. Consider a language with only four symbols: A, B, C, D. Symbol A occurs one-half of the time, symbol B one-fourth of the time, and symbols C and D one-eighth of the time each. A direct method to encode these symbols uses two bits per symbol, for example, A: 00, B: 01, C: 10, and D: 11. Alternatively, one can use the probabilities of when these symbols occur to calculate the uncertainty (the entropy) of the source as one and three-quarter bits per symbol, then devise a code to match that entropy, giving A: 0, B: 10, C: 110, D: 111. The new code requires only one and three-quarter bits per symbol to encode messages in this language, versus two bits per symbol, a compression ratio of seven to eight. No method uses fewer bits per symbol to encode this source.

32. David Kahn, *The Codebreakers: The Story of Secret Writing*, rev. ed. (New York: Macmillan, 1996), 741.

33. Dirk J. Struik, "The MIT Department of Mathematics during Its First Seventy-Five Years: Some Recollections," in *A Century of Mathematics in America*, 3 vols., ed. Peter Duren (Providence, RI: American Mathematical Society, 1988–1989), 3:163–177; Steve J. Heims, *John von Neumann and Norbert Wiener: From Mathematics to the Technologies of Life and Death* (Cambridge, MA: MIT Press, 1980), chap. 3; and Masani, *Norbert Wiener*.

34. Norbert Wiener, "Fourier Analysis and Asymptotic Series," appendix B to Vannevar Bush, *Operational Circuit Analysis* (New York: John Wiley, 1929), 366–379; Susann Puchta, "On the Role of Mathematics and Mathematical Knowledge in the Invention of Vannevar Bush's Early Analog Computers," *AHC* 18 (October–December 1996): 49–59; Larry Owens, "Vannevar Bush and the Differential Analyzer: The Text and Context of an Early Computer," *Technology and Culture* 27 (1986): 63–95; and Mindell, *Between Human and Machine*, chap. 5.

35. Yuk-Wing Lee, "Synthesis of Electric Networks by Means of the Fourier Transforms of Laguerre's Function" (Ph.D. diss., MIT, 1930); E. V. Griggs to Norbert Wiener, September 4, 1936; Wiener to Lee, November 2, 1936; J. G. Roberts to Wiener, February 10, 1937; E. W. Adams to Wiener, March 24, 1937; Wiener to W. E. Beatty, March 31, 1937; Wiener to E. W. Adams, May 24, 1937, NW, boxes 3-45, 3-46, 3-47; Lee to Vannevar Bush, July 25, 1941 (vita), NW, box 4-60; Wiener, *I Am a Mathematician: The Later Life of a Prodigy* (New York: Doubleday, 1956), 134–135; Conway and Siegelman, *Dark Hero of the Information Age*, 76, 360; and Masani, *Norbert Wiener*, 163–165, 177–179.

36. Steve J. Heims, "Introduction," in Norbert Wiener, *Invention: The Care and Feeding of Ideas* (rpt. of 1954 mss.; Cambridge, MA: MIT Press, 1993), ix–xxi.

37. Mindell, *Between Human and Machine*, chaps. 7, 10.

38. Ibid., 328. On Bernstein and Mooney, see Norbert Wiener to Warren Weaver, December 21, 1942 (quotation), NW, box 4-63; Wiener, *I Am a Mathematician*, 248–249; and Wiener to Weaver, February 22, 1942, NW, box 4-61. The term *computer* denoted a human calculator before it was applied to machines; see David A. Grier, *When Computers Were Human* (Princeton, NJ: Princeton University Press, 2005).

39. Masani, *Norbert Wiener*, chap. 14; Peter Galison, "The Ontology of the Enemy: Norbert Wiener and the Cybernetic Vision," *Critical Inquiry* 21 (1994): 228–266; Edwards, *Closed World*, 180–187; and Mindell, *Between Human and Machine*, chap. 8. In contrast to these authors, Mindell argues that Wiener "was reacting to and building on an evolving understanding, pervasive among engineers and psychologists, that the boundary between humans and machines affected the performance of dynamic systems and was a fruitful area of research" (285–286).

40. Segal, *Le zéro et le un*, 93–98, gives the fullest account to date of Wiener's theory of information. My account differs from Segal's in that I stress the inter-

actions between Shannon and Wiener, the differences in their theories, and how those differences relate to their wartime projects.

41. *NWCyb*, 18, his emphasis.

42. Galison, "Ontology of the Enemy," 233–245; Masani, *Norbert Wiener*, 181–185; and Mindell, *Between Human and Machine*, 277–283.

43. That is, they assumed that an "ensemble" of such time series was "stationary" (its statistical properties were independent of the time axis) and "ergodic" (statistically homogeneous).

44. Galison, "Ontology of the Enemy," 238–239; and Masani, *Norbert Wiener*, 185–187.

45. Mindell, *Between Human and Machine*, 243–244, 280–283; Masani, *Norbert Wiener*, 189–190; and Galison, "Ontology of the Enemy," 244–245. Mindell speculates that the termination may have been caused by friction between Weaver and Wiener. Masani and Galison say Wiener recognized major problems with his statistical approach in solving the antiaircraft problem.

46. Mindell, *Between Human and Machine*, 6–7, 277.

47. Thornton Fry to Norbert Wiener, July 22, 1936, NW, box 3-45.

48. Mindell, *Between Human and Machine*, 383n.4.

49. Norbert Wiener to Warren Weaver, February 22, 1942, NW, box 4-61.

50. Norbert Wiener, "The Extrapolation, Interpolation, and Smoothing of Stationary Time Series with Engineering Applications," NDRC report, February 1, 1942, copy no. 136 of 300 copies, NW, box 28A-561A, which does, indeed, have a yellow cover; and Wiener, *The Extrapolation, Interpolation, and Smoothing of Stationary Time Series with Engineering Applications* (Cambridge, MA, and New York: Technology Press and John Wiley, 1949). In *I Am a Mathematician*, 255, Wiener noted that before the war, "Yellow Peril" commonly referred to the yellow-bound mathematical books published by Springer Verlag.

51. Transcript of interview with Julian Bigelow by Steve J. Heims, November 12, 1968, SH, box 1-5, pp. 4, 7, 10. Wiener thought enough of Bigelow's mathematical abilities to mention him as a possible candidate for an applied mathematics position at the University of Pennsylvania. See Norbert Wiener to J. R. Kline, April 10, 1941, NW, box 4-59.

52. Mindell, *Between Human and Machine*, 279–280; and Masani, *Norbert Wiener*, 193. On the history of time series up to Wold, see Judy L. Klein, *Statistical Visions in Time: A History of Time Series Analysis, 1662–1938* (Cambridge: Cambridge University Press, 1997).

53. Wiener, "Extrapolation, Interpolation, and Smoothing of Stationary Time Series," 3 (quotation), 15. Segal, *Le zéro et le un*, 98, translates the quotation into French and credits Shannon with developing the essential idea into a measure of information. He does not analyze Wiener's use of "statistical information" in the NDRC report.

54. Rosenblueth, Wiener, and Bigelow, "Behavior, Purpose and Teleology," 21; and Segal, *Le zéro et le un*, 181–182.

55. Norbert Wiener to Arturo and Virginia [Rosenblueth], June 29, 1944, NW, box 4-66.

56. Hartley, "Transmission of Information." The term gained currency in 1946 when Denis Gabor in Britain published, "Theory of Communication," *Journal of the IEE*, pt. 3, 93 (1946): 429–459. Gabor redefined "amount of information," as used by Hartley, in terms of a "quantum of information," which he called a "logon," as the product of uncertainties in the time and frequency of an electrical signal.

57. P. C. Mahalanobis, "Biography," in Ronald A. Fisher, *Contributions to Mathematical Statistics* (New York: Wiley, 1950), 265–272, on 270; Segal, *Le zéro et le un*, 32–44; and Fisher, *The Design of Experiments* (London: Oliver & Boyd, 1935).

58. MIT Department of Mathematics, Program of Education and Research in Applied Mathematics, "Report to the Visiting Committee," January 15, 1944, MITMath, folder 1/3, 8-11, on 10.

59. On the Teleological Society, see Heims, *Cybernetics Group*, 40–46; and Heims, *John von Neumann and Norbert Wiener*, 184–186.

60. Norbert Wiener to E. H. Vestine, December 4, 1944, NW, box 4-66; John von Neumann to Howard Aiken et al., January 12, 1945, NW, box 4-67; and Wiener to von Neumann, January 24, 1945, NW, box 4-67.

61. Norbert Wiener to Warren McCulloch, August 30, 1943; and Wiener to N. Rashevsky, August 30, 1943 (quotation), both in NW, box 4-65.

62. Warren McCulloch to Arturo Rosenblueth, March 29, 1943, WSM, Lettvin folder.

63. Norbert Wiener to John von Neumann, January 12, 1945, NW, box 4-67; Wiener to E. J. McShane, January 7, 1946, NW, box 5-70; Wiener to Arturo Rosenblueth, January 6, 1947, NW, box 5-73; and Wiener to Joseph Doob, April 8, 1947, NW, box 5-77. The title of the paper was "Prediction Theory and Time Series."

64. Warren McCulloch, "To the Members of the Conference on Teleological Mechanisms—October 23 & 24, 1947," n.d., WSM, Macy Meetings I–VII, folder "IV 1947 October," 4. The transcription of a noisy SoundScriber recording verifies McCulloch's summary of Wiener's remarks. See "Conference on Feedback Mechanisms and Circular Causal Systems . . . , March 8 and 9, 1946," SH, 154. Reliable transcripts of the conferences were not made until the sixth meeting in 1949, when they began to be published. See Heims, *Cybernetics Group*, 18.

65. "Conference on Teleological Mechanisms," September 20, 1946, Sound-Scriber transcript, Log. Bet, 7.4 and 9.4, Disc. 1, WSM, Cybernetics Conferences.

66. Margaret Mead, notebook, October 18, 1946, folder "Macy Foundation . . . Conference on Feedback, 1946"; and Mead, notebook, March 13, 1947, folder "Macy Foundation . . . Conference on Feedback, 1947 (March)," MM, box F42. This is corroborated by McCulloch, "To the Members of the Conference on Teleological Systems," 7–8.

67. Norbert Wiener to J. Gordon Deutsch, February 26, 1947, NW, box 5-75; Warren Weaver diary, May 19, 1947, RFA, RG 12.1; and notes on interview with

Robert Fano by Steve J. Heims, April 19, 1978, SH, box 1-3. See, also, Fano, *Transmission of Information*, vii.

68. Filmer S. C. Northrop to Norbert Wiener, January 10 and February 12, 1947; Warren McCulloch to Wiener, April 8 and April 18, 1947; McCulloch to Northrop, April 15, 1947 (quotation); Wiener to McCulloch, April 16 and May 2, 1947, Wiener to Northrop, May 6, 1947, all in NW, box 5, folders 73, 75, 77, 78; program for "Conference on Teleological Mechanisms," New York Academy of Sciences, October 21–22, 1946; McCulloch to Northrop, March 18, 1947, Northrop to Wiener, May 5, 1947, McCulloch to Janet Freed, August 14, 1947, and Wiener and Arturo Rosenblueth to McCulloch, August 9, 1947, all in WSM—Macy II, III, IV, and WSM, ser. I, Rosenblueth correspondence; and Wiener, "Time, Communication, and the Nervous System," *Annals of the New York Academy of Sciences* 50 (1948): 197–219.

69. References are to the version McCulloch and Pitts edited: Norbert Wiener, "Time, Communication and the Nervous System," n.d. [April 1947], in WSM—Macy III. It was probably attached to McCulloch to Wiener, April 18, 1947, NW, box 5-77.

70. Warren Weaver used the same example to popularize Shannon's theory. Compare ibid., 7–8, with Weaver, "Recent Contributions to the Mathematical Theory of Communication," 101.

71. Wiener, "Time, Communication and the Nervous System," mss., 8, 12, 22 (quotations).

72. Norbert Wiener to E. Freymann, April 2, 1948, NW, box 5-82.

73. Erwin Schrödinger, *What Is Life?* (Cambridge: Cambridge University Press, 1944), 76; *NWCyb*, 19; and Kay, *Who Wrote the Book of Life?*, 64, 85. The reference was added at the galley proof stage. Compare Wiener, "Introduction," typescript, n.d., ca. November 1947, NW, box 28A-577, p. 16; with galley proof, Hermann et Cie, February 27, 1948, NW, box 28B-592.

74. Warren McCulloch, "Recollections of the Many Sources of Cybernetics," 1974; rpt. in *WSMCW*, 1:21–49, on 39; and unsigned [Julian Bigelow] to Warren Weaver, November 5, 1941, NW, box 4-61.

75. MIT Department of Mathematics, "Report to the Visiting Committee," January 15, 1944, MITMath, folder 1/3, 6–7.

76. Norbert Wiener to Giorgio de Santillana, October 16, 1945, NW, box 4-69; Wiener to Rosenblueth, January 24, 1946, NW, box 5-70; and *NWCyb*, 22.

77. Wiener, *I Am a Mathematician*, 179, 263.

78. N. J. A. Sloane and Aaron D. Wyner, "Biography of Claude Elwood Shannon," in *CESCP*, xi–xvii, on xi; and Anthony Liversidge, "Profile of Claude Shannon," 1987, rpt. in *CESCP*, xix–xxxiii, on xxv.

79. Owens, "Vannevar Bush and the Differential Analyzer," 63; and G. Pascal Zachary, *Endless Frontier: Vannevar Bush, Engineer of the American Century* (Cambridge, MA: MIT Press, 1997), 49–53, 73–74.

80. Liversidge, "Profile of Claude Shannon," xxv.

81. Sloane and Wyner, "Biography of Claude Elwood Shannon," xi–xii; Claude Shannon, "A Symbolic Analysis of Relay and Switching Circuits" (M.S. thesis, MIT, 1940), MITA.

82. For earlier examples in electrical power engineering, see Ronald R. Kline, *Steinmetz: Engineer and Socialist* (Baltimore: Johns Hopkins University Press, 1992), chaps. 3, 5.

83. Charles Rich to Vannevar Bush, May 4, 1938, CESLC, box 1; and Bush recommendation letter for Claude Shannon, December 18, 1939, VB, box 102-2401. Shannon thanks Hitchcock, Bush, and Samuel Caldwell for advice on the published paper; see Shannon, "A Symbolic Analysis of Relay and Switching Circuits," 713. On Bush's working on the Rapid Selector, see Colin Burke, *Information and Secrecy: Vannevar Bush, Ultra, and the Other Memex* (Metuchen, NJ: Scarecrow Press, 1994), 185.

84. Vannevar Bush to E. B. Wilson, December 15, 1938, VB, box 102-2401. On Bush's going to Washington, see Zachary, *Endless Frontier*, 89, 93.

85. Vannevar Bush to Barbara Burks, January 5, 1939, Burks to Bush, January 20, 1939, both in VB, box 18-403; and *American Men of Science*, 6th ed. (New York: Science Press, 1938), s.v., "Burks, Barbara S." On the checkered history of the Eugenics Record Office, see Garland E. Allen, "The Eugenics Record Office at Cold Spring Harbor, 1910–1940: An Essay in Institutional History," *Osiris*, 2nd ser., 2 (1986): 225–264.

86. Barbara Burks to Vannevar Bush, January 20, 1939, VB, box 18-403.

87. MIT Department of Mathematics, "Report to the Visiting Committee," May 15, 1939, MITMath, folder 1/3, 8; and Struik, "The MIT Department of Mathematics," 168 (quotation).

88. Liversidge, "Profile of Claude Shannon," xxvii.

89. Wiener, *I Am a Mathematician*, 179.

90. Norbert Wiener to J. R. Kline, April 10, 1941, NW, box 4-59.

91. Claude Shannon to Vannevar Bush, December 13, 1939, VB, box 102-2401; and Shannon, "An Algebra for Theoretical Genetics" (Ph.D. diss., MIT, 1940), rpt. in *CESCP*, 891–920.

92. On Bush's conservatism, see Nathan Reingold, "Vannevar Bush's New Deal for Research: Or the Triumph of the Old Order," *Historical Studies in the Physical and Biological Sciences* 17 (1987): 299–344. On patriarchal attitudes toward women scientists at this time, see Margaret W. Rossiter, *Women Scientists in America: Strategies and Struggles to 1940* (Baltimore: Johns Hopkins University Press, 1982).

93. Sloane and Wyner, "Biography of Claude Elwood Shannon," xii.

94. Claude Shannon to Vannevar Bush, February 16, 1939, VB, box 102-2401; rpt. in Friedrich-Wilhelm Hagemeyer, "Die Entstehung von Informationskonzepten in der Nachrichtentechnik" (Ph.D. diss., Freie Universität Berlin, 1979), 504–505; extracted in *CESCP*, 455–456. On the importance of this letter, see Hagemeyer, 423–425; and Segal, *Le zéro et le un*, 84–85.

95. Hartley, "Transmission of Information." On the use of the term *intelligence*, see e.g., Harry Nyquist, "Certain Factors Affecting Telegraph Speed," *BSTJ* 3 (1924): 324–346.

96. Claude Shannon to Vannevar Bush, December 15, 1939; Bush to Shannon, December 18, 1939; Shannon to Bush, March 8, 1940; Bush to Shannon, March 11, 1940, Bush to Shannon, April 30, 1940, all in VB, box 102-2401; and William Aspray, "The Emergence of Princeton as a World Center for Mathematical Research, 1896–1939," in *History and Philosophy of Modern Mathematics*, ed. Aspray and Philip Kitcher (Minneapolis: University of Minnesota Press, 1988), 346–366.

97. Hermann Weyl to Claude Shannon, April 11, 1940, CESLC, box 1.

98. Claude Shannon to Vannevar Bush, June 5, 1940, and Bush to Shannon, June 7, 1940, VB, box 102-2401.

99. Warren Weaver to Ross G. Harrison, November 1, 1940, and Harrison to Weaver, November 5, 1940, RFA, RG 1.1, series 200E, box 171-2080.

100. Mindell, *Between Human and Machine*, 289, 328.

101. Warren Weaver to Vannevar Bush, October 24, 1949, VB, box 117-2801. Shannon wrote five reports analyzing fire-control systems under this contract. See Mindell, *Between Human and Machine*, 196–197, 289–290.

102. Claude Shannon to Vannevar Bush, March 8, 1940, VB, box 102-2401, his question mark.

103. Warren Weaver to Vannevar Bush, October 24, 1949, VB, box 117-2801; Sloane and Wyner, "Biography of Claude Elwood Shannon," xiv; Norma Barzman, *The Red and the Blacklist: The Intimate Memoir of a Hollywood Expatriate* (New York: Thunder's Mouth Press, 2003), 378–379; and William Poundstone, *Fortune's Formula: The Untold Story of the Scientific Betting System That Beat the Casinos and Wall Street* (New York: Hill and Wang, 2005), 22–23, 27. In 1949, Shannon married Mary Elizabeth (Betty) Moore, a mathematician and human "computer" in his department at Bell Labs, a marriage that lasted until his death from Alzheimer's disease in 2001.

104. Warren Weaver to John Atanasoff, October 15, 1941, NW, box 4-61. On Stibitz and Fry witnessing the demonstration, see Mindell, *Between Human and Machine*, 280.

105. Conway and Siegelman, *Dark Hero of the Information Age*, 126. For a similar, briefer comment, see transcript of an interview with Julian Bigelow by Steve J. Heims, November 12, 1968, SH, box 1-5, p. 45. I have found no archival evidence to support Bigelow's recollections.

106. Shannon, "A Mathematical Theory of Communication," 652. See also 626–627n.4.

107. S. Millman, ed., *A History of Engineering and Science in the Bell System: Communication Sciences, 1925–1980* (New York: BTL, 1984), 104, 405; and M. D. Fagan, ed., *A History of Engineering and Science in the Bell System: National Service in War and Peace, 1925–1975* (New York: BTL, 1978), 296–313.

108. Fagan, ed., *History of Engineering and Science in the Bell System*, 316, 317.

109. Claude E. Shannon, "Analogue of the Vernam System for Continuous Time Series," Memorandum 43-110-44, BTL, May 10, 1943; rpt. in *CESCP*, 144–147, on 144. On the Vernam system, see Kahn, *Codebreakers*, 394–402.

110. Jonathan Sterne, *MP3: The Meaning of a Format* (Durham, NC: Duke University Press, 2012), 78–88. In stressing the commercial origins of Shannon's theory, Sterne does not note the tie between PCM and wartime cryptography. For the relationship between PCM and Shannon's concept of channel capacity, see B. M. Oliver, J. R. Pierce, and C. E. Shannon, "The Philosophy of PCM," *Proceedings of the IRE* 36 (1948): 1324–1331.

111. Andrew Hodges, *Alan Turing: The Enigma* (1983; rpt., New York: Walker, 2000), 235–237, 242–244.

112. Ibid., 249–251, quotations on 249, 250; Claude Shannon to Vannevar Bush, March 8, 1940, VB, box 102-2401; and Robert Price, "A Conversation with Claude Shannon," *IEEE Communications Magazine* 22, no. 5 (1984): 123–126. Hodges said Turing's "*ban* of weight of evidence made something ten times as likely," Shannon's "*bit* of information made something twice as definite" (250, his emphasis).

113. Kahn, *Codebreakers,* 744.

114. Price, "A Conversation with Claude Shannon," 124; and Claude Shannon, "Communication Theory of Secrecy Systems," *BSTJ* 28 (1949): 656–715.

115. The declassified version of the report published in 1949 omitted this sentence. Compare Claude Shannon, "A Mathematical Theory of Cryptography," Memorandum 45-110-92, BTL, September 1, 1945, ATTA, Reel FL-4620, vol. F, 2, with Shannon, "Communication Theory of Secrecy Systems."

116. Shannon, "A Mathematical Theory of Cryptography," 3.

117. Ibid., 3, 5; and Kahn, *Codebreakers*, 748.

118. Shannon, "A Mathematical Theory of Cryptography," 7, 9, 10. He also uses the phrase "total amount of information" on 10.

119. Compare ibid., 7–24, with Shannon, "A Mathematical Theory of Communication," 384–396.

120. Shannon, "A Mathematical Theory of Cryptography," 10–11, 18–23, 56–64.

121. Shannon, "A Mathematical Theory of Communication," 393.

122. Warren Weaver to Norbert Wiener, December 21, 1948, NW, box 6-87. I have found no answer to the letter in the archives.

123. Norbert Wiener to Francis Bello, October 13, 1953, NW, box 12-179.

124. Norbert Wiener, "[Review of] *The Mathematical Theory of Communication* . . . ," *Physics Today* 3 (September 1950): 31–32, on 31.

125. Wiener, *I Am a Mathematician*, 263, 264.

126. "Annual Meeting," *Proceedings of the IRE* 36 (1948): 366–367, on 366. On the organization of the meeting, see L. V. Berkner to Norbert Wiener, December 26, 1947, NW, box 5-79.

127. Henry Wallman to Claude Shannon, March 29, 1948, CESLC, box 1.

## Chapter 2: Circular Causality

1. "Conference on Teleological Mechanisms, September 20, 1946," n.d. (ca. October 1946), transcript, uneven pagination, SH, box 1, folder 8; and Steve J. Heims, *The Cybernetics Group* (Cambridge, MA: MIT Press, 1991), 28, 183–187, 193. Wiener later raised these concerns in *NWCyb*, 33–35, 189–191. In addition to Bateson, the following social scientists attended the special meeting in September 1946: sociologists Paul Lazarsfeld and Robert Merton, from Columbia, and Talcott Parsons, from Harvard; anthropologist Clyde Kluckholm, from Harvard; and psychologist Theodore Schneirla, from the American Museum of Natural History. Lazarsfeld was a permanent member of the cybernetics group, who resigned in 1948. Schneirla became a permanent member. See Heims, 285–286.

2. Gregory Bateson to Warren McCulloch, April 2, 1947, WSM—Macy IV; and Heims, *Cybernetics Group*, 87–89, 237–238.

3. Gregory Bateson to Warren McCulloch, November 27, 1946, WSM—Macy III.

4. Gregory Bateson to Warren McCulloch, January 19, 1948, WSM—Macy IV (quotations); Margaret Mead to Alex Bavelas, February 10, 1948, MM, box F41; and McCulloch, "To the Members of the Feedback Conference on Teleological Mechanisms," n.d. (ca. February 15, 1948), MM, box F41. On their separation in 1946 and divorce in 1950, see David Lipset, *Gregory Bateson: The Legacy of a Scientist* (Boston: Beacon Press, 1982), 175–176; Mary Catherine Bateson, *With a Daughter's Eye: A Memoir of Margaret Mead and Gregory Bateson* (New York: William Morrow, 1984), 49–50; Jane Howard, *Margaret Mead: A Life* (New York: Simon and Schuster, 1984), 265; and Hilary Lapsley, *Margaret Mead and Ruth Benedict: The Kinship of Women* (Amherst: University of Massachusetts Press, 1999), 295–297.

5. Margaret Mead to Dorothy Lee, January 22, 1948, MM, box F41.

6. Jakobson became an enthusiast for cybernetics and information theory; see chapter 5.

7. Margaret Mead, remarks at "Business Session, Eighth Conference on Cybernetics," WSM—Macy VIII, 26.

8. For the definition of *interdisciplinarity* as "communication and collaboration across academic disciplines," see Jerry A. Jacobs and Scott Frickel, "Interdisciplinarity; A Critical Assessment," *Annual Review of Sociology* 35 (2009): 43–65, quotation on 44. I use *interdisciplinary* as an adjective denoting the same meaning, rather than limiting it to the integration of the knowledge and practices of two or more disciplines, a common usage today. See, e.g., National Academy of Sciences, *Facilitating Interdisciplinary Research* (Washington, DC: National Academies Press, 2005), 26.

9. On the historical importance of the Macy conferences on cybernetics, especially in the social sciences, see Heims, *Cybernetics Group*; Jean-Pierre Dupuy, *The Mechanization of the Mind: On the Origins of Cognitive Science* (Princeton, NJ: Princeton University Press, 2000); and Paul N. Edwards, *The Closed World:*

*Computers and the Politics of Discourse in Cold War America* (Cambridge, MA: MIT Press, 1996), chap. 6.

10. Julie T. Klein, *Interdisciplinarity: History, Theory, and Practice* (Detroit: Wayne State University Press, 1990), chap. 1; and Robert Kohler, *Partners in Science: Foundation Managers and Natural Scientists, 1920–1945* (Chicago: University of Chicago Press, 1991), chaps. 10–12.

11. Frank Fremont-Smith, in "Introductory Discussion," *HVFCyb*, 6:9–26, on 9–10.

12. Dupuy, *Mechanization of the Mind*, 84–87.

13. Warren McCulloch, "To the Members of the Feedback Conference on Teleological Mechanisms," n.d. (ca. February 15, 1948), MM, box F41; and McCulloch to F. S. C. Northrop, March 18, 1947, WSM—Macy IV.

14. Warren McCulloch, "Summary of the Macy Meetings," n.d. (ca. June 1948), WSM—Macy VI, no. 2.

15. Frank Fremont-Smith to Margaret Mead, February 8, 1946, MM, box F41.

16. Unsigned [Frank Fremont-Smith to Margaret Mead], May 11, 1942, MM, box F42; Arturo Rosenblueth, Norbert Wiener, and Julian Bigelow, "Behavior, Purpose and Teleology," *Philosophy of Science* 10 (1943): 18–24; Warren McCulloch and Walter Pitts, "A Logical Calculus of the Ideas Immanent in Nervous Activity," *Bulletin of Mathematical Biophysics* 5 (1943): 115–133; Josiah Macy, Jr. Foundation, *The Josiah Macy, Jr. Foundation, 1930–1955: A Review of Activities* (New York: Macy Foundation, 1955), 20; Heims, *Cybernetics Group*, 14; and George P. Richardson, *Feedback Thought in Social Science and Systems Theory* (Philadelphia: University of Pennsylvania Press, 1991), 96–97.

17. Margaret Mead, "Cybernetics of Cybernetics," in *Purposive Systems: Proceedings of the First Annual Symposium of the American Society for Cybernetics*, ed. Heinz von Foerster et al. (New York: Spartan Books, 1968), 1–11, on 1.

18. Macy Foundation, *Josiah Macy, Jr. Foundation*, xi; Margaret Mead and Paul Beyers, *The Small Conference: An Innovation in Communication* (Paris: Mouton, 1968), 10; and Lapsley, *Margaret Mead and Ruth Benedict*, 233.

19. Lipset, *Gregory Bateson*, 174–175; Howard, *Margaret Mead*, chap. 15; and Bateson, *With a Daughter's Eye*, chap. 3. Mead was executive secretary of the National Research Council's Committee on Food Habits in Washington, DC. Bateson went on a long assignment in psychological warfare for the Office of Strategic Services (the predecessor of the CIA) to Southeast Asia in 1943.

20. Dennis Bryson, "Lawrence K. Frank, Knowledge, and the Production of the Social," *Poetics Today* 19 (1998): 401–421; Heims, *Cybernetics Group*, 170–178; and Dupuy, *Mechanization of the Mind*, 82–84.

21. Howard Aiken, John von Neumann, and Norbert Wiener to E. H. Vestine, December 4, 1944, NW, box 4-66; and Peter Galison, "Ontology of the Enemy: Norbert Wiener and the Cybernetic Vision," *Critical Inquiry* 21 (1994): 228–266, on 247–248.

22. Howard Aiken, John von Neumann, and Norbert Wiener to H. Goldstein, December 28, 1944, NW, box 4-66; and Galison, "Ontology of the Enemy," 248n.50.

23. Norbert Wiener to Arturo Rosenblueth, January 24, 1945, NW, box 4-67, quoted in Steve Heims, *John von Neumann and Norbert Wiener: From Mathematics to the Technologies of Life and Death* (Cambridge, MA: MIT Press, 1980), 185; and Galison, "The Ontology of the Enemy," 248.

24. John von Neumann to Howard Aiken et al., January 12, 1945, NW, box 4-67.

25. See, e.g., John von Neumann to Norbert Wiener, February 3, 1945, NW, box 4-67.

26. Heims, *John von Neumann and Norbert Wiener*, 185–189.

27. Ibid., 189–190; and William Aspray, *John von Neumann and the Origins of Modern Computing* (Cambridge, MA: MIT Press, 1990), 50–51.

28. "Conference on Teleological Mechanisms," September 20, 1946, Sound-Scriber transcript, Log. Bet, 7.4 and 9.4, disc 1, WSM, ser. II, Josiah Macy Jr. Foundation #1. On institutional support, see Henry Moe, "Awards of Guggenheim Post-Service Fellowships," press release, October 22, 1945, APS; Heims, *Cybernetics Group*, 49–51; and Lily Kay, *Who Wrote the Book of Life? A History of the Genetic Code* (Stanford, CA: Stanford University Press, 2000), 82–84.

29. A neurologist by training, Fremont-Smith had strong ties with the neurologists and social scientists who attended the conferences, including Rosenblueth, Bateson, Mead, and Frank. See Heims, *Cybernetics Group*, 170–178.

30. Lily E. Kay, "From Logical Neurons to Poetic Embodiments of Mind: Warren S. McCulloch's Project in Neuroscience," *Science in Context* 14 (2001): 591–614; and Tara Abraham, "Physiological Circuits: The Intellectual Origins of the McCulloch-Pitts Neural Networks," *Journal of the History of the Behavioral Sciences* 38 (2002): 3–25, on 10–11.

31. Warren McCulloch to Frank Fremont-Smith, June 24, 1942, October 23, 1943, November 10, 1943, and January 23, 1946. Macy funding acted as seed money that enabled McCulloch to get further grants in schizophrenia, EEG, and electric shock. See McCulloch to Fremont-Smith, July 1, 1948. All in WSM, ser. I, Fremont-Smith correspondence.

32. Warren McCulloch to Frank Fremont-Smith, January 23, 1946, WSM, ser. I, Fremont-Smith correspondence; and McCulloch to Walter Pitts, February 2, 1946, WSM, ser. I, Pitts correspondence.

33. Warren McCulloch to Frank Fremont-Smith, January 23, 1946, WSM, ser. I, Fremont-Smith correspondence; Gregory Bateson to McCulloch, November 27, 1946, WSM—Macy III; and Batseon, "Physical Thinking and Social Problems," *Science*, n.s., 103 (June 21, 1946): 717–718.

34. Norbert Wiener to Warren McCulloch, February 15, 1946; Frank Fremont-Smith to McCulloch, February 18, 1946; and Fremont-Smith to Wiener, February 28, 1946, all in WSM—Macy I.

35. Norbert Wiener to John von Neumann, February 5, 1946, NW, box 5-70; and Wiener to Warren McCulloch, February 15, 1946, WSM—Macy I.

36. Heims, *Cybernetics Group*. For criticism, see, e.g., Arturo Rosenbleuth to Warren McCulloch, July 8, 1946, WSM, ser. I, Rosenblueth correspondence; and Rosenblueth to McCulloch, February 3, 1947, WSM—Macy III.

37. Warren McCulloch, "Recollections of the Many Sources of Cybernetics," 1974; rpt. in *WSMCW*, 1:21–49, on 41; and Stewart Brand, ed., "For God's Sake, Margaret: Conversation with Gregory Bateson and Margaret Mead," *CoEvolution Quarterly* 10 (Summer 1976): 32–44, on 33.

38. According to Dupuy, *Mechanization of the Mind*, 79, these themes account for about one-half of the fifty-three published "units of discussion." The other half includes neuroses and pathology of mental life, human and social communication, and abnormal communication, which is covered in Heims, *Cybernetics Group*, chaps. 6–7, 9–10.

39. Discussion of Alex Bavelas, "Communication Patterns in Problem-Solving Groups," in *HVFCyb*, 8:1–44, on 42.

40. On the history and philosophy of scientific models, see Mary S. Morgan and Margaret Morrison, eds., *Models as Mediators: Perspectives on Natural and Social Science* (Cambridge: Cambridge University Press, 1999).

41. Arturo Rosenblueth and Norbert Wiener, "The Role of Models in Science," *Philosophy of Science* 12 (1945): 316–332; and Hans Lukas Teuber to Warren McCulloch, November 10, 1947, MM, box F41.

42. Margaret Mead to Erik Erikson, February 26, 1948, MM, box F41.

43. Warren McCulloch to Hans Lukas Teuber, December 10, 1947, WSM—Macy IV, 2.

44. Ralph W. Gerard, "Some of the Problems Concerning Digital Notions in the Central Nervous System," in *HVFCyb*, 7:11–57, on 11, 12.

45. The debate is discussed in Edwards, *Closed World*, 192; and Dupuy, *Mechanization of the Mind*.

46. Abraham, "Physiological Circuits"; Aspray, *John von Neumann*, chap. 3; and George Dyson, *Turing's Cathedral: The Origins of the Digital Universe* (New York: Random House, 2012).

47. Discussion of Gerard, "Problems Concerning Digital Notions in the Central Nervous System," in *HVFCyb*, 7:19, 20, 22.

48. Ibid., 47.

49. Ibid., 19, 43.

50. Ibid., 13.

51. Ibid., 26–27.

52. Ibid., 27.

53. Ibid., 32 (quotation), 33–36. Wiener made the distinction that *analog* referred to measuring; *digital*, to counting (43).

54. Ibid., 36, 43, 44. The origin of the term *analog* in reference to computers is still unclear; see David Mindell, *Between Human and Machine: Feedback, Con-*

*trol, and Computing before Cybernetics* (Baltimore: Johns Hopkins University Press, 2002), 387n.57.

55. Discussion of Gerard, "Problems Concerning Digital Notions in the Central Nervous System," in *HVFCyb*, 7:37.

56. Ibid., 50.

57. Claude Shannon, "Presentation of a Maze-Solving Machine," in *HVFCyb*, 8:173–180, on 175, 176; BTL, Press Release, May 2, 1952, ATTA, location 196 03 01; John Pfeiffer, "This Mouse Is Smarter Than You Are," *Popular Science* 160 (March 1952): 99–101; "Electrical Mouse, New Memory Prodigy . . . ," *NYT*, May 6, 1952, 31; "Mouse with a Memory," *Time*, May 19, 1952, 59–60; and Roberto Cordeschi, *The Discovery of the Artificial: Behavior, Mind and Machines before and beyond Cybernetics* (Dordrecht: Kluwer, 2002), 158–160.

58. Warren McCulloch, remarks in "Business Session, Eighth Conference on Cybernetics," WSM—Macy VIII, 24; and Shannon, "Presentation of a Maze-Solving Machine," 175, 179 (quotation).

59. See, e.g., N. Katharine Hayles, *How We Became Posthuman: Virtual Bodies in Cybernetics, Literature, and Informatics* (Chicago: University of Chicago Press, 1999).

60. Heinz von Foerster, Margaret Mead, and Hans Teuber, "A Note by the Editors," in *HVFCyb*, 8:xi–xx, on xvii–xix, their emphasis.

61. W. Grey Walter to Warren McCulloch, September 22, 1949, NW, box 7-104; Philip Husbands and Owen Holland, "The Ratio Club: A Hub of British Cybernetics," in *The Mechanical Mind in History*, ed. Husbands, Holland, and Michael Wheeler (Cambridge, MA: MIT Press, 2008), chap. 6, on 90, 93; and Andrew Pickering, *The Cybernetic Brain: Sketches of Another Future* (Chicago: University of Chicago Press, 2010), chap. 4.

62. Warren McCulloch to Margaret Mead, January 14, 1952; and McCulloch "To the Members of the Feedback Conference on Teleological Mechanisms," January 16, 1952, both in MM, box F42.

63. W. Ross Ashby, *Design for a Brain: The Origin of Adaptive Behaviour* (New York: John Wiley, 1952), chaps. 1–7. Ashby also used the term *black box* at the 1952 Macy meeting; see Discussion of Ashby, "Homeostasis," in *HVFCyb*, 9:73–108, on 79. Wiener used the term in commenting on a draft of Ashby's book; see Norbert Wiener to Ashby, March 7, 1951, NW, box 9-135.

64. Ashby, *Design for a Brain*, chap. 8. The random search process was mechanized by step switches selecting different resistors, whose values were chosen randomly, to change the field of the search.

65. Ibid., 130.

66. Ashby, "Homeostasis," 73.

67. Discussion of Ashby, "Homeostasis," in *HVFCyb*, 9:95.

68. Ibid., 104, 105.

69. Ibid., 106, 107.

70. Ibid., 106.

71. Discussion of John F. Young, "Discrimination and Learning in Octopus," in *HVFCyb*, 9:109–119, on 111, 114, 117, 118, 119.

72. Hayles, *How We Became Posthuman*, 16, 65–67.

73. Warren McCulloch to Donald MacKay, April 23, 1952, WSM, ser. I, MacKay correspondence.

74. *NWHUHB*, 2nd ed., 38. For Wiener's comments on the mathematics in a draft of Ashby's book manuscript, see Wiener to W. Ross Ashby, March 7, 1951, NW, box 9-135. Warren McCulloch agreed with Warren Weaver about "Ashby's mathematical naivete," but he admired him as a psychiatrist for learning so much mathematics. See McCulloch to Weaver, March 12, 1951, WSM, ser. I, Rockefeller Foundation correspondence. Ashby asked for help from the mathematicians at the Macy meetings; see Discussion of Ashby, "Homeostasis," in *HVFCyb*, 9:75, 90.

75. Discussion of Ralph W. Gerard, "Central Excitation and Inhibition," in *HVFCyb*, 9:127–150, on 147–148.

76. Max Black, *Models and Metaphors: Studies in Language and Philosophy* (Ithaca, NY: Cornell University Press, 1962), chap. 13.

77. Norbert Wiener and F. S. C. Northrop, the only philosopher at the Macy conferences on cybernetics, also served on the editorial board of *Philosophy of Science*. On scientists as philosophers of science in this period, see George A. Reisch, *How the Cold War Transformed Philosophy of Science: To the Icy Slopes of Logic* (Cambridge: Cambridge University Press, 2005).

78. McCulloch, "Recollections of the Many Sources of Cybernetics," 31, 36; Heims, *Cybernetics Group*, 122–123; and Richardson, *Feedback Thought*, 46–53, 59–90.

79. "Introductory Discussion," *HVFCyb*, 6:25.

80. Discussion of John Stroud, "The Psychological Moment in Perception," in *HVFCyb*, 6:27–63, on 37, 47, 53–54, 62. When McCulloch wrote to Stroud to invite him to the seventh conference, he hoped Stroud was "keeping up on Claude Shannon's work" because McCulloch wanted to invite him. See McCulloch to Stroud, November 22, 1949, WSM—Macy VII.

81. Kubie had switched from neurophysiology to psychiatry after the war, much to the chagrin of McCulloch, who distrusted psychiatry. See Heims, *Cybernetics Group*, chap. 6.

82. Discussion of Lawrence S. Kubie, "The Neurotic Potential and Human Adaptation," in *HVFCyb*, 6:64–111, on 107. Wiener had made a similar point at the first Macy conference; see Heims, *Cybernetics Group*, 25.

83. Warren McCulloch to Claude Shannon, May 24, 1949; and Shannon to McCulloch, June 2, 1949, both in WSM, ser. I, Shannon correspondence.

84. Warren McCulloch to Claude Shannon, March 11, 1950, CESLC, box 1.

85. This terminology comes from semiotics. See Charles W. Morris, *Foundations of the Theory of Signs* (Chicago: University of Chicago Press, 1938).

86. Claude Shannon, "The Redundancy of English," in *HVFCyb*, 7:123–158, on 123–125. For a more extensive account, see Shannon, "Prediction and Entropy

of Printed English," *BSTJ* 30 (1951): 350–364. The experiments were conducted because of the immense difficulty in calculating the redundancy of a natural language using entropy equations.

87. Discussion of Shannon, "The Redundancy of English," in *HVFCyb*, 7:129, 135.

88. Ibid., 146 (quotation), 149–154.

89. Ibid., 154–155.

90. Ibid., 155–158. Licklider's paper was "The Manner in Which and Extent to Which Speech Can Be Distorted and Remain Intelligible," in *HVFCyb*, 7:58–122.

91. Warren Weaver diary, April 14, 1950, RFA, RG 12.1.

92. Donald MacKay to Warren McCulloch, May 2, 1950; McCulloch to Warren Weaver, May 5, 1950; Weaver to McCulloch, May 22, 1950; MacKay to McCulloch, May 24, 1950; McCulloch to MacKay, November 27, 1950; and MacKay, "Report on Fellowship of the Rockefeller Foundation in the Natural Sciences, January–December 1951," n.d., attached to MacKay to McCulloch, January 22, 1952. Before he came over, MacKay had met Shannon and Wiesner at the London Symposium and at a dinner at the Ratio Club in September 1950. See MacKay to McCulloch, October 5, 1950. All documents in WSM, ser. I, MacKay correspondence. On their friendship, see WSM, ser. I, MacKay correspondence, 1951 to 1969, the year of McCulloch's death.

93. For more on MacKay's theory, see chapter 4.

94. Donald MacKay, "In Search of Basic Symbols," in *HVFCyb*, 8:181–221, on 181, 183, 185, 187, his emphasis.

95. Ibid., 189–191, 195–197.

96. See, e.g., Hayles, *How We Became Posthuman*, 54–56.

97. Discussion of MacKay, "In Search of Basic Symbols," in *HVFCyb*, 8: 207–208.

98. Ibid., 219.

99. Ibid., 220.

100. Von Foerster, Mead, and Teuber, "Note by the Editors," xiii, xiv.

101. Warren McCulloch's remarks at "Business Session, Ninth Conference on Cybernetics . . . March 21, 1952," n.d., SH, box 1-8. Those employing information theory included Ashby, Bateson, Bigelow, McCulloch, Mead, Pitts, Quastler, and Young. See *HVFCyb*, vol. 9, passim.

102. Warren McCulloch, "Summary of the Points of Agreement Reached in the Previous Nine Conferences on Cybernetics," in *HVFCyb*, 10:69–80, on 70, 75, 80. The term *negentropy* was coined by physicist Léon Brillouin; see Jérôme Segal, *Le zéro et le un: Histoire de la notion scientifique d'information au 20ᵉ siècle* (Paris: Éditions Syllepse, 2003), 358–359.

103. Yehoshua Bar-Hillel, "Semantic Information and Its Measures," in *HVFCyb*, 10:33–48, on 47; and W. Grey Walter, "Studies on Activity of the Brain," in *HVFCyb*, 10:19–31, on 19. On Grey Walter, see Pickering, *Cybernetic Brain*, chap. 3.

104. Geoffrey C. Bowker, "How to Be Universal: Some Cybernetic Strategies, 1943–1970," *Social Studies of Science* 23 (1993): 107–127; Bowker, *Memory Practices in the Sciences* (Cambridge, MA: MIT Press, 2005), 77–87; and Slava Gerovitch, *From Newspeak to Cyberspeak: A History of Soviet Cybernetics* (Cambridge, MA: MIT Press, 2002), chap. 2.

105. Scott Frickel, "Building an Interdiscipline: Collective Action Framing and the Rise of Genetic Toxicology," *Social Problems* 51 (2004): 269–287. On cybernetics and other postwar fields as interdisciplines, see Peter Galison, "The Americanization of Unity," *Daedalus* 127 (1998): 45–71, on 66. Cybernetics shares several of the characteristics of an interdiscipline discussed by Frickel (a hybrid form of knowledge and permeable boundaries), but it did not focus on solving specific problems, and it claimed to be a universal science.

106. Von Foerster, Mead, and Teuber, "A Note by the Editors," xi–xiii.

107. Discussion of Stroud, "The Psychological Moment in Perception," *HVF-Cyb*, 6:27–63, on 28, 31, 32, 48. Stroud drew much of his inspiration from British psychologist Kenneth Craik, whom McCulloch admired. See Edwards, *Closed World*, 182, 185–186, 196–199.

108. Hans Lukas Teuber to Warren McCulloch, November 10, 1947, MM, box F41. Ironically, Teuber himself engaged in robot-talk in the letter by using metaphors from digital computing: "Many of the issues [at the meeting] are still reverberating in my mind, even though this mind still oscillates ceaselessly (and undecidedly) 0, 1, 0, 1, 0, 1, on the chances of your computing robot to become an acceptable twin brother of the nervous system."

109. Although Peter Galison, "The Americanization of Unity," *Daedalus* 127 (1998): 45–71, argues that cybernetics was a continuation of the European Unity of Science Movement in the United States, that discourse was not common in the proceedings of the Macy conferences on cybernetics.

110. Warren McCulloch, "A Recapitulation of the Theory, with a Forecast of Several Extensions," *Annals of the New York Academy of Sciences* 50 (1948): 259–271, on 259.

111. McCulloch, "Summary of the Points of Agreement Reached in the Previous Nine Conferences on Cybernetics," 69.

112. Von Foerster, Mead, and Teuber, "A Note by the Editors," xi.

113. Warren McCulloch to Molly Harrower, February 5, 1948, WSM—Macy V. While McCulloch aligned himself with experimentalists in this letter, he looked up to the mathematicians and prided himself on his ability to collaborate with Pitts.

114. Discussion of Gerard, "Central Excitation and Inhibition," in *HVFCyb*, 9:137–138.

115. Norbert Wiener, *I Am a Mathematician: The Later Life of a Prodigy* (New York: Doubleday, 1956), 285.

116. Discussion of Kubie, "The Relationship of Symbolic Function in Language Formation and in Neurosis," in *HVFCyb*, 7:209–235, on 234. On McCulloch's criticism of psychoanalysis, see Heims, *Cybernetics Group*, 129–132.

117. Warren McCulloch to Frank Fremont-Smith, November 18, 1954; and Margaret Mead to McCulloch, November 22, 1954, both in WSM—Macy X. For a satirical review, see George A. Miller, "[Review of] *Cybernetics: Circular Causal and Feedback Mechanisms in Biological and Social Sciences*," *American Journal of Psychology* 66 (1953): 661–663. Miller became a proponent of cybernetics and information theory; see chapter 5.

118. [Cybernetic Group] to Norbert Wiener, April 24, 1953, NW, box 11-171. Northrop was the only one of the twenty-five attendees who did not sign the letter.

119. Norbert Wiener to James Killian, December 2, 1951; and Killian to Wiener, December 11, 1951, both in NW, box 10-144. The letters are partially quoted in Flo Conway and Jim Siegelman, *Dark Hero of the Information Age: In Search of Norbert Wiener, the Father of Cybernetics* (New York: Basic Books, 2005), 221. On McCulloch's lifestyle, see ibid., 168–169.

120. Norbert Wiener to John Young, December 14, 1951, NW, box 10-144.

121. Norbert Wiener to Josiah Macy, Jr., Foundation, February 11, 1952, NW, box 10-146; and Wiener to Fremont-Smith, April 2, 1953, NW, box 11-168. The latter letter was probably not sent, because there is an original in NW.

122. On Wiener's disputes that led to letters of resignation, see Conway and Siegelman, *Dark Hero of the Information Age*, 208–210, 221.

123. Ibid., chap. 11. Heims, *Cybernetics Group*, 138, refers to the breakup as "the specific result of some personal matters involving McCulloch and members of Wiener's family."

124. See, e.g., Conway and Siegelman, *Dark Hero of the Information Age*, 234.

125. Norbert Wiener to W. Grey Walter, April 2, 1953, NW, box 11-168.

### Chapter 3: The Cybernetics Craze

1. Muriel Rukeyser to Norbert Wiener, April 18, 1949; and Wiener to Rukeyser, April 26, 1949, both in NW, box 6-96. On the wartime meeting, see Wiener to Giorgio de Santillana, June 7, 1943, NW, box 4-64. On Wiener's admiration of Gibbs, see *NWHUHB*, 2nd ed., preface.

2. "The Thinking Machine," *Time*, January 23, 1950, 54ff., on 55; and Max Born to Denis Gabor, February 6, 1951, DG, box MB, 10/3.

3. Beverly Brooks to Norbert Wiener, February 21, 1949, NW, box 6-92; and Brooks to Wiener, December 19, 1949, NW, box 7-108. For complaints about the poor proofreading, see Claude Shannon, "[Review of] *Cybernetics . . . ,*" 1949, rpt. in *CESCP*, 872–873.

4. "Salesmen as Critics," *NYT Book Review*, February 13, 1949, 5; and "Machines That Think," *Business Week*, February 19, 1949, 38ff., on 40.

5. "In Man's Image," *Time*, December 27, 1948, 45; Harrison Smith, "The Machine in Man's Image," *Saturday Review of Literature* 32 (January 8, 1949): 22; John Pfeiffer, "The Stuff That Dreams Are Made On," *NYT Book Review*,

January 23, 1949, 27; and Norfolk, VA, *Ledger-Dispatch*, April 14, 1949, in NW, box 25c-378.

6. Will Feller to Norbert Wiener, November 18, 1948, NW, box 6-86; and Joseph Doob to Wiener, December 9, 1948, NW, box 6-87.

7. T. J., Jr., "Cybernetics," *American Speech*, 24 (February 1949): 78; Norbert Wiener to Storm Whaley, Siloam Springs, AK, February 4, 1949, NW, box 6-91 (quotation); and G & C Merriam & Co. to Wiener, June 20, 1949, NW, box 7-100.

8. Diana Bennett to Norbert Wiener, August 6, 1952; Wiener to Bennett, August 15, 1952; and Bennett to Wiener, August 20, 1952, all in NW, box 10-154.

9. N. Katharine Hayles, *How We Became Posthuman: Virtual Bodies in Cybernetics, Literature, and Informatics* (Chicago: University of Chicago Press, 1999); Christina Dunbar-Hester, "Listening to Cybernetics: Music, Machines, Nervous Systems, 1950–1980," *Science, Technology, and Human Values* 35 (2010): 113–139; Andrew Pickering, *Cybernetic Brain: Sketches of Another Future* (Chicago: University of Chicago Press, 2010), 78–89; and James Baldwin, *No Name in the Street* (1972), rpt. in Baldwin, *Collected Essays* (New York: Library of America, 1998), 349–476, on 384.

10. On these blurred boundaries between "upstream and downstream science," see Stephen Hilgartner, "The Dominant View of Popularization: Conceptual Problems, Political Uses," *Social Studies of Science* 20 (1990): 519–539.

11. C. Diane Martin, "The Myth of the Awesome Thinking Machine," *Communications of the ACM* 36 (April 1993): 120–133; and Jennifer S. Light, "When Computers Were Women," *Technology and Culture* 40 (1999): 455–483.

12. *NWCyb*, introduction, chaps. 5–7; and H. Smith, "The Machine in Man's Image."

13. Pfeiffer, "The Stuff That Dreams Are Made On"; and Pfeiffer, "Mechanical Logicians," *NYT Book Review*, December 11, 1949, 19.

14. David Dietz, "Three Scientists Produced Brilliant Books in '49," December 6, 1949, unidentified newspaper clipping, NW, box 28c-597.

15. H. Smith, "The Machine in Man's Image"; "Machines That Think"; and "Chess by Machine," Charleston, WV, *Gazette*, May 21, 1949, clipping in NW, box 25c-378.

16. Barbara Wiener to "Ron," October 13, 1948; and Science Service, "Future Release: Friday Afternoon, October 22," n.d. (revised on October 18, 1948), both in NW, box 6-85.

17. "The Brain Is a Machine," *Newsweek*, November 15, 1948, 89; H. Smith, "The Machine in Man's Image"; and "Machines That Think."

18. Kenneth Smith, "Devaluing Brains in Industry," *Philadelphia Inquirer*, April 10, 1949; and Rennie Taylor, "Mechanical Slaves Forecast," New Orleans, *Item*, May 25, 1949, both in NW, box 25c-378. On these phrases, see *NWCyb*, 37.

19. Norbert Wiener, "World of Robot Brains," *Science Digest* 26 (September 1949): 56–58.

20. Harry M. Davis, "An Interview with Norbert Wiener," *NYT Book Review*, April 10, 1949, 23; and "Revival of R.U.R. with New Prologue," *NYT*, May 7, 1950, E 13.

21. William L. Laurence, "Cybernetics, a New Science, Seeks the Common Elements in Human and Mechanical Brains," *NYT*, December 19, 1948, E 9; T. C. Simmons, "Industrial L.A.," *Los Angeles [Daily] News*, June 3, 1949, NW, box 25c-378; H. Smith, "The Machine in Man's Image"; and Victor Riesel, "Inside Labor," Wilmington, DE, *Star*, May 15, 1949, NW, box 25c-378.

22. K. Smith, "Devaluing Brains in Industry."

23. *NWCyb*, 37–38. For quotations of this passage, see, e.g., H. Smith, "The Machine in Man's Image"; K. Smith, "Devaluing Brains in Industry"; and John B. Thurston, "Devaluing the Human Brain," *Saturday Review of Literature* 32 (April 23, 1949): 24–25, on 24.

24. Amy Sue Bix, *Inventing Ourselves Out of Jobs? America's Debate over Technological Unemployment, 1929–1981* (Baltimore: Johns Hopkins University Press, 2000), 237–250, on 237.

25. "A Key to the Automatic Factory," *Fortune* 40 (November 1949): 139ff.; Norbert Wiener to Congressman John Marsalis, May 22, 1950, NW, box 8-118; and "Come the Revolution," *Time*, November 27, 1950, 66–68, on 66. On Wiener's meeting with Reuther, see Flo Conway and Jim Siegelman, *Dark Hero of the Information Age: In Search of Norbert Wiener, the Father of Cybernetics* (New York: Basic Books, 2005), 245–247.

26. Science Service, "Future Release," 6; and Norbert Wiener, "Cybernetics," *Scientific American* 179 (November 1948): 14–19, on 15–16.

27. See, for example, Laurence, "Cybernetics"; Editor, "Machine-Rule Turns Out Human Misfits," Los Angeles, *Daily News*, December 13, 1949, NW, box 25c-378; and *Time*, "Thinking Machine." For the exception, see John D. Trimmer, "Instrumentation and Cybernetics," *Scientific Monthly* 69 (1949): 328–331, on 330.

28. Harry M. Davis, "Mathematical Machines," *Scientific American* 180 (April 1949): 29–38, on 29 (quotations); and Davis, "Interview with Norbert Wiener." The closest formulations I have found in Wiener's writings are that he wrote about an "age of communication and control," claimed that "information is information, not matter or energy," and contrasted power and communications engineering on the basis of energy versus information. See *NWCyb*, 50, 155; and Wiener, "A New Concept of Communication Engineering," *Electronics* 22 (January 1949): 74–76, on 74.

29. On these mechanisms in the United States and Britain, see John F. Young, *Cybernetics* (London: Iliffe Books, 1969), chap. 7.

30. *NWHUHB*, 196; and Mara Mills, "On Disability and Cybernetics: Helen Keller, Norbert Wiener, and the Hearing Glove," *Differences* 22 (2011): 74–111, 84–85, 92–93, 96–99, quotation on 92.

31. *NWHUHB*, 196–203. For photographs of Wiener using the device, see Mills, "Disability and Cybernetics," 86, 95.

32. "Code Device May Aid Deaf," *NYT*, February 19, 1949, 9; Norbert Wiener, "Sound Communication with the Deaf," *Philosophy of Science* 16 (1949): 260–262; and Jerome B. Wiesner, Wiener, and Leon Levine, "Some Problems in Sensory Prosthesis," *Science*, n.s., 110 (1949): 512.

33. Norbert Wiener, "Problems of Sensory Prosthesis" (1951), rpt. in *NWCW*, 4:413–421; "'Hearing Fingers' May Soon Aid Deaf," *NYT*, December 29, 1949, 3; Alton L. Blakesley, "Electronic Device Enables Hearing via Finger Tips," *Salt Lake City Tribune* (AP story), December 29, 1949, NW, box 7-110; and "U.S. Science Holds Its Biggest Powwow," *Life* 28 (January 9, 1950): 17–23, on 17. On the negative stereotype, see Mills, "Disability and Cybernetics," 84.

34. MIT, RLE, *Quarterly Progress Report*, October 15, 1949, 54–55.

35. Ibid., January 15, 1950, 67 (quotation); April 15, 1950; and January 15, 1951, 75–76.

36. Helen Keller to Norbert Wiener, February 12, 1950, NW, box 7-111. For examples of letters to parents and others, see Wiener to B. M. Frost, March 2, 1949, NW, box 6-93; Jerome Wiesner to K. L. Raheja, September 16, 1949, NW, box 7-104; and Wiener to Peter L. McLaughlin, January 5, 1950, NW, box 7-109. On Keller's interest and visit, see Thornton Fry to Wiener, December 30, 1949, NW, box 7-108; Wiener to Keller, February 11, 1950, NW, box 7-111; and Mills, "Disability and Cybernetics," 83–84.

37. Norbert Wiener to Jerome Wiesner, November 17, 1952, NW, box 11-159, and Wiesner to Wiener, December 1, 1952, NW, box 11-160.

38. Norbert Wiener to Edward Kern, March 8, 1949, NW, box 6-93 (quotation); Kern to Wiener, March 15, 1949, and April 7, 1949, NW, box 6-93.

39. *NWHUHB*, 190–194; and Norbert Wiener to Edward Kern, April 11, 1949, NW, box 6-95.

40. Edward Kern to Norbert Wiener, March 15, 1949, NW, box 6-93; and Jerome Wiesner to Norbert Wiener, November 1, 1949, NW, box 7-106.

41. *Time*, "Thinking Machine." On the exchange about the cartoonist, see Edward Kern to Norbert Wiener, April 7, 1949; and Wiener to Kern, April 11, 1949, both in NW, box 6-95. For a cartoon by Artzybasheff on cybernetics, titled "Executive of the Future," see Herbert Kulby, "Cybernetics: Doom or Destiny?," *Esquire*, April 1952, 65, 132–134; and Domenic J. Iacono, "The Art of Boris Artzybasheff," *Scientific American* 269 (November 1993): 72, 76–80, on 79.

42. Edward Kern to Norbert Wiener, March 9, 1950, NW, box 7-113; *NWHUHB*, 193; John Pfeiffer, "A Machine-Eye View of Why We Behave like Human Beings," *NYT Book Review*, August 20, 1950, 3; and Stuart Chase, "The Next Industrial Revolution: An Age of Machines That Think," *New York Herald Tribune Book Review*, August 20, 1950, 5. Chase called the moth/bedbug a "monster," a term he also applied to ENIAC and other gigantic computing machines.

43. W. Grey Walter, *The Living Brain* (London: Gerald Duckworth, 1953), chap 5; Rhodri Hayward, "The Tortoise and the Love Machine: Grey Walter and

the Politics of Electroencephalography," *Science in Context* 14 (2001): 615–641; and Pickering, *Cybernetic Brain*, chap. 3.

44. W. Grey Walter, "An Imitation of Life," *Scientific American* 182 (May 1950): 42–45; Grey Walter, "A Machine That Learns," *Scientific American* 185 (August 1951): 60–63; Waldemar Kaempffert, "Two Electro-Mechanical 'Tortoises' Exercise Something That Resembles Free Will," *NYT*, March 19, 1950, E 9; John Pfeiffer, "This Mouse Is Smarter Than You Are," *Popular Science* 160 (March 1952): 99–101; "Electrical Mouse, New Memory Prodigy . . . ," *NYT*, May 6, 1952, 31; and "Mouse with a Memory," *Time*, May 19, 1952, 59–60. AT&T supplied a press release on the mouse; see "For Immediate Release," May 2, 1952, ATTA, Microfiche, 1945–1984, at 945–196,03,01. The only popular account I have found of the homeostat in the United States is in Serge Fliegers, "Will Machines Replace the Human Brain?," *American Mercury* 76 (January 1953): 53–61, on 58.

45. Fliegers, "Will Machines Replace the Human Brain?," 53, 58, 60; *Time*, "Thinking Machine," 56; and Serge Fliegers to Mrs. George Baldwin, n.d. (ca. December 1, 1952), NW, box 11-160.

46. "How U.S. Cities Can Prepare for Atomic War," *Life* 29 (December 18, 1950): 77ff., on 81; George Stibitz to Norbert Wiener, March 18, 1951, NW, box 9-135; and Wiener to Stibitz, April 16, 1951, NW, box 9-136. For a reappraisal of Wiener's contributions to computing during the war, see Pesi R. Masani, *Norbert Wiener, 1894–1964* (Basel: Birkhauser, 1990), 171–175.

47. Robert Morison diary, February 21, 1949, RFA, RG 12.1, 47–48. Wiener wanted to make money from his books, telling Warren McCulloch, "I am an author and I eat my royalties. So don't ask for [copies of] *Cybernetics*." See Wiener to McCulloch, November 20, 1950, NW, box 9-129.

48. Norbert Wiener to Arturo Rosenblueth, March 4, 1949, NW, 6-93.

49. Paul Brooks to Norbert Wiener, August 10, 1949, NW, box 7-102; Wiener to Brooks, October 16, 1949, NW, box 7-105; and Wiener to Brooks, November 19, 1949, NW, box 7-107.

50. Paul Brooks to Norbert Wiener, November 10, 1949; Wiener to Brooks, November 14, 1949, both in NW, box 7-106; and *NWHUHB*, 15–16, 187–189, quotation on 16. On Wiener's suggesting the titles "Pandora" and "Cassandra," see Wiener to Frances Ancker, September 9, 1949, NW, box 7-104; and Wiener to Arthur Blanchard, October 25, 1949, NW, box 7-105.

51. *NWHUHB*, quotations on 8, 16. Although the term *second industrial revolution* is only implied in *Cybernetics* (1948), the media attributed it to Wiener before he used it in *NWHUHB*. See, e.g., H. Smith, "The Machine in Man's Image"; "Machines That Think," 38; and Louis Ridenour, "Mechanical Brains," *Fortune* 39 (May 1949): 109ff., on 109, 117. The term dates to the 1920s, when it referred to electrification and the growth of the chemical industry, which has been its subsequent usage in the history of technology. See I. Bernard Cohen, *Revolution in Science* (Cambridge, MA: Harvard University Press, 1985), 268.

52. Chase, "Next Industrial Revolution."

53. "[Review of] *The Human Use of Human Beings . . . ,*" *Scientific American* 183 (December 1950): 60.

54. Irwin Edman, "Mind in Matter," *New Yorker* 26 (October 14, 1950): 124–126, on 124 (quotation); "The New Automatic Age," *Atlantic Monthly* 186 (September 1950); "Automatic Age," *Parade Magazine*, March 4, 1951, copy in NW, box 25c-379; and Mark Starr, "The Cybernetic Way of Life," *Saturday Review of Literature* 33 (August 19, 1950): 15–16. See, also, Charles Noyes, "Men and Machines," *Nation* 171 (August 26, 1950): 190–191; Carleton Beals, "Cybernetics," *Rotarian*, September 1953, 14–16; and "Are We Men or Machines?," *Quick*, June 6, 1949, 19.

55. Robert Peel, "Man among Animals and Machinery," *Christian Science Monitor*, August 26, 1950, 7.

56. On his psychological problems (manic-depression), see Conway and Siegelman, *Dark Hero of the Information Age*, 93–99.

57. Toni Strassman to Norbert Wiener, September 12, 1950, NW, box 8-124; Barbara Welles, WOR, to Wiener, September 26, 1950, NW, box 8-125; and Louis Adamic, WNYC, to Secretary, October 27, 1950, NW, box 8-127.

58. "How U.S. Cities Can Prepare for Atomic War," 85; Jennifer S. Light, *From Warfare to Welfare: Defense Intellectuals and Urban Problems in Cold War America* (Baltimore: Johns Hopkins University Press, 2003), chap. 2; and Robert Kargon and Arthur Molella, "The City as Communications Net: Norbert Wiener, the Atomic Bomb, and Urban Dispersal," *Technology and Culture* 45 (2004): 764–777.

59. Allan Morris to Norbert Wiener, March 13, 1953, NW, box 11-166; and Wiener to Morris, April 21, 1953, NW, box 11-170.

60. Norbert Wiener to J. J. Smith, June 14, 1951, NW, box 9-141; Wiener, "Cybernetics," *Scientia* (1952), rpt. in *NWCW*, 4:203–205; and Lavina Dudley to Wiener, September 2, 1954, NW, box 14-200.

61. Jason Epstein to Norbert Wiener, May 19, 1953, NW, box 12-172; Epstein to Wiener, September 25, 1953, NW, box 12-174; and *NWHUHB*, 2nd ed., especially preface and chaps. 1–2, 12.

62. Jason Epstein to Norbert Wiener, February 25, 1958, NW, box 15-227; and Lynwood Bryant to Wiener, January 18, 1960, NW, box 19-272.

63. Norbert Wiener to Walter Yust, August 30, 1954, NW, box 14-199; and Claude Shannon, "Cybernetics," in *Encyclopedia Britannica*, 13th ed., 1957, 6:916.

64. Walter LaFeber, *America, Russia, and the Cold War, 1945–1996*, 8th ed. (New York: McGraw-Hill, 1997), chaps. 4–6; John L. Gaddis, *The Cold War: A New History* (New York: Penguin Press, 2005), chap. 2; Paul Forman, "Behind Quantum Electronics: National Security as Basis for Physical Research in the United States, 1940–1960," *Historical Studies in the Physical and Biological Sciences* 18 (1987): 149–229; and Stuart W. Leslie, *Cold War and American Science:*

*The Military-Industrial-Academic Complex at MIT and Stanford* (New York: Columbia University Press, 1993), chaps. 2–3.

65. For newspaper coverage in the *NYT*, see, e.g., William M. Blair, "Harvard Unveils Huge Calculator," January 8, 1947, 21; William G. Weart, "Army Soon to Get Super-Calculator," March 3, 1947, 22; and Austin Stevens, "Air Force Unveils Fastest Computer," June 21, 1950, 5.

66. Frank Rockett, "The Transistor," *Scientific American* 179 (September 1948): 53–55; Louis Ridenour, "A Revolution in Electronics," *Scientific American* 185 (August 1951): 13–17; Wiener, "Cybernetics" (1948); Davis, "Mathematical Machines"; Warren Weaver, "The Mathematics of Communication," *Scientific American* 181 (July 1949): 11–15; Claude Shannon, "A Chess-Playing Machine," *Scientific American* 182 (February 1950): 48–51; Grey Walter, "An Imitation of Life"; Grey Walter, "A Machine That Learns"; and Horace C. Levinson and Arthur A. Brown, "Operations Research," *Scientific American* 184 (March 1951): 15–17. On the Cold War support for this research, see Leslie, *Cold War and American Science*; Paul N. Edwards, *Closed World: Computers and the Politics of Discourse in Cold War America* (Cambridge, MA: MIT Press, 1996); and Philip Mirowski, *Machine Dreams: Economics Becomes a Cyborg Science* (Cambridge: Cambridge University Press, 2002).

67. On this point, see Slava Gerovitch, *From Newspeak to Cyberspeak: A History of Soviet Cybernetics* (Cambridge, MA: MIT Press, 2002), 129. Cybernetics had a Cold War image inside military-funded projects; see, e.g., Thomas P. Hughes, *Rescuing Prometheus: Four Monumental Projects That Changed the Modern World* (New York: Pantheon Books 1998), 21–22.

68. Norbert Wiener, "A Scientist Rebels," *Atlantic Monthly* 179 (January 1947): 179; and Wiener to George E. Forsythe, Boeing, December 2, 1946, NW, box 5-72.

69. "Wiener Denounces Devices 'for War,'" *NYT*, January 9, 1947, 4; and Norbert Wiener, "A Rebellious Scientist after Two Years," *Bulletin of the Atomic Scientists* 14 (1948): 338–339, on 338. On Wiener's stance against military funding, see Steve J. Heims, *John von Neumann and Norbert Wiener: From Mathematics to the Technologies of Life and Death* (Cambridge, MA: MIT Press, 1980), chap. 13.

70. Norbert Wiener to Warren McCulloch, January 8, 1947, NW, box 5-73; and Wiener to Milton Greenberg, February 24, 1950, NW, box 7-112.

71. Jerry Lettvin, a researcher under Warren McCulloch at MIT, said in an interview that Wiener's position was "disingenuous" because he knew of the military support of the RLE. See Conway and Siegelman, *Dark Hero of the Information Age*, 223.

72. Wiener, "Rebellious Scientist after Two Years," 338.

73. Press release quoted in Conway and Siegelman, *Dark Hero of the Information Age*, 242; and *NWCyb*, 38–39.

74. Norbert Wiener to W. Grey Walter, November 21, 1951, NW, box 10-143.

75. *NWCyb*, 31–32, 35, 162–163.

76. Norbert Wiener to Ted Martin, August 14, 1950, NW, box 8-122.

77. My argument differs from that in Fred Turner, *From Counterculture to Cyberculture: Stewart Brand, the Whole Earth Network, and the Rise of Digital Utopianism* (Chicago: University of Chicago Press, 2006), 16–28, which argues that the interdisciplinary research that characterized cybernetics helped to create a "forgotten openness of the closed world."

78. Edwards, *Closed World*, chap. 1; Norbert Wiener to J. R. Kline, October 5, 1948, NW, box 6-85; and Wiener "Too Damn Close," *Atlantic Monthly* 186 (July 1950): 50–52.

79. Norbert Wiener to Paul Brooks, December 8, 1949, NW, box 7-108; and *NWHUHB*, 123–143, 206–208, quotation on 206.

80. Norbert Wiener to James Killian, September 13, 1951, NW, box 9-141; Warren Weaver diary, August 15, 1951, RFA, RG 12.1 (quotation); and Conway and Siegelman, *Dark Hero of the Information Age*, chap. 13.

81. Norbert Wiener to Duncan E. Macdonald, April 16, 1953, NW, box 11-170; and Wiener to Rear Admiral W. McL. Hogue, January 12, 1953, NW, box 11-162.

82. Lily Kay, "From Logical Neurons to Poetic Embodiments of Mind: Warren S. McCulloch's Project in Neuroscience," *Science in Context* 14 (2001): 591–614, on 603n.9.

83. Norbert Wiener to Warren McCulloch, January 8, 1947, NW, box 5-73.

84. Transcript of Sixth Macy Conference, 1949, NW, box 36-620.

85. *NWHUHB*, 111; and Norbert Wiener, *I Am a Mathematician: The Later Life of a Prodigy* (New York: Doubleday, 1956), 270. In contrast, Hayles, *How We Became Posthuman*, praises science fiction as a pathway between cybernetics and the broader culture.

86. John Diebold, *Automation: The Advent of the Automatic Factory* (New York: D. van Nostrand, 1952), 154; and George A. Miller, "[Review of] *Cybernetics; Circular Causal and Feedback Mechanisms . . . 1952*," *American Journal of Psychology* 66 (1953): 661–663, on 661.

87. Special Issue on Automatic Control, *Scientific American* 187 (September 1952): 166, 170, 172, 177.

88. Isaac Asimov, *I, Robot* (New York: Doubleday, 1950).

89. Bernard Wolfe, *Limbo* (New York: Random House, 1952), 104, 134, 141, 149; Barbara J. Lewis, "Human Machines: The Treatment of the Cyborg in American Popular Fiction and Film" (Ph.D. diss., New York University, 1997), 108–119; and Hayles, *How We Became Posthuman*, chap. 5.

90. Erik Fennel to *Time* magazine, December 26, 1948, NW, box 6-87.

91. Norbert Wiener to Erik Fennel, January 4, 1949, NW, box 6-89. On Wiener's writing of science fiction and mysteries, some of which was published, see Conway and Siegelman, *Dark Hero of the Information Age*, 288–290.

92. Clyde Beck, "Cybernetics, Science Fiction, and Survival," copy in NW, box 29b-653.

93. John E. Burchard to Martin Matheson, John Wiley, September 22, 1948, NW, box 6-85.

94. Kurt Vonnegut, Jr., *Player Piano* (New York: Delacorte Press, 1952), 12–13; and Norbert Wiener to Hope English, July 17, 1952, NW, box 10-153.

95. C. Noyes, "Men and Machines"; and Starr, "Cybernetic Way of Life," 15.

96. Donald Bruce to Norbert Wiener, September 15, 1950, NW, box 8-124; and the attached clipping, Joseph Levine, "Machines May Ruin Us, If Reds Don't," *Vancouver Sun*, n.d., dateline, August 25, [1950].

97. CHQ, Technocracy, Inc., "The Problem of Human Governance," *Technocracy Digest*, August 1951, 15–18, 35–41, on 15, 41, copy in NW, box 25c-379.

98. Norbert Wiener to Donald Bruce, September 18, 1950, NW, box 8-125; and Wiener to *Technocracy Digest*, November 16, 1954, NW, box 14-204.

99. Martin Gardner, *Fads and Fallacies in Science*, 2nd ed. (New York: Dover, 1957), chaps. 22–23.

100. L. Ron Hubbard, *Dianetics: The Modern Science of Mental Health* (New York: Hermitage House, 1950); and Hubbard, "The New Science of the Mind," *Astounding Science Fiction*, April 1950.

101. Yvette Gittleson, "Sacred Cows in Collision," *American Scientist* 38 (1950): 603–609, on 604, 605, 608. Hubbard said in "Terra Incognita: The Mind," *Explorers Journal* 28 (Winter–Spring 1950): 1–4, that dianetics merely drew an analogy between the brain and the computer. But commentators such as Gittelson pointed to statements in the book to the contrary. See an exchange of letters between another reviewer and Hubbard in the *NYT Book Review*, August 6, 1950, 12.

102. John Campbell to Norbert Wiener, March 9, 1950, NW, box 7-113; and Wiener to Edward Alexander, September 19, 1950, NW, box 8-123. On Campbell as a dianetics patient, see Gardner, *Fads and Fallacies*, 264; and as a student of Wiener's, see Patricia S. Warrick, *The Cybernetic Imagination in Science Fiction* (Cambridge, MA: MIT Press, 1980), 54.

103. Norbert Wiener to Waldo Frank, November 21, 1950, NW, box 9-130. In 1957, Gardner, *Fads and Fallacies*, chaps. 3, 5, 22, categorized all of these topics as scientific fads.

104. Norbert Wiener to Waldo B. Noyes, February 25, 1951, NW, box 9-134 (quotation); Noyes to Hubbard Dianetic Research Foundation, March 3, 1951, NW, box 9-135; and Wiener to Noyes, March 19, 1951, NW, box 9-135.

105. Claude Shannon to Warren McCulloch, August 23, 1949, WSM, ser. I, Shannon correspondence.

106. Warren McCulloch to Claude Shannon, October 20, 1949, WSM, ser. I, Shannon correspondence; L. Ron Hubbard to Claude Shannon, December 6, 1949, CESLC, box 1; and Hubbard, "Terra Incognita," 4.

107. "Psychologists Act against Dianetics," *NYT*, September 9, 1950, 19; "Dianetic Funds Attached," *NYT*, May 15, 1951, 29; Gardner, *Fads and Fallacies*, 264; and Warren McCulloch to Bernard W. Joseph, March 12, 1951, WSM, ser. I, Detroit Dianetics Assoc. correspondence.

108. Gardner, *Fads and Fallacies*, chap. 23; Alfred Korzybski to Norbert Wiener, January 19, 1949, NW, box 6-90; and Anatol Rapoport and Alfonso Shimbel, "Mathematical Biophysics, Cybernetics and General Semantics," *ETC: A Review of General Semantics* 6 (1949): 145–159, on 147, 155.

109. Paul Huntsinger to Norbert Wiener, March 16, 1949, NW, box 6-94; and Wiener to Huntsinger, April 26, 1949, NW, box 6-96.

110. These included John Cage in experimental music, William Burroughs in Beat poetry, and acid-guru Timothy Leary in "flicker" (strobe lights producing visual effects). See Dunbar-Hester, "Listening to Cybernetics"; and Pickering, *Cybernetic Brain*, 78–89.

111. On this form of boundary work, see Thomas F. Gieryn, "Boundary Work and the Demarcation of Science from Non-Science: Strains and Interests in Professional Ideologies of Scientists," *American Sociological Review* 48 (1983): 781–795.

112. Geoff Bowker, "How to Be Universal: Some Cybernetic Strategies, 1943–1970," *Social Studies of Science* 23 (1993): 107–127; and Peter Galison, "The Americanization of Unity," *Daedalus* 127 (1998): 45–71.

113. Churchill Eisenhart, "Cybernetics: A New Discipline," *Science* 109 (1949): 398–399; John von Neumann, "[Review of] *Cybernetics* . . . ," *Physics Today* 2 (May 1949): 33–34, on 34; and Léon Brillouin, "Life, Thermodynamics, and Cybernetics," *American Scientist* 37 (1949): 554–568.

114. Evelyn Fox-Keller, *Refiguring Life: Metaphors of Twentieth-Century Biology* (New York: Columbia University Press, 1995), chap. 3; and Lily E. Kay, *Who Wrote the Book of Life? A History of the Genetic Code* (Stanford, CA: Stanford University Press, 2000).

115. Although Henry Quastler, ed., *Essays on the Use of Information Theory in Biology* (Urbana: University of Illinois Press, 1953), 2, barely mentioned cybernetics, he told McCulloch that his group at the University of Illinois was "trying to use cybernetics in fields like cellular biology, chemical communication, etc." See Quastler to Warren McCulloch, December 24, 1952, WSM, ser. I, Quastler correspondence.

116. Walter A. Rosenblith, "[Review of] *Cybernetics* . . . ," *Annals of the American Academy of Political and Social Science* 264 (1949): 187–188, on 187; unsigned [Margot Zemurray] to Nikolai Tucci, June 13, 1953, NW, box 8-119; and Rosenblith, "Brothers," in *Jerry Wiesner: Scientist, Humanist, Memories and Memoirs*, ed. Rosenblith (Cambridge, MA: MIT Press, 2003), 3–17, on 3, 7, 9.

117. Peter J. Taylor, "Technocratic Optimism, H. T. Odum, and the Partial Transformation of the Ecological Metaphor after World War II," *Journal of the History of Biology* 26 (1988): 213–244.

118. Warren McCulloch, *Embodiments of Mind* (Cambridge, MA: MIT Press, 1965).

119. Ralph Bown to R. K. Honaman, February 4, 1949, ATTA, FC 3106, vol. 5; Bown to Honaman, November 9, 1949, ATTA, vol. K; and Shannon, "[Review of] *Cybernetics* . . . "

120. Eugene F. Coleman to Norbert Wiener, May 12, 1950, NW, box 8-117. John Pfeiffer, the science writer who touted *Cybernetics* in the newspapers, was also a member of the group.

121. Hsue Shen Tsien, *Engineering Cybernetics* (New York: McGraw-Hill, 1954).

122. "[Review of] *Cybernetics* . . . ," *American Sociological Review* 14 (1949): 332.

123. Dixon Wecter, "In Defense of Talk," *Saturday Review of Literature* 32 (November 5, 1949): 9–11, 36–38, on 11; and Lewis Mumford to Norbert Wiener, November 4, 1950, NW, box 8-125, reproduced in Masani, *Norbert Wiener*, 336–337.

124. H. M. McLuhan to Norbert Wiener, March 28, 1951, NW, box 9-135; McLuhan to Wiener, November 10, 1951, NW, box 10-143; and McLuhan, *The Mechanical Bride: Folklore of Industrial Man* (New York: Vanguard Press, 1951), 34.

125. Max Lerner, *America as a Civilization: Life and Thought in the United States Today* (New York: Simon and Schuster, 1957), 238–249, 767 (quotation).

126. See, e.g., Bowker, "How to Be Universal"; and Peter Galision, "Ontology of the Enemy: Norbert Wiener and the Cybernetic Vision," *Critical Inquiry* 21 (1994): 228–266.

127. Alphonse Chapanis, "[Review of] *Cybernetics* . . . ," *Quarterly Review of Biology* 24 (1949): 266.

128. Robert M. Fano to Colin Cherry, February 12, 1953, CC, box 12/3/1.

129. Joseph Doob to Norbert Wiener, July 23, 1948, NW, box 5-84; Doob to Wiener, October 5, 1948, NW, box 6-85; and Esther P. Potter to Wiener, December 2, 1949, NW, box 7-108.

130. Donald MacKay, "Report on Fellowship of the Rockefeller Foundation in Natural Sciences, January–December 1951," n.d., attached to MacKay to Warren McCulloch, January 22, 1952, WSM, ser. I, MacKay correspondence.

131. Warren Weaver to Norbert Wiener, December 21, 1948, NW, box 6-87; and Weaver to Wiener, March 24, 1949, NW, box 6-94.

132. J. B. S. Haldane to Wiener, January 11, 1949, NW, box 6-89. On their friendship, see Conway and Siegelman, *Dark Hero of the Information Age*, 78–79, 90, 171.

133. Daniel Bell, "Work and Its Discontents," in Bell, *The End of Ideology: On the Exhaustion of Political Ideas in the Fifties*, rev. ed. (New York: Free Press, 1962), 227–272, on 267.

134. Russell L. Ackoff, "[Review of] *Cybernetics* . . . ," *Philosophy of Science* 16 (1949): 159–160.

135. Norbert Wiener, "Cybernetics" (1950), rpt. in NWCW, 4:790–792; Gerald Holton to Wiener, October 23, 1950, NW, box 8-127 (quotation); Holton to Wiener, November 19, 1950, NW, box 9-129; Program Committee, Institute for Unity of Science, to Wiener, November 22, 1950, NW, box 9-130; Holton, "Ernst Mach and the Fortunes of Positivism in America," *Isis* 83 (1992): 27–60, on 55–56; Galison, "Americanization of Unity," 48, 54 (quotation), 56, 59; and George A. Reisch, *How the Cold War Transformed Philosophy of Science: To the Icy Slopes of Logic* (Cambridge: Cambridge University Press, 2005), 294–305.

136. Norbert Wiener, "[Review of] *Modern Science and Its Philosophy*, by Philipp Frank . . . ," *NYT Book Review*, August 14, 1949, BR-3.

137. Steering Committee, Robert Fano chair, to The Cybernetics Communications Group of the Institute for the Unity of Science, December 18, 1952, NW, box 11-161. Yehoshua Bar-Hillel was also on the steering committee at this time. Karl Deutsch and Walter Rosenblith were on it in 1951; see Steering Committee to Dear Sir, October 17, 1951, NW, box 9-142.

138. Karl W. Deutsch, "Some Notes on Research on the Role of Models in the Natural and Social Sciences," *Synthese* 7 (1948/1949): 506–533. On Deutsch's attending the institute's seminars, see Holton, "Ernst Mach and the Fortunes of Positivism in America," 56.

139. Charles W. Morris, "The Science of Man and Unified Science," *Proceedings of the American Academy of Arts and Sciences* 80, no. 1 (July 1951): 37–44, on 39; and F. Dermont Barrett and Herbert A. Shepard, "A Bibliography of Cybernetics," ibid., no. 3 (March 1953): 204–222.

140. Norbert Wiener, "The Role of the Observer," *Philosophy of Science* 3 (1936): 307–319; Arturo Rosenblueth, Wiener, and Julian Bigelow, "Behavior, Purpose and Teleology," *Philosophy of Science* 10 (1943): 18–24; and Rosenblueth and Wiener, "The Role of Models in Science," *Philosophy of Science* 12 (1945): 316–322.

141. E. N. [Ernest Nagel], "[Review of] *Cybernetics* . . . ," *Journal of Philosophy* 46 (1949): 736–737. Nagel attended the institute's seminar in Cambridge as a visitor; see Holton, "Ernst Mach and the Fortunes of Positivism in America," 56. On logical empiricism and the philosophy of science in the United State, see Reisch, *How the Cold War Transformed Philosophy of Science.*

142. C. West Churchman and Russell L. Ackoff, "Purposive Behavior and Cybernetics," *Social Forces* 29 (October 1950–May 1951): 32–39, on 33. On their involvement with the Unity of Science movement, see Churchman and Ackoff, "Varieties of Unification," *Philosophy of Science* 13 (1946): 287–300.

143. On cybernetics outside the United States and Britain, see David A. Mindell, Jérôme Segal, and Slava Gerovitch, "From Communications Engineering to Communications Science: Cybernetics and Information Theory in the United States, France, and the Soviet Union," in *Science and Ideology: A Comparative History*, ed. Mark Walker (London: Routledge, 2003), 66–96.

144. Michael J. Apter, "Cybernetics: A Case Study of a Scientific Subject-

Complex," in *The Sociology of Science*, ed. Paul Halmos, Sociological Review Monograph, no. 18 (Keele, UK: University of Keele, 1972), 93–115, on 112.

145. Compare McCulloch, *Embodiments of Mind* with W. Ross Ashby, *Introduction to Cybernetics* (London: Chapman and Hall, 1956), 4.

146. Norbert Wiener, "Cybernetics," *Encyclopedia Americana*, 1952, rpt. in *NWCW*, 4:804; and Wiener, "Cybernetics," *Scientia*, 203.

147. Leslie, *Cold War and American Science*, 25–32; and Karl Wildes and Nilo A. Lindgren, *A Century of Electrical Engineering and Computer Science at MIT, 1882–1982* (Cambridge, MA: MIT Press, 1985), chap. 16.

148. Albert Mueller and Karl Mueller, eds., *An Unfinished Revolution? Heinz von Foerster and the Biological Computer Laboratory, BCL, 1958–1976* (Vienna: Echoraum, 2007).

### Chapter 4: The Information Bandwagon

1. "Professional Group Notes," *Proceedings of the IRE* 39 (1951): 703; and Willis Jackson, "Information Theory," *Nature* 167 (1951): 20–23, on 20.

2. Nelson Blachman, "The Third London Symposium," *IRE Trans IT* 2 (1956): 17–23. Macy veterans Henry Quastler, J. C. R. Licklider, and Yehoshua Bar-Hillel also attended.

3. L. A. de Rosa, "In Which Fields Do We Graze?," *IRE Trans IT* 1, no. 3 (December 1955): 2.

4. Claude E. Shannon, "The Bandwagon," *IRE Trans IT* 2 (1956): 3.

5. Peter Elias to Norbert Wiener, July 17, 1956, NW, box 15-217.

6. Norbert Wiener, "What Is Information Theory?," *IRE Trans IT* 2 (1956): 48.

7. "'In Which Fields Do We Graze' [letters to the editor]," *IRE Trans IT* 2 (1956): 96–97. Blachman published the report on the London symposium that led to the debate. See note 2 above.

8. On concepts of boundary work and interpretative flexibility in the sciences, see Thomas F. Gieryn, "Boundaries of Science," in *Handbook of Science and Technology Studies*, ed. Sheila Jasanoff et al. (London: Sage, 1999), 393–443; and Harry Collins, *Changing Order: Replication and Induction in Scientific Practice* (London: Sage, 1985).

9. Yehoshua Bar-Hillel, "An Examination of Information Theory," *Philosophy of Science* 22 (1955): 86–105, on 97, 104.

10. Colin Cherry, *On Human Communication: A Review, a Survey, and a Criticism* (Cambridge, MA: MIT Press, 1957), 216 (quotation), 247, his emphasis.

11. Warren S. McCulloch, "Recollections of the Many Sources of Cybernetics" (1974), rpt. in *WSMCW*, 1:21–49, on 38. On their careers, see T. E. Allibone, "Denis Gabor," *Biographical Memoirs of the Fellows of the Royal Society* 26 (1980): 107–147; Marvin J. McDonald, "Mind and Brain, Science and Religion: Belief and Neuroscience in Donald M. MacKay and Roger W. Sperry," in *The Role of Beliefs in the Natural Sciences*, ed. Jitse M. van der Meer (Lanham, MD: Uni-

versity Press of America, 1996), 199–226; and Carol Wilder, "A Conversation with Colin Cherry," *Human Communication Research* 3 (1977): 354–362.

12. Denis Gabor to Warren Weaver, February 12, 1951; and Gabor to Lord Cherwell, February 12, 1951, both in DG, EN 2/1.

13. Ralph Bown to F. S. Barton, February 24, 1950; Mervin Kelly to Bown, March 24, 1950; and Willis Jackson to Claude Shannon, March 28, 1950; all in CESLC, box 1.

14. Unsigned, "Minutes of Informal Meeting of the 'Communications Theory Symposium' Committee . . . 30th March 1950," n.d., CC, 10/1/3; Willis Jackson to Claude Shannon, March 31, 1950, CESLC, box 1; and Colin Cherry, "Minutes of Informal Meeting of the Committee on 'Symposium on Communications Theory' . . . 28th April 1950," n.d., CC, 10/1/3, item 10.

15. "Symposium on Information Theory . . . [26–29] September, 1950 . . . Programme," n.d. CC, 10/3/3; Colin Cherry, "A History of the Theory of Information," *Trans PGIT* 1, no. 1 (February 1953): 22–43; and Cherry to W. E. Braser, n.d. (ca. December 1950), CC, 11/4. For the use of history by participants to form the field of information theory, see Bernard Geoghegan, "The Historiographic Conceptualization of Information: A Critical Survey," *AHC* 30 (2008): 69–81.

16. Denis Gabor, "Communication Theory, Past, Present and Prospective," *Trans PGIT* 1, no. 1 (February 1953): 2–4, on 4.

17. W. Grey Walter, "Possible Features of Brain Function and Their Imitation," *Trans PGIT* 1, no. 1 (February 1953): 134–136; A. M. Uttley, "Information Machines and Brains," *Trans PGIT* 1, no. 1 (February 1953):143–149; and Philip Husbands and Owen Holland, "The Ratio Club: A Hub of British Cybernetics," in *The Mechanical Mind in History*, ed. Husbands, Holland, and Michael Wheeler (Cambridge, MA: MIT Press, 2008), chap. 9.

18. Colin Cherry, "Minutes of Informal Meeting of the Committee on 'Symposium on Communications Theory' . . . 28th April 1950," n.d., CC, 10/1/3, item 15; Donald MacKay "The Nomenclature of Information Theory," *Trans PGIT* 1, no. 1 (February 1953): 9–21; J. A. V. Bates, "Glossary of Physiological Terms," *Trans PGIT* 1, no. 1 (February 1953): 5–8; Cherry, "Meeting of the Speakers at the Proposed Symposium on Information Theory . . . 16th June [1950]," n.d., CC, 10/1/3, item 3 (quotation); and MacKay to Denis Gabor, July 21, 1950, DG, MM-2.

19. Cherry, "A History of the Theory of Information" (1953), 28 (quotation), 39–40; and Donald MacKay to Denis Gabor, July 21, 1950, DG, MM-2.

20. Willis Jackson, "Foreword," *Trans PGIT* 1, no. 1 (February 1953): n.p.; and Jackson, "Information Theory," quotations on 20, 21.

21. Warren McCulloch to Mina Rees and Joachim Weyl, March 12, 1951, WSM, ser. II, U.S. Navy, Office of Naval Research #1. MacKay is not referenced, for example, in McCulloch's *Embodiments of Mind* (Cambridge, MA: MIT Press, 1965).

22. *WJCT*, 96, 383 (quotation), 475, 512.

23. Willis Jackson, "Opening Address," in *WJCT*, x; and "Applications of Communication Theory," *Nature* 170 (1952): 1051–1052.

24. Donald MacKay to Warren McCulloch, August 6, 1952, WSM, ser. I, MacKay correspondence.

25. Denis Gabor, "A Summary of Communication Theory," in *WJCT*, 1–24, on 1.

26. On U.S. researchers using the term, see *WJCT*, 22, 82, and 531. On non-U.S. users, in addition to MacKay and Bell noted above, see x, 503, and 528.

27. Yehoshua Bar-Hillel, discussion in *WJCT*, 512.

28. *CCIT*, 3, 33, 46, 215ff., 354, 366.

29. "Extracts from Referees' Observations on a Paper . . . by E. C. Cherry . . . ," n.d. attached to Secretary of IEE to Colin Cherry, May 16, 1951, CC, 11/4, p. 1.

30. Denis Gabor to Earl [Bertrand] Russell, February 14, 1951, DG, EN 2/1.

31. Denis Gabor to André Gabor, April 28, 1953, DG, E 0/1.

32. Denis Gabor, "Information Theory: From Telegraphy to a New Science," *Times Science Review* (Spring 1953): 11–12, on 12.

33. Denis Gabor, "Photons and Waves," *Nature* 166 (1950): 724–727, on 726; and Gabor, "Communication Theory and Cybernetics," *Transactions of the IRE Professional Group on Circuit Theory* 1, no. 1 (1954): 19–31, on 19, 21.

34. Colin Cherry, "A History of the Theory of Information," *Proceedings of the IEE*, Part III, 98 (1951): 383–393, on 392; Cherry, "A History of the Theory of Information" (1953), 39–40; Cherry, "The Communication of Information," *American Scientist* 40 (1952): 640–664, on 663; and Cherry, *On Human Communication*, 64–65, 216, 219, 303, 306.

35. Cherry, *On Human Communication*, 216 (quotation, his emphasis), 246–247.

36. Donald MacKay, "Operational Aspects of Some Fundamental Concepts of Human Communication," *Synthese* 9 (1955): 182–198, on 198.

37. Donald MacKay, "The Place of 'Meaning' in the Theory of Information," in *CCIT*, 223.

38. Cherry, *On Human Communication*, 219–226, 236–241, quotations on 221, 231.

39. Denis Gabor to Warren Weaver, October 17, 1951, RFA, RG 1.2, series 401D, box 18; and Gabor to Jerome Wiesner, November 28, 1951, DG, EN/1.

40. Colin Cherry to W. Locke, May 5, 1952, CC 10/u/3; Cherry to Claude Shannon, July 15, 1952, CESLC, box 1; Cherry to Roman Jakobson, November 21, 1952, RJ, box 40-44; Cherry to Jakobson, January 6, 1953, RJ, box 40-44; Robert Fano to Cherry, February 12, 1953, CC, 12/3/1; Cherry to Yehoshua Bar-Hillel, November 17, 1954, CC, 12/3/1; and Bar-Hillel, "Introduction," in Bar-Hillel, *Language and Information: Selected Essays on Their Theory and Application* (Reading, MA: Addison-Wesley, 1964), 6–7. On Jakobson's affiliation

with the RLE, see Robert Morison diary, November 25, 1955, RFA, RG 12-1, box 47, p. 204. On Cherry at the RLE, see MIT, RLE, *Quarterly Progress Report*, April 15, 1952, vi; July 15, 1952, vi; and Wilder, "Conversation with Colin Cherry," 355.

41. Claude Shannon, "The Lattice Theory of Information," *Trans PGIT* 1, no. 1 (February 1953): 105.

42. Robert Fano, "Concluding Discussion," in *WJCT*, 531–532.

43. Robert Fano to Colin Cherry, February 12, 1953, CC, 12/3/1. Fano did not have a high opinion of MacKay; he could not recommend him for a university professorship at Keele (which MacKay did receive). See Fano to D. S. Jones, October 29, 1959, NW, box 18-266.

44. Robert Fano, "The Information Theory Point of View in Speech Communication," *Journal of the Acoustical Society of America* 22 (1950): 691–696, on 691.

45. Colin Cherry to Yehoshua Bar-Hillel, July 3, 1953, CC, 12/3/1, his emphasis.

46. At BTL, see, e.g., Brockway McMillan, "Basic Theorems of Information Theory," *Annals of Mathematical Statistics* 24 (1953): 196–219, on 196.

47. David Slepian, "Report on the Third London Symposium on Information Theory," Technical Memorandum, BTL, October 17, 1955, MM-55-114-41, ATTA, reel FC-4618, vol. T, p. 1.

48. Warren Weaver diary, July 15, 1955, RFA, RG 12.1, his emphasis; and Jerome Wiesner to Donald Price, Ford Foundation, July 18, 1955. On Wiesner as liaison and the British wanting Mooers, see Wiesner to Colin Cherry, January 6, 1955; and Cherry to Wiesner, January 31, 1955, all in JW, box 1-30.

49. Jerome Wiesner to Arnold Shostak, ONR, April 26, 1955; and Calvin Mooers to Wiesner, May 13, 1955, both in JW, box 1-30. On his master's thesis, see Mooers, "Choice and Coding in Information Retrieval Systems," *Trans PGIT* 4, no. 4 (September 1954): 112–118, on 118.

50. Jerome Wiesner to Donald Price, July 18, 1955; Price to Margaret Mead, July 27, 1955; and Mead to Warren McCulloch, August 5, 1955 (quotation), all in JW, box 1-30. Wiesner found funding (source not mentioned) for Mooers while waiting for an answer from Price at the Ford Foundation, which decided not to fund Mead.

51. "Professional Groups," *Proceedings of the IRE* 40 (1952): 114; "William G. Tuller," *Proceedings of the IRE* 42 (1954): 1710; de Rosa, "In Which Fields Do We Graze?"; and Michael J. di Toro, Jr., "Applications of Information Theory," *IRE Trans IT* 2 (1956): 101.

52. Karl Wildes and Nilo A. Lindgren, *A Century of Electrical Engineering and Computer Science at MIT, 1882–1982* (Cambridge, MA: MIT Press, 1985), 253, 255; and William G. Tuller, "Theoretical Limitations on the Rate of Transmission of Information," *Proceedings of the IRE* 37 (1949): 468–476.

53. William Tuller to Willis Jackson, September 19, 1952; and Jackson to Tuller, October 1, 1952; both in CC, box 10.1.3.

54. F. Louis H. M. Stumpers, "A Bibliography of Information Theory (Com-

munication Theory—Cybernetics)," *Trans PGIT* 2, no. 2 (November 1953): ii–60, on ii. The PGIT published three supplements to this bibliography with the same title in 1955, 1957, and 1960.

55. William Tuller to L. G. Abraham, January 11, 1954; Robert Fano to Mrs. W. G. Tuller, September 15, 1954, both in RMF, box 3, IRE—AIEE folder; and "William G. Tuller."

56. Howard Gardner, *The Mind's New Science: A History of the Cognitive Revolution* (New York: Basic Books, 1985), 28–29; and Noam Chomsky, "Three Models for the Description of Language," *IRE Trans IT* 2, no. 3 (September 1956): 113–124, on 113.

57. I omit the 1953 volume from my survey because it consisted only of the papers from the first London symposium, which the PGIT did not organize, and Stumpers's bibliography.

58. Authors referred to Norbert Wiener's *The Extrapolation, Interpolation, and Smoothing of Stationary Time Series with Engineering Applications* (Cambridge, MA, and New York: Technology Press and John Wiley, 1949), rather than to *NWCyb*. Only one author referred to *NWCyb*, for a definition of information.

59. See John R. Pierce, *Symbols, Signals, and Noise: The Nature and Process of Communication* (New York: Harper & Brothers, 1961), 42.

60. Peter Elias, "Predictive Coding—Part I," *IRE Trans IT* 1 (1955): 16–24.

61. Peter Elias, "Two Famous Papers," *IRE Trans IT* 4 (1958): 99.

62. One target of Elias's sarcasm may have been physicist Léon Brillouin's "Life, Thermodynamics, and Cybernetics," *American Scientist* 37 (1949): 554–568.

63. Robert Price, "The Search for Truth," *IRE Trans IT* 5 (1959): 39–40.

64. Peter Elias et al., "Progress in Information Theory in the U.S.A., 1957–1960," *IRE Trans IT* 7 (1961): 128–144, on 128; and L. A. Zadeh, ed., "Report on Progress in Information Theory in the U.S.A., 1960–1963," *IEEE Trans IT* 9 (1963): 221–264, on 221.

65. Elias, "Coding and Information Theory" (1959), rpt. in *Cybernetics*, ed. C. R. Evans and A. D. J. Robertson (Baltimore: University Park Press, 1968), 252–266, on 253–254. Aaron Wyner at Bell Labs was associate editor for Shannon theory from 1970 to 1972.

66. Robert M. Fano, *Transmission of Information: A Statistical Theory of Communications* (New York: MIT Press and John Wiley, 1961), 1; and Yuk-Wing Lee, *Statistical Theory of Communication* (New York: John Wiley, 1960), 2–3.

67. See, e.g., Fazlollah M. Reza, *Introduction to Information Theory* (New York: McGraw-Hill, 1961); and Robert Ash, *Information Theory* (New York: Interscience, 1965).

68. Joseph Doob to Norbert Wiener, October 5, 1948; and Wiener to Doob, October 8, 1948, NW, box 6-85. On Doob's career, see J. Laurie Snell, "A Conversation with Joe Doob," *Statistical Science* 12 (1997): 301–311.

69. Norbert Wiener to Doob, April 16, 1951, NW, box 9-136; Doob, *Stochas-*

*tic Processes* (New York: Wiley, 1953), vi; and Snell, "Conversation with Joe Doob," 306.

70. G. A. Barnard, "The Theory of Information," *Journal of the Royal Statistical Society*, ser. B, 13 (1951): 46–64, on 46 (quotation), 58–61; M. S. Bartlett, "The Statistical Approach to the Analysis of Time Series," *Trans PGIT* 1, no. 1 (February 1953): 81–101, on 87; and *NWCyb*, 76. For a fuller account of the discussion of Barnard's paper, see Jérôme Segal, *Le zéro et le un: Histoire de la notion scientifique d'information au 20ᵉ siècle* (Paris: Éditions Syllepse, 2003), 297–299.

71. Joseph L. Doob, "[Review of] Shannon, C. E., A mathematical theory of communication . . . ," *Mathematical Reviews* 10 (1949): 133. For quotations of Doob's remark, see, e.g., John R. Pierce, "The Early Days of Information Theory," *IEEE Trans IT* 19 (1973): 3–8, on 5; and Solomon W. Golomb et al., "Claude Elwood Shannon (1916–2001)," *Notes of the American Mathematical Society* 49 (January 2002): 8–15, on 11.

72. Anthony Liversidge, "Profile of Claude Shannon" (1987), rpt. in *CESCW*, xix–xxxiii.

73. McMillan, "Basic Theorems of Information Theory," 196. On McMillan as a student of Wiener, see Pesi R. Masani, *Norbert Wiener, 1894–1964* (Basel: Birkhauser, 1990), 375.

74. A. I. Khinchin, *Mathematical Foundations of Information Theory*, trans. R. A. Silverman and M. D. Friedman (New York: Dover, 1957), 30–33; Herman H. Goldstine, "Information Theory," *Science* 133 (1961): 1395–1399, on 1397; David Slepian, "Information Theory in the Fifties," *IEEE Trans IT* 19 (1973): 145–148, on 147; and Segal, *Le zéro et le un*, chap. 9.

75. Joseph L. Doob, "Editorial," *IRE Trans IT* 5 (1959): 3; and Denis Gabor, "Guest Editorial," *IRE Trans IT* 5 (1959): 97.

76. Edmund C. Berkeley, *Giant Brains: Or Machines That Think* (New York: John Wiley, 1949), xiii, 1–5, 10–15, 20–21, 229, 248, 254. On his career, see Atsushi Akera, "Edmund Berkeley and the Origins of ACM," *Communications of the ACM* 50, no. 5 (May 2007): 30–35.

77. "Office Robots," *Fortune* 29 (January 1952): 82ff., on 84.

78. Herbert B. Nichols, "New Machine Gives Promise of Sifting Rumor from Fact," *Christian Science Monitor*, March 7, 1950, copy in NW, box 7-114.

79. Randall L. Dahling, "Shannon's Information Theory: The Spread of an Idea," in Stanford University, Institute for Communication Research, *Studies of Innovation and of Communication to the Public* (Stanford, CA, 1962), 117–139, on 134–135.

80. Warren Weaver, "The Mathematics of Communication," *Scientific American* 181 (July 1949): 11–15, on 13.

81. Warren Weaver to Claude Shannon, July 7 and July 23, 1948, CESLC, box 1; and Weaver diary, September 10, 1948, RFA, RG 12.1. On the translation

project, see Yehoshua Bar-Hillel, "The State of Machine Translation in 1951," *American Documentation* 2 (1951): 229–237.

82. Warren Weaver to Claude Shannon, January 27, 1949, CESLC, box 1. On Weaver's interest in the popularization of science and in interdisciplinarity, see Robert Kohler, *Partners in Science: Foundation Managers and Natural Scientists, 1920–1945* (Chicago: University of Chicago Press, 1991), chap. 10

83. Louis Ridenour to Warren Weaver, November 8, 1948, RF, RG 2, ser. 216, box 416, folder 2802; and Weaver diary, February 11, 1949, RFA, RG 12.1. On Ridenour, see Robert Buderi, *The Invention That Changed the World: How a Small Group of Radar Pioneers Won the Second World War and Launched a Technological Revolution* (New York: Simon and Schuster, 1996), 109–111, 132, 250.

84. Warren Weaver to Louis Ridenour, February 24, 1949; Ridenour to Wilbur Schramm, March 1, 1949; Weaver to Ridenour, March, 18, 49, all in JAC, box 3, Loomis folder; Dahling, "Shannon's Information Theory," 133; and Wilbur Schramm, *The Beginnings of Communication Study in America*, ed. Steven H. Chaffee and Everett M. Rodgers (London: Sage, 1997), 140.

85. Louis Ridenour to Warren Weaver, March 21, 1949, RF, RG 2, ser. 216, box 453, folder 3045; and Ridenour to Mervin Kelly, April, 12, 1949, JAC, box 3, Loomis folder.

86. Warren Weaver to Claude Shannon, March 28, 1949; Weaver to Louis Ridenour, March 28, 1949; Ridenour to Weaver, April 12, 1949; and Ridenour to Mervin Kelly, April 12, 1949, all in JAC, box 3, Loomis folder.

87. For examples, see G. D. Gull, "Implications for the Storage and Retrieval of Knowledge," *Library Quarterly* 25 (1955): 333–343, on 339–340; Andreas Koutsoudas, "Mechanical Translation and Zif's Law," *Language* 33 (1957): 545–552, on 547n.6; D. F. Kerridge, "Inaccuracy and Inference," *Journal of the Royal Statistical Society*, ser. B, 23 (1961): 184–194, on 184; and Debora Shaw and Charles H. Davis, "Entropy and Information: A Multidisciplinary Overview," *Journal of the American Society for Information Science* 34 (1983): 67–74, on 68.

88. Warren Weaver to Louis Ridenour, November 17, 1949, CESLC, box 1; and Weaver to Vannevar Bush, October 24, 1949, VB, box 117, folder 2801.

89. "Proposed Copy for Dust Cover of Shannon and Weaver," attached to Louis Ridenour to Warren Weaver, November 23, 1949, CESLC, box 1; and Weaver to Ridenour, February 24, 1949, JAC, box 3, Loomis folder.

90. Francis Bello, "Information Theory," *Fortune* 48 (December 1953): 136ff., on 136.

91. Ibid.

92. "Comments on Mr. Bello's Article by W. E. Kock, B. McMillan, C. E. Shannon, and D. Slepian," October 14, 1953, attached to Claude Shannon to Francis Bello, October 15, 1953 (quotations), ATTA, Reel FC-4617; and Bello, "Information Theory," 137.

93. Francis Bello to Warren McCulloch, November 3, 1953, WSM, ser. I, Fortune Magazine correspondence.

94. Francis Bello to Norbert Wiener, October 9, 1953, NW, box 12-179; Marjorie Jack to Wiener, October 25, 1953, NW, box 12-180; and Bello to Wiener, November 3, 1953, NW, box 12-181.

95. Fritz Leiber, "Tomorrow's Electronic Messages," *Science Digest* 35 (March 1954): 41–43, on 41.

96. "Information: Now It's the Realm," *Business Week*, July 30, 1955, 58ff.; and Mark D. Bowles, "Crisis in the Information Age: How the Information Explosion Threatened Science, Democracy, the Library, and the Human Body, 1945–1999" (Ph.D. diss., Case Western Reserve University, 1999).

97. Claude Shannon, "A Chess-Playing Machine," *Scientific American* 182 (February 1950): 48–51.

98. Claude Shannon, "Information Theory," in *Encyclopedia Britannica*, 14th ed., 1957, vol. 12, pp. 350–350B, on 350.

99. Claude Shannon, "A Mathematical Theory of Communication," *BSTJ* 27 (1948): 379–423, 623–656, on 393; Shannon to G. W. Gilman, September 28, 1948, ATTA, reel FC-4616, vol. E; Shannon to Warren McCulloch, August 23, 1949, WSM, ser. I, Shannon correspondence; Shannon, "[Review of] *Cybernetics . . .*" (1949), rpt. in *CESCP*, 872–873, on 872; Ralph Bown to R. K. Honaman, November 19, 1949, two letters with same date, ATTA, film, vol. K; Shannon, "Some Topics in Information Theory" (1952), rpt. in *CESCP*, 458–459; and Shannon, "Information Theory," seminar notes, MIT, 1956 and succeeding years, outline in *CESCP*, xlii.

100. He did not revise this statement in "Information Theory" (1968), rpt. in *CESCP*, 212–220.

101. Compare, e.g., Leonard Engel, "Sees Study of Robot Brains Helping to Solve Riddle of the Human Mind," *Toronto Star*, June 25, 1949, NW, box 25C-378; "The Thinking Machine," *Time*, January 23, 1950, 54–56, 58–60; Stuart Chase, "The Next Industrial Revolution," *New York Herald Tribune Book Review*, August 20, 1950, 5; John Pfeiffer, "This Mouse Is Smarter Than You Are," *Popular Science* 160 (March 1952): 99–101; and J. L., "Information on a Theorist: C. E. Shannon," *Saturday Review of Literature* 39 (December 1, 1956): 73.

102. Waldemar Kaempffert, "What Next in the Attributes of Machines? It Might Be Power to Reproduce Themselves," *NYT*, May 3, 1953, E 9; J. L., "Information on a Theorist"; "Bright Spectrum," *Time*, November 18, 1957, 25–26, on 26; and Brock Brower, "The Man-Machines May First Talk to Dr. Shannon," *Vogue*, April 15, 1963, 88–89, 138–139.

103. Vannevar Bush to E. B. Wilson, January 3, 1939, VB, box 119-2920.

104. Warren Weaver diary, November 2, 1950, RFA, RG 12.1.

105. John R. Pierce, *Electrons, Waves, and Messages* (Garden City, NY: Hanover House, 1956), preface, chaps. 14–15; and William Baker, Edward David,

and A. Michael Noll, "John R. Pierce," *Proceedings of the American Philosophical Society* 148 (2004): 146–149.

106. Pierce, *Symbols, Signals, and Noise,* quotations on ix, 210, 227–228, his emphasis.

107. The term *glamor science* comes from Pierce's colleague, E. N. Gilbert, at Bell Labs; see Gilbert, "Information Theory after 18 Years," *Science* 152 (1966): 320–326, on 320.

108. Pierce, *Symbols, Signals, and Noise,* 268–269.

109. Segal, *Le zéro et le un,* part 2, covers many of these areas.

110. Dahling, "Shannon's Information Theory," 123–125.

111. Ibid., 127–133. His thesis contains the full bibliography used for his analysis; see Dahling, "Shannon's Information Theory: The Spread of an Idea" (Master's thesis, Stanford University, 1957), 36–57.

112. Lily E. Kay, *Who Wrote the Book of Life? A History of the Genetic Code* (Stanford, CA: Stanford University Press, 2000), chap. 4; Evelyn Fox-Keller, *Refiguring Life: Metaphors of Twentieth-Century Biology* (New York: Columbia University Press, 1995), chap. 3; and Donna J. Haraway, "The High-Cost of Information in Post–World War II Evolutionary Biology: Ergonomics, Semiotics, and the Sociobiology of Communication Systems," *Philosophical Forum* 13, nos. 2–3 (Winter–Spring 1981–1982): 244–278.

113. Ronald R. Kline, "What Is Information Theory a Theory of? Boundary Work among Information Theorists and Information Scientists in the United States and Britain during the Cold War," in *The History and Heritage of Scientific and Technical Information Systems: Proceedings of the 2002 Conference, Chemical Heritage Foundation,* ed. W. Boyd Rayward and Mary Ellen Bowden (Medford, NJ: Information Today, 2004), 15–28.

114. Murray Eden, "Human Information Processing," *IRE Trans IT* 9 (1963): 253–256; Pierce, *Symbols, Signals, and Noise,* chap. 12; Kay, *Who Wrote the Book of Life?,* 297–307; and chapter 5 below.

115. Philip Mirowski, *Machine Dreams: Economics Becomes a Cyborg Science* (Cambridge: Cambridge University Press, 2002), 370–390.

116. Segal, *Le zéro et le un,* chap. 5; Denis Gabor to Norbert Wiener, February 14, 1951, NW, box 9-134; Wiener to Jerome Rothstein, March 19, 1951, NW, box 9-135; Wiener to Gabor, April 6, 1951, NW, box 9-136; and Masani, *Norbert Wiener,* 111.

117. Andrew J. Viterbi, "Information Theory in the Sixties," *IEEE Trans IT* 19 (1973): 257–262, on 259; and Joel West, "Commercializing Open Science: Deep Space Communication as the Lead Market for Shannon Theory, 1960–1973," *Journal of Management Studies* 45 (2008): 1506–1532.

118. Jonathan Sterne, *MP3: The Meaning of a Format* (Durham, NC: Duke University Press, 2012), 20–21, 64.

119. Henry Quastler, ed., *Information Theory in Psychology: Problems and*

*Methods* (Glencoe, IL: Free Press, 1955), 2. On Quastler's work in molecular biology, see Kay, *Who Wrote the Book of Life?*, 115–126.

120. Yehoshua Bar-Hillel and Rudolf Carnap, "Semantic Information," in *WJCT*, 503–512, on 503; George A. Miller, "What Is Information Measurement?," *American Psychologist* 8 (1953): 3–11, on 3; and Bar-Hillel, "Examination of Information Theory," 94.

121. Warren Weaver, "Recent Contributions to the Mathematical Theory of Communication," in *The Mathematical Theory of Communication,* ed. Claude E. Shannon and Warren Weaver (Urbana: University of Illinois Press, 1949), 93–117, on 114, 116. On the sales figure, see Dahling, "Shannon's Information Theory" (1962), 122.

122. See Stephen Hilgartner, "The Dominant View of Popularization: Conceptual Problems, Political Uses," *Social Studies of Science* 20 (1990): 519–539.

123. *NWHUHB*, 6, 8.

124. Donald MacKay, "In Search of Basic Symbols," in *HVFCyb*, 8:181–221, on 192.

125. Bar-Hillel, "Examination of Information Theory," 97–98.

126. Ibid., 100, 103; and Gabor, "Summary of Communication Theory," 4.

127. Shaw and Davis, "Entropy and Information."

128. Cherry, *On Human Communication*, 214.

129. Weaver, "Recent Contributions to the Mathematical Theory of Communication," 103.

130. See, e.g., Kay, *Who Wrote the Book of Life?*; Mirowski, *Machine Dreams*; and Hunter Crowther-Heyck, *Herbert A. Simon: The Bounds of Reason in Modern America* (Baltimore: Johns Hopkins University Press, 2005).

131. Robert Fano to Colin Cherry, February 12, 1953, CC, 12/3/1.

132. Shannon, "The Bandwagon."

133. For the case of information science, see Kline, "What Is Information Theory a Theory Of?" On users as innovators, see Nellie Oudshoorn and Trevor Pinch, eds., *How Users Matter: The Co-construction of Users and Technologies* (Cambridge, MA: MIT Press, 2003).

134. Robert Fano, "Concluding Discussion," in *WJCT*, 531.

135. For molecular biology and economics, see Kay, *Who Wrote the Book of Life?*, and Mirowski, *Machine Dreams*, 370–390.

### Chapter 5: Humans as Machines

1. Charles C. Holt to Norbert Wiener, February 22, 1954, NW, box 13-188; and Wiener to Holt, April 14, 1954, NW, box 13-191. Wiener's letter may have reinforced Simon's preference for Ashby's version of cybernetics; see Hunter Crowther-Heyck, *Herbert A. Simon: The Bounds of Reason in Modern America* (Baltimore: Johns Hopkins University Press, 2005), 189. On Holt, see ibid., 163, 210. For examples of Wiener's replying in a similar manner to other economists, see Wiener to William B. Simpson, March 20, 1950, NW, box 7-114; Wiener to

Andrew Pikler, May 22, 1950, NW, box 8-118; and Oskar Morgenstern to Wiener, December 21, 1950, NW, box 9-132.

2. George P. Richardson, *Feedback Thought in Social Science and Systems Theory* (Philadelphia: University of Pennsylvania Press, 1991), 105–108, notes that because the mathematical analysis of control systems had been around since the 1930s, Wiener's skepticism must refer to the inapplicability of his autocorrelation prediction theory to the social sciences because of the spotty runs of social science data.

3. Dorothy Ross, "Changing Contours of the Social Science Disciplines," in *The Modern Social Sciences*, vol. 7 of *The Cambridge History of Science,* ed. Theodore M. Porter and Dorothy Ross (Cambridge: Cambridge University Press, 2003), 205–237; Mary S. Morgan, "Learning from Models," in *Models as Mediators: Perspectives on Natural and Social Science*, ed. Morgan and Margaret Morrison (Cambridge: Cambridge University Press, 1999), 347–388; Richardson, *Feedback Thought*, chap. 2; and Sharon E. Kingsland, "Economics and Evolution: Alfred James Lotka and the Economy of Nature," in *Natural Images in Economic Thought: "Markets Red in Tooth and Claw,"* ed. Philip Mirowski (Cambridge: Cambridge University Press, 1994), 231–246.

4. Max Black, *Models and Metaphors: Studies in Language and Philosophy* (Ithaca, NY: Cornell University Press, 1962), 223–228, on 223, based on a chapter published in 1960.

5. Kenneth J. Arrow, "Mathematical Models in the Social Sciences," in *The Policy Sciences*, ed. Daniel Lerner and Harold D. Lasswell (Stanford, CA: Stanford University Press, 1951), 129–154, on 129.

6. Philip Mirowski, *Machine Dreams: Economics Becomes a Cyborg Science* (Cambridge: Cambridge University Press, 2002), refers to these modeling methods as Cold War "cyborg sciences," which he defines as ones that relied on the digital computer, breached the boundaries between the natural and the social, as well as reality and simulation, and used "information" as a physical concept related to thermodynamics (12–18). Although cybernetics shares many of these characteristics, Mirowski says that "cybernetics itself never attained the status of a fully fledged cyborg science but instead constituted the philosophical overture to a whole phalanx of cyborg sciences" (11).

7. Bernard Berelson, "Behavioral Sciences," in *International Encyclopedia of the Social Sciences*, 2nd ed. (New York: Macmillan, 1968), 2:41–47; Hunter Crowther-Heyck, "Patrons of the Revolution: Ideas and Institutions in Postwar Behavioral Science," *Isis* 97 (2006): 420–446; and Crowther-Heyck, *Herbert A. Simon*, 170.

8. Andrew Pickering, *The Cybernetic Brain: Sketches of Another Future* (Chicago: University of Chicago Press, 2010), 93 (Ashby); Richard L. Merritt, Bruce M. Russett, and Robert A. Dahl, "Karl Wolfgang Deutsch," *Biographical Memoirs of the National Academy of Sciences* 80 (2001): 58–79, on 59; Claude Shannon to Hendrik Bode, October 30, 1957, CESLC, box 2; and George A. Miller, Eugene

Galanter, and Karl H. Pribram, *Plans and the Structure of Behavior* (New York: Henry Holt, 1960), 2.

9. This point is also made, in a different context, by Mirowski, *Machine Dreams*, 373n.5. In this regard, George Miller suggested to Herbert Simon that because Shannon "seems to be on a social science kick this year [at the Center for Advanced Study in the Behavioral Sciences], it might be a good time to get him thinking about our problems." See Miller to Simon, September 18, 1957. They wanted Shannon to serve on a Committee on Cognition of the Social Science Research Council, but Miller suggested waiting to see "how Claude reacts to the interaction with social scientists at the Center." See Miller to Simon, October 21 and October 22, 1957 (quotation), all in HAS, box 46, folder 6286.

10. Irwin Pollack, "Information Theory," in *International Encyclopedia of the Social Sciences*, 2nd ed., 7:331–337, on 331.

11. George Miller, "What Is Information Measurement?," *American Psychologist* 8 (1953): 3–11; and Yehoshua Bar-Hillel, "Examination of Information Theory," *Philosophy of Science* 22 (1955): 86–105, on 101.

12. Paul N. Edwards, *Closed World: Computers and the Politics of Discourse in Cold War America* (Cambridge, MA: MIT Press, 1996), 222–230; and Hunter Crowther-Heyck, "George A. Miller, Language, and the Computer Metaphor of Mind," *History of Psychology* 2 (1999): 37–64.

13. Fred Attneave, *Applications of Information Theory to Psychology: A Summary of Basic Concepts, Methods, and Results* (New York: Henry Holt, 1959), 26; Wendell R. Garner, "The Contribution of Information Theory to Psychology," in *The Making of Cognitive Science: Essays in Honor of George A. Miller*, ed. William Hirst (Cambridge: Cambridge University Press, 1988), chap. 2, on 21; Edwards, *Closed World*, 225; and Crowther-Heyck, "George A. Miller," 45.

14. George Miller and Frederick C. Frick, "Statistical Behavioristics and Sequences of Responses," *Psychological Review* 56 (1949): 311–324, on 320, 323; and Frick and Miller, "A Statistical Description of Operant Conditioning," *American Journal of Psychology* 64 (1951): 20–36, on 24.

15. Miller, "What Is Information Measurement?," popularized the method; Miller, "The Magical Number Seven, Plus or Minus Two: Some Limits on Our Capacity for Processing Information," *Psychological Review* 63 (1956): 81–97, summarized and compared experiments using the method.

16. Miller, "The Magical Number Seven, Plus or Minus Two," 93–95.

17. Attneave, *Applications of Information Theory to Psychology*, chaps. 2–3; and Garner, "Contribution of Information Theory to Psychology," 24–31.

18. Garner, "Contribution of Information Theory to Psychology," 32–33; and Ulric Neisser, "Cognitive Recollections," in *Making of Cognitive Science*, ed. Hirst, chap. 6, on 84.

19. Lily E. Kay, *Who Wrote the Book of Life? A History of the Genetic Code* (Stanford, CA: Stanford University Press, 2000), 297–307; and Bernard Geoghe-

gan, "From Information Theory to French Theory: Jakobson, Lévi-Strauss, and the Cybernetics Apparatus," *Critical Inquiry* 38 (2011): 96–126.

20. Roman Jakobson to Norbert Wiener, February 24, 1949, NW, box 6-96, partially quoted in Kay, *Who Wrote the Book of Life?*, 297–298.

21. Roman Jakobson to Warren Weaver, July 30, 1950; and Jakobson to Charles Fahs, February 22, 1950, both in RJ, box 6-37. Quoted in Kay, *Who Wrote the Book of Life?*, 300; and Geoghegan, "From Information Theory to French Theory," 109–110. On the grant, see Geoghegan, 111.

22. E. Colin Cherry, Morris Halle, and Roman Jakobson, "Toward the Logical Description of Languages in Their Phonemic Aspect," *Language* 29 (1953): 34–46, on 39, 42. See also Kay, *Who Wrote the Book of Life?*, 301–302; and Geoghegan, "From Information Theory to French Theory," 112.

23. A decade later, Jakobson used Donald MacKay's version of information theory; see Roman Jakobson, "Linguistics and Communication Theory" (1960), rpt. in Jakobson, *On Language*, ed. Linda R. Waugh and Monique Monville-Burston (Cambridge, MA: Harvard University Press, 1990), chap. 28; and Kay, *Who Wrote the Book of Life?*, 302–303.

24. For a later argument in this regard, see Walter F. Buckley, *Sociology and Modern Systems Theory* (Englewood Cliffs, NJ: Prentice Hall, 1967), 3. Richardson, *Feedback Thought*, 92–93, includes Deutsch and Simon in the list of postwar social scientists who introduced engineering control theory into the social sciences. While I analyze the varieties of cybernetic modeling under one label, Richardson distinguishes between a homeostatic, nonmathematical "cybernetics thread," and a dynamic, mathematical "servomechanism thread" in the social sciences. In his scheme, Deutsch, Miller, and Parsons come under the cybernetics thread, Simon is under the servomechanisms thread, and Bateson is a precursor to the cybernetics thread.

25. Merritt, Russett, and Dahl, "Karl Wolfgang Deutsch"; and Karl Deutsch, "Some Memories of Norbert Wiener: The Man and His Thoughts," *IEEE Trans SMC* 3 (1975): 368–372. On Deutsch's attending the Wiener seminar, see unsigned [Margot Zemurray] to Nikolai Tucci, June 13, 1953, NW, box 8-119.

26. Unsigned, "How U.S. Cities Can Prepare for Atomic War . . . ," *Life*, December 18, 1950, 77–82, on 85; Karl Deutsch to Dana Munro, January 23, 1951, attached to Deutsch to Norbert Wiener, January 29, 1951, NW, box 9-133; Wiener to Deutsch, February 28, 1951, NW, box 9-134; Deutsch to Wiener, June 7, 1956, NW, box 15-217; Karl Deutsch and Norbert Wiener, "The Lonely Nationalism of Rudyard Kipling," *Yale Review* (1963), rpt. in *NWCW*, 4:868–886; and Steve J. Heims, *John von Neumann and Norbert Wiener: From Mathematics to the Technologies of Life and Death* (Cambridge, MA: MIT Press, 1980), 301, 376–377.

27. Karl Deutsch, *The Nerves of Government: Models of Political Communication and Control* (New York: Free Press of Glencoe, 1963), 77.

28. Karl Deutsch, "Some Notes on Research on the Role of Models in the Natural and Social Sciences," *Synthese* 7 (1948/1949): 506–533; Deutsch, "Mechanism, Organism, and Society: Some Models in Natural and Social Science," *Philosophy of Science* 18 (1951): 230–252; and Deutsch, "On Communication Models in the Social Sciences," *Public Opinion Quarterly* 16 (1952): 356–380, on 358.

29. Karl Deutsch, *Nationalism and Social Communication: An Inquiry into the Foundations of Nationality* (Cambridge, MA: Technology Press, 1953), chaps. 4, 6, appendix 5. On the Deutsch-Solow model, see Richardson, *Feedback Thought*, 222–226.

30. Deutsch, *Nationalism and Social Communication*, 189–190.

31. Ibid., 258–261.

32. Anatol Rapoport, "[Review of] Karl W. Deutsch, *The Nerves of Government* . . . ," *Annals of the American Academy of Political and Social Science* 358 (March 1965): 197–198, on 197.

33. Deutsch, *Nerves of Government*, 42.

34. Crowther-Heyck, *Herbert A. Simon*, chap 9, on 185; and Herbert A. Simon, *Models of My Life* (New York: Basic Books, 1991), 114.

35. Herbert A. Simon, "On the Application of Servomechanism Theory in the Studies of Production Control," *Econometrica* 20 (1952): 247–268, on 258.

36. Herbert A. Simon, "Some Strategic Considerations in the Construction of Social Science Models," in *Mathematical Thinking in the Social Sciences*, ed. Paul Lazarsfeld (Glencoe, IL: Free Press, 1954), 398, 399.

37. Herbert A. Simon, *Models of Man, Social and Rational: Mathematical Essays on Rational Human Behavior in a Social Setting* (New York: John Wiley, 1957), 196–206.

38. Herbert A. Simon, *The Sciences of the Artificial* (Cambridge, MA: MIT Press, 1969), 5n.2, 6 (quotation), 85n.2, 96n.12. Crowther-Heyck, *Herbert A. Simon*, 189–193, states that Simon "was not enthralled by Wiener's formulation of cybernetic theory" but was more influenced by traditional servo theory, which he termed "b.c," that is, "before cybernetics," and by Ashby.

39. Miller, Galanter, and Pribram, *Plans and the Structure of Behavior*, chaps. 2–3, quotations on 26–27, 44. Crowther-Heyck, "George A. Miller," 54–55, notes that the TOTE represents Miller's development of the computer metaphor of mind for psychology, with the informational calculus as an intermediate step. See also Edwards, *Closed World*, 230–233.

40. Robert C. Bannister, "Sociology," in *Modern Social Sciences*, ed. Porter and Ross, 329–353; and Adam Kuper, "Anthropology," in ibid., 354–378.

41. Talcott Parsons, *The Social System* (Glencoe, IL: Free Press, 1951); and Parsons, "Social Systems," in *International Encyclopedia of the Social Sciences*, 2nd ed., 15:458–473.

42. Talcott Parsons, "On Building Social System Theory: A Personal History," *Daedalus* 99 (1970): 826–881, on 831, 850, quoted in Steve J. Heims, *The Cyber-*

*netics Group* (Cambridge, MA: MIT Press, 1991), 184; and Crowther-Heyck, *Herbert A. Simon*, 188.

43. Talcott Parsons, "Cause and Effect in Sociology," in *Cause and Effect: The Hayden Colloquium on Scientific Method and Concept*, ed. Daniel Lerner (New York: Free Press, 1965), 51–64, on 53; Parsons, *Politics and Social Structure* (New York: Free Press, 1969), 10–11, 32; and Parsons, "Social Systems," 466, 468. On this relationship between Simon and Parsons, see Crowther-Heyck, *Herbert A. Simon*, 174–175, 189.

44. Jurgen Ruesch and Gregory Bateson, *Communication: The Social Matrix of Psychiatry* (New York: W. W. Norton, 1951), 177 (in a chapter written by Bateson); Bateson, "From Versailles to Cybernetics," unpub. 1966 lecture, rpt. in Bateson, *Steps to an Ecology of Mind: Collected Essays in Anthropology, Psychiatry, Evolution, and Epistemology* (1972; rpt., Northdale, NJ: Jason Aronson, 1987), 477–485, on 477; and Bateson, "Preface," in ibid., xiv.

45. Bateson, "The Science of Mind and Order," in Bateson, *Steps to an Ecology of Mind*, xvii–xxvi, on xxvi.

46. David Lipset, *Gregory Bateson: The Legacy of a Scientist* (Boston: Beacon Press, 1982), chaps. 9–11.

47. Norbert Wiener to Lawrence Frank, October 26, 1951, NW, box 9-142, quoted in Heims, *John von Neumann and Norbert Wiener*, 305. In this letter, Wiener criticized a paper by Frank for being too vague.

48. Ruesch and Bateson, *Communication*, chap. 7.

49. Gregory Bateson to Norbert Wiener, September 22, 1952, NW, box 10-155; Bateson to Wiener, May 15, 1954, NW, box 13-193; Wiener to Bateson, June 16, 1954, NW, box 13-194 (quotation); Bateson to Wiener, November 29, 1954, NW, box 14-204 (quotation); Steve J. Heims, "Gregory Bateson and the Mathematicians: From Interdisciplinary Interactions to Societal Functions," *Journal of the History of the Behavioral Sciences* 13 (1977): 141–159, on 147–151; Lipset, *Gregory Bateson*, chap. 12; and Max Visser, "Gregory Bateson on Deutero-Learning and Double Bind: A Brief Conceptual History," *Journal of the History of the Behavioral Sciences* 39 (2003): 269–278.

50. Gregory Bateson, *Naven: A Survey of the Problems Suggested by a Composite Picture of the Culture of a New Guinea Tribe Drawn from Three Points of View*, 2nd ed. (Stanford, CA: Stanford University Press, 1958), 289–290.

51. Gregory Bateson, "Form, Substance, and Difference" (1970); rpt. in Bateson, *Steps to an Ecology of Mind*, 448–465, on 461.

52. Bateson, "Bali: The Value of a Steady State" (1949), rpt. in Bateson, *Steps to an Ecology of Mind*, 107–127, on 109–111.

### Chapter 6: Machines as Human

1. Norbert Wiener, *God and Golem, Inc.: A Comment on Certain Points Where Cybernetics Impinges on Religion* (Cambridge, MA: MIT Press, 1964), 95.

2. Ibid., chaps. 2–5, quotation on 71.

3. See, e.g., James Moor, "The Dartmouth College Artificial Intelligence Conference: The Next Fifty Years," *AI Magazine* 27, no. 4 (2006): 87–91; Pamela McCorduck, *Machines Who Think* (San Francisco: W. H. Freeman, 1979), chap. 5; and Paul N. Edwards, *The Closed World: Computers and the Politics of Discourse in Cold War America* (Cambridge, MA: MIT Press, 1996), 239. This section comes from Ronald R. Kline, "Cybernetics, Automata Studies, and the Dartmouth Conference on Artificial Intelligence," *AHC* 33, no. 4 (2011): 5–16.

4. Howard Gardner, *The Mind's New Science: A History of the Cognitive Revolution* (New York: Basic Books, 1985).

5. Flo Conway and Jim Siegelman, *Dark Hero of the Information Age: In Search of Norbert Wiener, the Father of Cybernetics* (New York: Basic Books, 2005), 175; McCorduck, *Machines Who Think*, 84; and "An Interview with Marvin L. Minsky," conducted by Arthur L. Norberg, November 1, 1989, CBIOH, 10–12.

6. Marvin Minsky, ed., *Semantic Information Processing* (Cambridge, MA: MIT Press, 1968), 7; and Allen Newell, "Intellectual Issues in the History of Artificial Intelligence," in *The Study of Information: Interdisciplinary Messages*, ed. Fritz Machlup and Una Mansfield (New York: John Wiley, 1983), 187–227, quotations on 191, 201. Many historians see cybernetics as an origin point for early AI. See, e.g., McCorduck, *Machines Who Think*, chap. 4; Edwards, *Closed World*, chap. 8; and Roberto Cordeschi, *The Discovery of the Artificial: Behavior, Mind, and Machines before and beyond Cybernetics* (London: Dordrecht, 2002), chap. 5.

7. Claude Shannon and John McCarthy to Norbert Wiener, May 19, 1953, NW, box 12-173; Donald MacKay to McCarthy, October 9, 1953; W. Grey Walter to Shannon and McCarthy, June 11, 1953; John von Neumann to Shannon and McCarthy, June 15, 1953; and W. Ross Ashby to McCarthy. All in CESLC, box 1.

8. Claude Shannon and John McCarthy to Norbert Wiener, May 19, 1953, NW, box 12-173; Donald MacKay to McCarthy, October 9, 1953; W. Grey Walter to Shannon and McCarthy, June 11, 1953; John von Neumann to Shannon and McCarthy, June 15, 1953; W. Ross Ashby to McCarthy, June 16, 1954; S. C. Kleene to Shannon and McCarthy, May 29, 1954; and Albert Uttley to McCarthy, October 20, 1953, all in CESLC, box 1.

9. McCorduck, *Machines Who Think*, 94.

10. Norbert Wiener to Crane Brinton, March 4, 1954, NW, box 13-189. On McCarthy at Bell Labs, see McCorduck, *Machines Who Think*, 94.

11. Alan Turing to W. Ross Ashby, ca. November 20, 1946, rpt. in Harry D. Huskey, "From ACE to the G-15," *AHC* 6, no. 4 (1984): 350–371, on 360; and Turing to Claude Shannon, June 3, 1953, CESLC, box 1.

12. John McCarthy to Claude Shannon, April 5, 1954, CESLC, box 1; and Donald MacKay, "The Epistemological Problem for Automata," in *Automata Studies*, ed. Claude Shannon and John McCarthy (Princeton, NJ: Princeton Uni-

versity Press, 1956): 235–251. On MacKay's work on automata, see McCorduck, *Machines Who Think*, 79–81; and Roberto Cordeschi, "Steps toward the Synthetic Method: Symbolic Information Processing and Self-Organizing Systems in Early Artificial Intelligence Modeling," in *The Mechanical Mind in History*, ed. Philip Husbands, Owen Holland, and Michael Wheeler (Cambridge, MA: MIT Press, 2008), chap. 10.

13. W. Ross Ashby, "Design for an Intelligence Amplifier," in *Automata Studies*, ed. Shannon and McCarthy, 215–234; Peter M. Asaro, "From Mechanisms of Adaptation to Intelligent Amplifiers: The Philosophy of W. Ross Ashby," in *Mechanical Mind in History*, ed. Husbands, Holland, and Wheeler, chap. 7; and Andrew Pickering, *The Cybernetic Brain: Sketches of Another Future* (Chicago: University of Chicago Press, 2010), chap. 4.

14. W. Ross Ashby to John McCarthy, June 16, 1954; Donald MacKay to McCarthy, October 9, 1953; and McCarthy to Claude Shannon, August 11, 1954, all in CESLC, box 1.

15. Shannon and McCarthy, eds., *Automata Studies*, ix; and Claude Shannon, "Von Neumann's Contributions to Automata Theory" (1958); rpt. in *CESCP*, 831–835, on 831.

16. John McCarthy to Claude Shannon, August 11, 1954; and Marvin Minsky to Shannon, January 18, 1954 [1955], both in CESLC, box 1.

17. Claude Shannon, "Computers and Automata" (1953); rpt. in *CESCP*, 703–710, on 709.

18. John McCarthy to Claude Shannon, November 24, 1954, CESLC, box 1.

19. John McCarthy and Claude Shannon, "Preface," in *Automata Studies*, ed. Shannon and McCarthy, v–viii, on v.

20. Claude Shannon to Irene Angus, August 8, 1952, CESLC, box 1.

21. John McCarthy to Claude Shannon, November 24, 1954, CESLC, box 1.

22. John McCarthy to Claude Shannon, February 18, 1955, CESLC, box 1.

23. Warren Weaver diary, April 4, 1955, RF-Dartmouth.

24. Warren Weaver to John McCarthy, June 7, 1955, RF-Dartmouth.

25. Warren Weaver to Robert Morison, June 14, 1955, RF-Dartmouth; and Weaver, "Recent Contributions to the Mathematical Theory of Communication," in Claude Shannon and Weaver, *The Mathematical Theory of Communication* (Urbana: University of Illinois Press, 1949), 93–117.

26. Robert Morison diary, June 17, 1955, RF-Dartmouth. On Minsky and Selfridge, see "An Interview with Oliver Selfridge," in *Mechanical Mind in History*, ed. Husbands, Holland, and Wheeler, chap. 17, on 399.

27. Robert Morison diary, June 17, 1955, RF-Dartmouth.

28. John McCarthy to Robert Morison, September 2, 1955, RF-Dartmouth. On McCarthy's working with Rochester at IBM, see "An Interview with John McCarthy," conducted by William Aspray, March 2, 1989, CBIOH, p. 8.

29. Edwards, *Closed World*, 253.

30. John McCarthy, Marvin Minsky, Nathan Rochester, and Claude Shannon,

"A Proposal for the Dartmouth Summer Research Project on Artificial Intelligence," August 31, 1955, RF-Dartmouth, pp. 1–3.

31. Ibid., 6–17, on 6, 7, 16, 17.

32. Robert Morison to John McCarthy, September 22, 1955, RF-Dartmouth; McCarthy to Claude Shannon, November 7, 1955, CESLC, box 1; and Morison to McCarthy, November 30, 1955, RF-Dartmouth.

33. On the Rosenbleuth-Wiener grant, see Lily E. Kay, *Who Wrote the Book of Life? A History of the Genetic Code* (Stanford, CA: Stanford University Press, 2000), 82–84, 349.

34. John McCarthy to Robert Morison, December 8, 1955. On the revised budget, see Donald Morrison to Robert Morison, December 19, 1955, both letters in RF-Dartmouth.

35. McCorduck, *Machines Who Think*, 132–135; and Hunter Crowther-Heyck, *Herbert A. Simon: The Bounds of Reason in Modern America* (Baltimore: Johns Hopkins University Press, 2005), chap. 10.

36. Herbert Simon to Claude Shannon, February 21, 1956, CESLC, box 2.

37. John McCarthy to Nathan Rochester, Claude Shannon, and Marvin Minsky, February 23, 1956, CESLC, box 2; and Allen Newell and Herbert Simon, "Plans for the Dartmouth Summer Research Project on Artificial Intelligence," March 6, 1956, HAS, box 38, Problem Solving Papers, # 9.

38. Claude Shannon to Herbert Simon, April 24, 1956, CESLC, box 2.

39. "An Interview with Allen Newell," conducted by Arthur Norberg, June 10–12, 1991, CBIOH, p. 12. I have found no documentation of these exchanges.

40. John McCarthy to Robert Morison, May 25, 1955, RF-Dartmouth.

41. Donald MacKay to Warren McCulloch, July 23, 1956, WSM, ser. I, MacKay correspondence. On the Rockefeller Foundation's paying his way to the United States, see MacKay to Warren Weaver, June 26, 1956, RF-MacKay.

42. "An Interview with John Holland," in *Mechanical Mind in History*, ed. Husbands, Holland, and Wheeler, 389; "Interview with Allen Newell," 12; and McCorduck, *Machines Who Think*, 94. Trenchard More thanks Bigelow in his master's thesis, but Herbert Simon does not recall Bigelow's being there. See Simon to Jacques Berleur, e-mail, November 19, 1999, HAS, box 77, Publications, Collected Papers on Early History of AI.

43. "[Interview with] Bernard Widrow," in *Talking Nets: An Oral History of Neural Networks*, ed. James A. Anderson and Edward Rosenfeld (Cambridge, MA: MIT Press, 1998), chap. 3, quotation on 49.

44. John McCarthy to [Robert Morison], November 18, 1956; and McCarthy to Morison, September 3, 1956 (quotation), both in RF-Dartmouth.

45. Interview of Herbert Simon by Pamela McCorduck, November 6, 1974, HAS, box 79, Publications, Interview by Pamela McCorduck . . . transcript, # 2, p. 17; and McCorduck, *Machines Who Think*, 108.

46. Gardner, *Mind's New Science*, 28–31; and Peter Elias, "The Rise and Fall

of Cybernetics in the US and the USSR," *Proceedings of the Symposia in Pure Mathematics* 60 (1997): 21–29, quoting Miller on 23–24.

47. Claude Shannon, "The Zero Capacity for a Noisy Channel" (1956); rpt. in *CESCP*, 221–238.

48. Stephen Welch, Dartmouth College, financial report to Rockefeller Foundation, February 7, 1957; and Rowe Steel to Welch, February 15, 1957, both in RF-Dartmouth.

49. Robert Morison diary, May 8, 1957; and Richmond Anderson to Oliver Selfridge, June 10, 1957, both in RF-Dartmouth. On the conference proposal, see Marvin Minsky and Selfridge to P. M. Woodward, February 11, 1957, attached to Morison to Warren Weaver, n.d., RF-Dartmouth.

50. Cordeschi, *Discovery of the Artificial*, 187.

51. Marvin Minsky, "Some Methods of Artificial Intelligence and Heuristic Programming," in *Mechanisation of Thought Processes: Proceedings of a Symposium Held at the National Physical Laboratory . . . 1958*, 2 vols. (London: HMSO, 1959), 1:3–26.

52. Daniel S. Halacy, Jr., *Cyborg—Evolution of the Superman* (New York: Harper & Row, 1965), 41; and Chris H. Gray, "An Interview with Jack E. Steele," in *The Cyborg Handbook*, ed. Chris H. Gray (New York and London: Routledge, 1995), 61–69, quotation on 64. This section and the next one come from Ronald Kline, "Where Are the Cyborgs in Cybernetics?," *Social Studies of Science* 39 (2009): 331–362.

53. Jerry Y. Lettvin, Humberto R. Maturana, Warren S. McCulloch, and Walter H. Pitts, "What the Frog's Eye Tells the Frog's Brain," *Proceedings of the IRE* 47 (1959): 1940–1959.

54. Heinz von Foerster, "Some Aspects of the Design of Biological Computers," in *Proceedings of the Second International Congress on Cybernetics, Namur, September 3–10, 1958* (Namur: Association Internationale de Cybernétique, 1960): 240–255; and Alan Corneretto, "Electronics Learns from Biology," *Electronic Design* 8 (September 14, 1960): 38–54.

55. Marshall Yovits, ONR, to Heinz von Foerster, December 30, 1957, HVF, box 7—McCulloch; von Foerster, "Final Report [on ONR contact Nonr 1834 (210)]," Electrical Engineering Research Laboratory, Engineering Experiment Station, University of Illinois, December 31, 1963, 1; Warren McCulloch to von Foerster, November 7, 1957, HVF, box 7—McCulloch; von Foerster to Jerome Lettvin, March 20, 1958, HVF, box 7—Lettvin; and Albert Müller, "A Brief History of the BCL," in *An Unfinished Revolution? Heinz von Foerster and the Biological Computer Laboratory, BCL, 1958–1976*, ed. Albert Müller and Karl Müller (Vienna: Echoraum, 2007), chap. 12.

56. Harvey E. Savely, "Air Force Research on Living Prototypes," in U.S. Air Force, Wright Air Development Division, *Living Prototypes—The Key to New Technology [Proceedings of the First] Bionics Symposium, September 13–15,*

*1960*, WADD Technical Report 60-600, December 1960 (Wright-Patterson Air Force Base, Ohio): 41–47. On reliability problems owing to complex aircraft systems, see John R. Downer, "The Burden of Proof: Regulating Reliability in Civil Aviation" (Ph.D. diss., Cornell University, 2006).

57. Gray, "Interview with Jack E. Steele," 62.

58. Leo E. Lipetz, "Bionics," *Science* 133 (1961): 588, 590, 592–593; USAF, *Living Prototypes*; and David R. Heinley, "The Role of Biology in the Instrumentation Industry," *American Institute of Biological Sciences Bulletin,* 13 (1963): 35–37.

59. John Keto, "Keynote Address," in USAF, *Living Prototypes*, 7–12, on 7; and Heinz von Foerster, "Bionics," in *McGraw-Hill Yearbook of Science and Technology* (New York: McGraw-Hill, 1963), 148–151, on 148.

60. Jack E. Steele, "How Do We Get There?," in USAF, *Living Prototypes*, 487–490.

61. See *Oxford English Dictionary*, online edition, s.v., "bionic."

62. Heinz Von Foerster, "Biological Ideas for the Engineer," *New Scientist* 15: 306–309; von Foerster, "Final Report," 5; and Peter Asaro, "Heinz von Foerster and the Bio-Computing Movements of the 1960s," in *An Unfinished Revolution?*, ed. Müller and Müller, chap. 11.

63. Keto, "Keynote Address," 10.

64. Heather M. David, "Symposium Points to More AF Emphasis on Hardware," *Missiles and Rockets* 12 (April 1, 1963): 41, 43, on 34, 35; and Daniel S. Halacy, Jr., *Bionics: The Science of "Living" Machines* (New York: Holiday House, 1965), 146, 172.

65. Donald R. Taylor, Jr., "A Bioelectric Pattern Recognition Control for Prosthesis," in *Bionics Symposium 1966, Cybernetic Problems in Bionics* (New York: Gordon and Breach): 885–893, on 885.

66. Halacy, *Bionics*, 146–147, 172.

67. Halacy, *Cyborg*, 145.

68. Edwin G. Johnsen and William R. Corliss (n.d.), "Teleoperators and Human Augmentation," NASA [Report] SP-5047, An AEC-NASA Technology Survey; rpt. in *Cyborg Handbook*, ed. Gray, 83–92, on 85, 87, their emphasis.

69. Halacy, *Bionics*, 146, 150–151.

70. Ibid., 181.

71. Chris H. Gray, Steven Mentor, and Heidi J. Figueroa-Sarriera, "Cyborgology: Constructing the Knowledge of Cybernetic Organisms," in *Cyborg Handbook*, ed. Gray, 1–14.

72. Nathan S. Kline and Manfred Clynes, "Drugs, Space, and Cybernetics: Evolution to Cyborgs," in *Psychophysiological Aspects of Space Flight*, ed. Bernard E. Flaherty (New York: Columbia University Press, 1961), 345–371, on 347–348. See also Clynes and Kline, "Cyborgs and Space," *Astronautics* 5, no. 9 (1960): 26–27, 74–76, on 27.

73. *NWCyb*, 19.

74. See, e.g., J. Law and I. Moser, "Cyborg," in *IESB*, 5:3202–3204.

75. Gray, Mentor, and Figueroa-Sarriera, "Cyborgology," 2.

76. Ibid.; Barbara J. Lewis, "Human Machines: The Treatment of the Cyborg in American Popular Fiction and Film" (Ph.D. diss., New York University, 1997), 5–7; and N. Katherine Hayles, *How We Became Posthuman: Virtual Bodies in Cybernetics, Literature, and Informatics* (Chicago: University of Chicago Press, 1999), 141.

77. Hayles, *How We Became Posthuman*, chap. 3. She mentions the hearing glove on 98–99. For a criticism of Hayles's interpretation of disembodiment in the hearing glove, see Mara Mills, "On Disability and Cybernetics: Helen Keller, Norbert Wiener, and the Hearing Glove," *Differences* 22 (2011): 74–111.

78. Norbert Wiener, "Sound Communication with the Deaf," *Philosophy of Science* 16 (1949): 260–262, on 261; *NWHUHB*, 201; and H. Marshall McLuhan, *Understanding Media: The Extensions of Man* (New York: McGraw-Hill, 1964), 43–46, 57, 68. On his admiration of Wiener, see McLuhan to Wiener, March 28, 1951, NW, box 9-135; and McLuhan, *The Mechanical Bride: Folklore of Industrial Man* (New York: Vanguard Press, 1951), 31, 34, 92.

79. Norbert Wiener, "Homeostasis in the Individual and Society," *Journal of the Franklin Institute* 251 (1951): 65–68, on 66; and Wiener, "The Concept of Homeostasis in Medicine," *Transactions and Studies of the College of Physicians of Philadelphia* 20 (1953): 87–93, on 92–93.

80. Norbert Wiener and J. P. Schadé, "Introduction to Neurocybernetics," *Progress in Brain Research* 2 (1963): 1–7, on 1.

81. Robert W. Mann, "Sensory and Motor Prostheses in the Aftermath of Wiener," *Proceedings of Symposia in Applied Mathematics* 52 (1997): 401–439, on 402–405; and Conway and Siegelman, *Dark Hero of the Information Age*, 322–324.

82. Norbert Wiener to Scott Allan, July 17, 1963, NW, box 23-328; John E. Mangelsdorf to Wiener, April 16, 1953, NW, box 23-325; Wiener to Mangelsdorf, April 22, 1963, NW, box 23-325; and Manfred Clynes to Wiener, November 13, 1961, NW, box 21-305.

83. Kline quoted in "Spaceman Is Seen as Man-Machine," *NYT*, May 21, 1960, 31.

84. Chris Gray, "An Interview with Manfred Clynes," in *Cyborg Handbook*, ed. Gray, 43–53.

85. Arthur W. Martin to Warren McCulloch, June 1, 1959; draft of letter from McCulloch to Martin, n.d. [ca. June 1959]; Manfred Clynes to McCulloch, January 4, 1960; "Curriculum Vitae [of Clynes]," n.d., ca. January 1960; McCulloch, recommendation for Clynes, n.d., ca. March 1960; McCulloch to Hudson Hoaglund, January 13, 1969, all in WSM, ser. I, Manfred Clynes correspondence; Manfred Clynes, "Unidirectional Rate Sensitivity: A Biocybernetic Law of Reflex and Humoral Systems as Physiologic Channels of Control and Communication," *Annals of the New York Academy of Sciences* 92 (1961): 946–969, on 946; and Steve J. Heims, *The Cybernetics Group* (Cambridge, MA: MIT Press, 1991), 129, 133.

86. Kline and Clynes, "Drugs, Space, and Cybernetics," 344, 345, 361.

87. "Spaceman Is Seen as Man-Machine."

88. "Man Remade to Live in Space," *Life*, July 11, 1960, 77–78.

89. Lewis, "Human Machines," 79–83, 142–148.

90. Joseph Shelley to the Editor, and Editor's response, "Man and Cyborgs," *Life*, August 1, 1960, 9.

91. William Beller, "Ways Sought to Ease Ordeal of Space," *Missiles and Rockets*, June 13, 1960, 38–40, on 38, 40.

92. Heather M. David, "Drugs May Halve Radiation Damage," *Missiles and Rockets*, September 26, 1960, 39–40, on 40.

93. G. Dale Smith to Richard J. Preston, April 12, 1962; and Smith and Alfred M. Mayo, "NASA Procurement Request," April 24, 1962, attached to Arthur B. Freeman to Mayo, April 25, 1962. All in NARCR, box RMO-3.

94. Heather M. David, "UAC Cyborg Study in Second Phase," *Missiles and Rockets*, May 13, 1963, 41, 43, on 43. On the OART, see John A. Pitts, *The Human Factor: Biomedicine in the Manned Space Program to 1980* (Washington, DC: NASA, 1985), 78, 80.

95. Robert W. Driscoll et al., "Engineering Man for Space: The Cyborg Study," Final Report [on NASA Contract] NASw-512, Corporate Systems Center, United Aircraft, Farmington, CT, May 15, 1963," part I-1, 3; part II-1 (quotation); part III-28.

96. Ibid., II-33 (his emphasis); III-17–18; IV-13; and V-1.

97. Ibid., VI-13.

98. David, "UAC Cyborg Study in Second Phase."

99. Driscoll, et al., "Engineering Man for Space," VII-3, 4 (quotation).

100. Ibid., VII-1, 4–5 (quotations).

101. I have found only three references to it after United Aircraft issued the final report on Phase I in May 1963: Unsigned, "Biotechnology Research Pushed by Big Companies," *Missiles and Rockets*, November 25, 1963, 83–84, 86, 88–89; Eugene Konecci, "Advanced Concepts in Man-Machine Control: A Review," speech, September 7–12, 1964, WSM, ser. II, NASA Committee on Biotechnology and Human Research, November 1964, # 1, pp. 42–45; and Halacy, *Bionics*, 173.

102. On this score, Gray, Mentor, and Figueroa-Sarriera, "Cyborgology," 8, say that "after this study the agency [NASA] seemed almost allergic to the term 'cyborg' and instead used more technical, and usually specific, locutions like teleoperators, human augmentation, biotelemetry, and bionics."

103. R. T. Allen to Frank Voris, August 27, 1963, NARCR, ser. 19, Central Files, 1959–1967, box RMO-5.

104. Frank B. Voris to James D. Hardy, September 9, 1963, NARCR, ser. 19, Central Files, 1959–1967, box RMO-5.

105. G. M. McDonnel to Warren McCulloch, December 14, 1964, WSM, ser. II, NASA Committee on Biotechnology and Human Research, March 1965, # 3; and Eugene Konecci to McCulloch, April 29, 1966, WSM, ser. II, NASA. On Konecci's resignation, see Pitts, *Human Factor*, 86–87.

106. Walter Pitts to Warren McCulloch, April 21, 1969, WSM, ser. I, Walter Pitts correspondence.

107. Neil R. Smalheiser, "Walter Pitts," *Perspectives in Biology and Medicine* 43 (2000): 217–226; and Juan de Dios Pozo-Olano, "Warren Sturgis McCulloch (1898–1969)," *Journal of the Neurological Sciences* 10 (1970): 414–416.

108. Wiener, *God and Golem*, 74, 76. On the earlier statements, see *NWHUHB*, 195.

109. Halacy, *Cyborg*, 15.

110. Arthur C. Clarke, *Profiles of the Future: An Inquiry into the Limits of the Possible* (New York: Harper & Row, 1962), 226–227, his emphasis.

111. See, e.g., Kathleen M. Krajewski, "Scientists and Social Responsibility: University of Delaware, July 1975 [A Conference Report]," *Technology and Culture* 18 (1977): 56–61; Lewis, "Human Machines," chap. 6; David M. Rorvik, *As Man Becomes Machine: The Evolution of the Cyborg* (Garden City, NY: Doubleday, 1971); and Thomas Stritch, "The Banality of Utopia," *Review of Politics* 34, no. 1 (1972): 103–106.

112. Harrison Smith, "The Machine in Man's Image," *Saturday Review of Literature* 32 (January 8, 1949): 22.

## Chapter 7: Cybernetics in Crisis

1. Walter A. Rosenblith, "Norbert Wiener: In Memoriam," *Kybernetik* 2 (1965): 195–196, on 196.

2. Peter Elias, "Cybernetics: Past and Present, East and West," in *The Study of Information: Interdisciplinary Messages*, ed. Fritz Machlup and Una Mansfield (New York: John Wiley, 1983), 441–444, on 442, 443.

3. Denis Gabor, "[Review of Gordon Pask's] *An Approach to Cybernetics . . . ,*" *British Journal for the Philosophy of Science* 9 (1958): 168–170, on 168.

4. Donald MacKay to Heinz von Foerster, May 25, 1959, HVF, box 7, MacKay correspondence; and MacKay, "Norbert Wiener: 'Catalytic Irritant,'" *Journal of Neurons and Mental Disease* 140 (1965): 9–10, on 9. On ARTORGA, see Michael J. Apter, *Cybernetics and Development* (Oxford: Pergamon Press, 1966), 25. MacKay tempered his position on using the term *cybernetics* in "What Is Cybernetics?," *Discovery* 23 (October 1962): 13–17.

5. W. Grey Walter, "Neurocybernetics (Communication and Control in the Living Brain)," in *Survey of Cybernetics*, ed. J. Rose (London: Iliffe, 1969), 93–108, on 94.

6. John McCarthy, "[Review of] Bloomfield, Brian, ed. *The Question of Artificial Intelligence . . . ,*" *AHC* 10 (1988): 224–229, on 227.

7. Charles R. Dechert, "The Development of Cybernetics," in *The Social Impact of Cybernetics*, ed. Dechert (New York: Simon and Schuster, 1966), 11–37, on 18.

8. Jerome Rothstein to T. C. Helvey, February 13, 1967, JR.

9. Satosi Watanabe, "Norbert Wiener and Cybernetical Concept of Time," *IEEE Trans SMC* 5 (1975): 372–375, on 373.

10. Warren McCulloch, draft of speech to the ASC inaugural dinner, n.d. [October 1964], WSM—ASC, By-laws, p. 18.

11. For examples of McCulloch's using the term *cybernetics* before the dinner, in 1955 and 1956, see McCulloch, *Embodiments of Mind* (Cambridge, MA: MIT Press, 1965), 158, 194.

12. Margaret Mead, "The Cybernetics of Cybernetics," in *Purposive Systems: Proceedings of the First Annual Symposium of the American Society for Cybernetics*, ed. Heinz von Foerster et al. (New York: Spartan Books, 1968), 1–11, on 2; and Mead, "Crossing Boundaries in Social Science Communication," *Social Science Information* 8 (1969): 7–15.

13. Julian Bigelow, interview with Steve Heims, November 12, 1968, SH, box 1-5, on 45, 46, 47.

14. See, e.g., William Aspray, *John von Neumann and the Origins of Modern Computing* (Cambridge, MA: MIT Press, 1990), 210.

15. Murray Eden, "Cybernetics," in *Study of Information*, ed. Machlup and Mansfield, 409–439, on 439. On the original aims of the journal, see Léon Brillouin, Colin Cherry, and Peter Elias, "A Statement of Editorial Policy," *Information and Control* 1 (1957): i–ii.

16. Fritz Machlup and Una Mansfield, "Cultural Diversity in Studies of Information," in *Study of Information*, ed. Machlup and Mansfield, 3–56, on 40.

17. Yehoshua Bar-Hillel, *Language and Information: Selected Essays on Their Theory and Application* (Reading, MA: Addison-Wesley, 1964), 11.

18. John F. Young, *Cybernetics* (London: Iliffe Books, 1969), preface, n.p.; and Michael J. Apter, "Cybernetics: A Case Study of a Scientific Subject-Complex," in *The Sociology of Science*, ed. Paul Halmos, Sociological Review Monograph, no. 18 (Keele: University of Keele, 1972): 93–115, on 111.

19. Paul Ryan, *Cybernetics of the Sacred* (New York: Doubleday, 1974), 31; and Eden, "Cybernetics," 416.

20. Geoffrey C. Bowker, "How to Be Universal: Some Cybernetic Strategies, 1943–1970," *Social Studies of Science* 23 (1993): 107–127, on 116; and Fred Turner, *From Counterculture to Cyberculture: Stewart Brand, the Whole Earth Network, and the Rise of Digital Utopianism* (Chicago: University of Chicago Press, 2006), 84–85.

21. In contrast, Warren McCulloch and Claude Shannon were friends with L. Ron Hubbard and corresponded with him about his research for *Dianetics*. See chapter 3.

22. Slava Gerovitch, *From Newspeak to Cyberspeak: A History of Soviet Cybernetics* (Cambridge, MA: MIT Press, 2002), chaps. 3, 5, 6. See a lengthy review of volume 1 by D. G. Malcolm in *Operations Research* 11 (1963): 1007–1012.

23. Dechert, "The Development of Cybernetics," 18; Bigelow interview with Heims, 44–47; Eden, "Cybernetics," 417; and Elias, "Cybernetics."

24. M. E. Maron, "Cybernetics," in *International Encyclopedia of the Social Sciences*, 2nd ed. (New York: Macmillan, 1968), 4:3–6, on 6; and Peter Elias,

"The Rise and Fall of Cybernetics in the US and the USSR," *Proceedings of the Symposia in Pure Mathematics* 60 (1997): 21–29, on 26.

25. Jessica Wang, *American Scientists in an Age of Anxiety: Scientists, Anticommunism, and the Cold War* (Chapel Hill: University of North Carolina Press, 1999).

26. Slava Gerovitch, "The Cybernetics Scare and the Origins of the Internet," *Baltic Worlds* 2 (2009): 32–38, on 38. On Wiener's trip to Russia, see Wiener to W. T. Martin, November 25, 1960, NW, box 20-288; and Gerovitch, *From Newspeak to Cyberspeak*, 244 (photo).

27. See, e.g., Philipp Aumann, "The Distinctiveness of a Unifying Science: Cybernetics' Way to West Germany," *AHC* 33, no. 4 (2011): 17–27.

28. Donald L. M. Blackmer, *The MIT Center for International Studies: The Founding Years, 1951–1969* (Cambridge, MA: MIT CENIS, 2002); Frances S. Saunders, *The Cultural Cold War: The CIA and the World of Arts and Letters* (New York: New Press, 1999), chap. 9; David Price, "Interlopers and Invited Guests: On Anthropology's Witting and Unwitting Links to Intelligence Agencies," *Anthropology Today* 18, no. 6 (December 2002): 16–26; and Price, "The AAA and the CIA," *Anthropology News* 41, no. 8 (November 2000): 13–14.

29. Ronald E. Doell and Allan A. Needell, "Science, Scientists, and the CIA: Balancing International Ideals, National Needs, and Opportunities," in *Eternal Vigilance? 50 Years of CIA*, ed. Rhodri Jeffreys-Jones and Christopher Andrew (London: Frank Cass, 1997), 59–81; and Jeffrey T. Richelson, "The Wizards of Langley: The CIA's Directorate of Science and Technology," in ibid., 82–103.

30. Doell and Needell, "Science, Scientists, and the CIA," 68.

31. John Marks, *The Search for the "Manchurian Candidate": The CIA and Mind Control* (New York: Times Books, 1979), 59, 64–65, 120; and Steve J. Heims, *Cybernetics Group* (Cambridge, MA: MIT Press, 1990), 167–168.

32. On his ties to the CIA, see Lily Kay, "From Logical Neurons to Poetic Embodiments of Mind: Warren S. McCulloch's Project in Neuroscience," *Science in Context* 14 (2001): 591–614, on 595.

33. John Ford to Arthur Schlesinger, Jr., October 17, 1962, cited in Gerovitch, "The Cybernetics Scare," 35. On Ford and the cybernetics scare, see also Benjamin Peters, "From Cybernetics to Cyber Networks: Norbert Wiener, the Soviet Internet, and the Cold War Dawn of Information" (Ph.D. diss., Columbia University, 2010), 183–187.

34. Zuoyue Wang, *In Sputnik's Shadow: The President's Science Advisory Committee and Cold War America* (New Brunswick, NJ: Rutgers University Press, 2008), 86.

35. Karl Wildes and Nilo A. Lindgren, *A Century of Electrical Engineering and Computer Science at MIT, 1882–1982* (Cambridge, MA: MIT Press, 1985), 262–266; and Walter Rosenblith, "Brothers," in *Jerry Wiesner: Scientist, Humanist, Memories and Memoirs*, ed. Rosenblith (Cambridge, MA: MIT Press, 2003), 3–17.

36. Unsigned, "Cybernetics Panel Membership," April 1, 1963, PE, box 8—PSAC folder. The members were Peter Elias at MIT, John Pierce at AT&T, John Tukey at Princeton, Leonid Hurwicz at the University of Minnesota, Willis Ware at RAND, Ross Adey at the University of California, Los Angeles, and George Miller at Harvard. Pierce and Tukey were the only regular members of PSAC; see Wang, *In Sputnik's Shadow*, 327.

37. "Cybernetic Panel Minutes," April 1, 1963, quotation on 4; and Stephen L. Aldrich (CIA) to Peter Bing, July 8, 1963, both in WR, box 50—PSAC folder. On the CIA and PSAC, see Wang, *In Sputnik's Shadow*, 83, 105, 108, 117.

38. John Ford, "Long-Range-Scientific Capabilities of the USSR (1963–'73): The 'Complex Scientific Problem, Cybernetics,'" May 1, 1963, working draft, WR, box 50, pp. i, 4, 10–11.

39. Peter S. Bing, "Summary," February 13, 1964, on 1, 2. A revised report was to be presented to PSAC in July 1964; see Bing to Cybernetics Panel members, June 10, 1964. At least one survey was conducted; see Willis H. Ware to Robert Fano, February 22, 1965. All documents in PE, box 8—PSAC folder. Wiesner later traveled to the Soviet Union, where he verified that it was far behind the United States in computer control and, thus, that there was no "cybernetics gap." See Spurgeon Keeny, Jr., "The Search for Soviet Cybernetics," in *Jerry Wiesner*, ed. Rosenblith, 81–88.

40. Even though the term *cybernetics* had multiple meanings in the USSR; see Gerovitch, *From Newspeak to Cyberspeak*, 246–251.

41. Flo Conway and Jim Siegelman, *Dark Hero of the Information Age: In Search of Norbert Wiener, the Father of Cybernetics* (New York: Basic Books, 2005), 317–320, 330–331, on 319.

42. Gerard L. Daniel to Walter Rosenblith, October 20, 1964, WR, box 50, folder with no title.

43. John Ford, "Soviet Cybernetics and International Development," in *Social Impact of Cybernetics*, ed. Dechert, 161–192. The Ad Hoc group became the Cybernetic Society of Washington, a chapter of the ASC, in 1965. See Heyward E. Canney, Jr., "Ad Hoc Group on the Larger Cybernetic Problem, First Meeting and Other Matters," February 3, 1965; Canney, "Cybernetics Society of Washington, First Meeting," April 4, 1965; both in WSM—ASC, Local Chapters.

44. Paul Henshaw to Roman Jakobson, January 25, 1965, and John Ford to Jakobson, September 28, 1965, both in RJ, box 5-74. The incorporators were Ford, neuroscientist Robert Livingston, and Walter Munster of the Atomic Energy Commission; see ASC, "Planning Conference," October 16–17, 1965, WSM—ASC, Bylaws, "Preamble." Ford was a director of the ASC as late as 1968; see Rudolph Constantine, "Minutes of the [ASC] Council Meeting Held on April 16, 1968," n.d., WSM—ASC, Board Meetings, no. 1.

45. Saunders, *Cultural Cold War*, chaps. 23–26; and Blackmer, *MIT Center for International Studies*, chaps. 5–6.

46. Conway and Siegelman, *Dark Hero of the Information Age*, chaps. 12–13.

47. Frank Fremont-Smith to Warren McCulloch, September 18, 1964, WSM—ASC, Board Meetings, no. 2. The other founders were F. S. C. Northrop, Hans Teuber, Yuk-Wing Lee, Hermann Goldstine, Oskar Morgenstern, and Francis Schmitt. Northrop and Teuber were veterans of the Macy conferences.

48. ASC, Attendees at Inaugural Dinner, October 16, 1964; and ASC, Inaugural Dinner, program, October 16, 1964, both in WSM—ASC, Board Meetings, no. 1. Among the attendees was Colonel Raymond Sleeper of the Foreign Technology Division of the Air Force Systems Command. He shared Ford's view of the threat of Soviet cybernetics; see Sleeper, "Cybernetics in the Service of Communism," *Air University Review*, March–April 1967.

49. Warren McCulloch to Paul Henshaw, April 22, 1964, WSM—ASC, Board Meetings, no. 2.

50. Warren McCulloch to John [Dixon], Jack [Ford], and Bob [Livingston], March 16, 1965, WSM—ASC, Board Meetings, no. 2. On Bigelow's election to the IAS, see George Dyson, *Turing's Cathedral: The Origins of the Digital Universe* (New York: Random House, 2012), 266.

51. Julian Bigelow to Paul Henshaw, March 30, 1965, WSM—ASC, Board Meetings, no. 2. The letter is quoted in Elizabeth Corona and Bradley Thomas, "A New Perspective on the Early History of the American Society for Cybernetics," *Journal of the Washington Academy of Sciences* 96, no. 2 (Summer 2010): 21–34, on 26–27, which is based on privately held ASC archives and does not discuss the ASC's ties to the CIA.

52. John Ford to Roman Jakobson, September 28, 1965, RJ, box 5-74.

53. ASC, "Planning Conference," October 16–17, 1965, WSM—ASC, Bylaws, "Discussion."

54. John Ford to Warren McCulloch, March 24, 1966, WSM—ASC, Board Meetings, no. 2.

55. Heinz von Foerster to Warren McCulloch, June 29, 1966, WSM—ASC, Board Meetings, no. 1. On von Foerster's election to the board, see Minutes of ASC meeting, June 24, 1966, attached to John Ford to McCulloch, June 27, 1966, WSM—ASC, Board Meetings, no. 2, p. 177.

56. Lawrence Fogel to Warren McCulloch, July 5, 1966, WSM—ASC, Board Meetings, no. 1.

57. Corona and Thomas, "New Perspective on the Early History of the American Society for Cybernetics," 29.

58. George Jacobi to Warren McCulloch, November 1, 1966; WSM—ASC, Local Chapters, no. 2; Michael Arbib to McCulloch, April 28, 1967, WSM, ser. I, Arbib correspondence; Lawrence Fogel to Arbib, January 4, 1968, WSM—ASC, Correspondence, no. 3; and Rudolph Constantine, "Minutes of the [ASC] Council Meeting Held on April 16, 1968," n.d., WSM—ASC, Board Meetings, no. 1.

59. *Purposive Systems*, ed. von Foerster et al. Macy veterans Yehoshua Bar-Hillel, Ralph Gerard, and J. C. R. Licklider also presented papers.

60. "John J. Ford Dies at 70: CIA, White House Official," *Washington Post*, July 28, 1993, B4.

61. Constantine, "Minutes of the [ASC] Council Meeting Held on April 16, 1968"; unsigned, "Purpose of the A.S.C.," n.d. (ca. 1969), WSM—ASC, Correspondence, no. 3; Douglas E. Knight to Lawrence Fogel, July 10, 1969, HVF, box 12, correspondence 1967–1969; *Purposive Systems*, ed. von Foerster et al., vi–xvi; *Cybernetics and the Management of Large Systems: Proceedings of the Second Annual Symposium of the American Society for Cybernetics*, ed. Edmond M. Dewan (New York: Spartan Books, 1969); *Cybernetics, Simulation, and Conflict Resolution: Proceedings of the Third Annual Symposium of the American Society for Cybernetics*, ed. Knight, Huntington W. Curtis, and Fogel (New York: Spartan Books, 1971); and "Cybernetics Meeting," *PS: Political Science and Politics* 3 (1970): 442.

62. W. D. Rowe, "Why Systems Science and Cybernetics?," *IEEE Trans SSC* 1 (1965): 2–3, on 2.

63. W. Ross Ashby, "The Cybernetic Viewpoint," *IEEE Trans SSC* 2 (1966): 7–8.

64. The three cyberneticians were Murray Babcock, a professor at the BCL, who had chaired the IEEE's Cybernetics Committee; von Foerster himself; and psychologist Milton Katz at President Johnson's Office of Economic Opportunity, who had also been a member of the Cybernetics Committee. See unsigned, "SSC Group ADCOM Members," *IEEE Trans SSC* 3 (1967): 1–5. Julian Reitman, who was chair of the SSC group in 1969, recalled that cybernetics was included in the SSC's name for mostly internal political reasons within the IEEE (conversation with the author, Ithaca, NY, May 23, 2009).

65. J. N. Warfield, "Editorial," *IEEE Trans SSC* 4 (1968): 85; Warfield, "Editorial," *IEEE Trans SSC* 5 (1969): 1; Julian Reitman, "Editorial," *IEEE Trans SSC* 5 (1969): 265; Stanley M. Altman, "A Future Direction for Systems Science and Cybernetics," *IEEE Trans SSC* 6 (1970): 257; and Jay W. Forrester, "Systems Analysis as a Tool for Urban Planning," *IEEE Trans SSC* 6 (1970): 258–264.

66. Denis Gabor, "Cybernetics and the Future of Our Industrial Civilization," *Journal of Cybernetics* 1, no. 2 (April–June 1971): 1–4, on 1, 2, his emphasis.

67. Heinz von Foerster, "Responsibilities of Competence," *Journal of Cybernetics* 2, no. 2 (April–June 1972): 1–6, on 2, his emphasis; and von Foerster, "Physics and Anthropology: A Personal Account by the New President of the Wenner-Gren Foundation," *Current Anthropology* 5 (1964): 330–331.

68. Howard Brick, *Age of Contradiction: American Thought and Culture in the 1960s* (New York: Twanye, 1998), chap. 6, on 124.

69. Thomas P. Hughes and Agatha C. Hughes, "Introduction," in *Systems, Experts, and Computers: The Systems Approach in Management and Engineering, World War II and After*, ed. Hughes and Hughes (Cambridge, MA: MIT Press, 2000), 1–26.

70. Stephen B. Johnson, "Three Approaches to Big Technology: Operations Research, Systems Engineering, and Project Management," *Technology and Culture* 38 (1997): 891–919; Thomas P. Hughes, *Rescuing Prometheus: Four Monumental Projects That Changed the Modern World* (New York: Pantheon, 1998), chap. 4; George P. Richardson, *Feedback Thought in Social Science and Systems Theory* (Philadelphia: University of Pennsylvania Press, 1991), 118–127, 149–159; Debora Hammond, *The Science of Synthesis: Exploring the Social Implications of General Systems Theory* (Boulder: University Press of Colorado, 2003), chap. 8; and Paul Erickson, "Mathematical Models, Rational Choice, and the Search for Cold War Culture," *Isis* 101 (2010): 386–392.

71. Hammond, *Science of Synthesis*, chap. 9. The other founding members were Anatol Rapoport and James Miller, faculty members of the Committee of Behavioral Sciences at the University of Chicago.

72. Jennifer S. Light, *From Warfare to Welfare: Defense Intellectuals and Urban Problems in Cold War America* (Baltimore: Johns Hopkins University Press, 2003), chaps. 2–3, quotations on 46, 47; and Hughes, *Rescuing Prometheus*, 166–171.

73. *Cybernetics and the Management of Large Systems*, ed. Dewan, sec. 3.

74. Edward E. David, Jr., "A Time for the Future," *Journal of Cybernetics* 1, no. 1 (January–March 1971): 1–2, on 2; and David, "Bionics or Electrology?," *IEEE Trans IT* 8 (1962): 74–77.

75. William D. McElroy, "Cybernetics and Biology," *Journal of Cybernetics* 1, no. 3 (July–September 1971): 1–4, on 3, 4.

76. H. Guyford Stever, "Science, Systems, and Society," *Journal of Cybernetics* 2, no. 3 (July–September 1972): 1–3, on 1.

77. Emanuel R. Piore, "Computers, Computation, and Problems," *Journal of Cybernetics* 1, no. 4 (October–December 1971): 1–3; and Harvey Brooks, "Cybernetics and Politics," *Journal of Cybernetics* 2, no. 1 (January–March 1972): 1–2. On Piore's trip to Russia, see Albert Parry, "The Russians Are—Computing," *NYT Magazine*, August 28, 1966, 25, 94, 96, 99–100, on 96.

78. William B. Gevarter, Washington Chapter of the ASC, *Newsletter*, July 1969, WSM—ASC, Correspondence, no. 3; and SSC and ASC, *Proceedings of the 1972 International Conference on Cybernetics and Society* (New York: IEEE, 1972).

79. Apter, "Cybernetics," 110. Brunel University did establish a Department of Cybernetics in 1968; see Andrew Pickering, *The Cybernetic Brain: Sketches of Another Future* (Chicago: University of Chicago Press, 2010), 237–240.

80. Andrew Abbott, *The System of Professions: An Essay on the Division of Expert Labor* (Chicago: University of Chicago Press, 1988), chap. 8 (on OR); Michael Fortun and Silvan S. Schweber, "Scientists and the Legacy of World War II: The Case of Operations Research (OR)," *Social Studies of Science* 23 (1993): 596–642; Stuart W. Leslie, *The Cold War and American Science: The Military-Industrial-Academic Complex at MIT and Stanford* (New York: Colum-

bia University Press, 1993), chap. 8 (on materials science); and Scott Frickel, "Building an Interdiscipline: Collective Action Framing and the Rise of Genetic Toxicology," *Social Problems* 51 (2004): 269–287, on 273.

81. Merrill Flood, "New Operations Research Potentials," *Operations Research* 10 (July–August 1962): 423–436, on 433. On Flood, see Philip Mirowski, *Machine Dreams: Economics Becomes a Cyborg Science* (Cambridge: Cambridge University Press, 2002), 353–359.

82. Ludwig von Bertalanffy, *General System Theory: Foundations, Development, Applications* (New York: George Braziller, 1969), 17, 23, 28–29, 44, 97, 149–150; Richardson, *Feedback Thought in Social Science and Systems Theory*, 122; and Hammond, *Science of Synthesis*, 64, 120–121. Although von Bertalanffy titled his book *General System Theory*, contemporaries, and even Bertalanffy in later years, adopted the term *general systems theory* for this expanding field. See, e.g., Kenneth E. Boulding, "General Systems Theory—The Skeleton of Science," *Management Science* 2 (1956): 197–208; W. Ross Ashby, "General Systems Theory as a New Discipline," *General Systems* 3 (1958): 1–6; and von Bertalanffy, "The History and Status of General Systems Theory," *Academy of Management Journal* 15 (1972): 407–426. On the problem of organized complexity, see Warren Weaver, "Science and Complexity," *American Scientist* 36 (1948): 536–544.

83. Boulding, "General Systems Theory"; and Boulding to Norbert Wiener, January 12, 1954, NW, box 13-186, and February 25, 1954, NW, box 13-188.

84. Jay W. Forrester, "Industrial Dynamics—A Response to Ansoff and Slevin," *Management Science* 14 (1968): 601–618, on 606; and Forrester, "Systems Analysis as a Tool for Urban Planning," 258 (abstract).

85. Jay W. Forrester, *Industrial Dynamics* (Cambridge, MA: MIT Press, 1961); Forrester, *Urban Dynamics* (Cambridge, MA: MIT Press, 1969); and Forrester, *World Dynamics* (Cambridge, MA: MIT Press, 1971). On his career and modeling methods, see David A. Mindell, *Between Human and Machine: Feedback, Control, and Computing before Cybernetics* (Baltimore: Johns Hopkins University Press, 2002), chap. 8; Richardson, *Feedback Thought in Social Science and Systems Theory*, 296–313; William Thomas and Lambert Williams, "The Epistemologies of Non-Forecasting Simulations, Part I: Industrial Dynamics and Management Pedagogy at MIT," *Science in Context* 22 (2009): 245–270; and Brian P. Bloomfield, *Modeling the World: The Social Construction of Systems Analysis* (London: Basil Blackwell, 1986).

86. Talcott Parsons, "Facilitation of Technological Innovation in Society," in *Purposive Systems*, ed. von Foerster et al., 153–161.

87. Walter F. Buckley, *Sociology and Modern Systems Theory* (Englewood Cliffs, NJ: Prentice Hall, 1967); and Robert Lilienfeld, *The Rise of Systems Theory: An Ideological Analysis* (New York: John Wiley, 1978), chap. 7.

88. Robert C. Bannister, "Sociology," in *The Modern Social Sciences*, vol. 7 of *The Cambridge History of Science*, ed. Theodore M. Porter and Dorothy Ross (Cambridge: Cambridge University Press, 2003), 329–353, on 352–353.

89. Hammond, *Science of Synthesis*, 22, 250, 252.

90. Jerry Y. Lettvin, Humberto R. Maturana, Warren S. McCulloch, and Walter H. Pitts, "What the Frog's Eye Tells the Frog's Brain" (1959), rpt. in McCulloch, *Embodiments of Mind*, 230–255; Maturana, "Introduction," in Maturana and Francisco Varela, *Autopoiesis and Cognition: The Realization of the Living* (Boston: D. Reidel, 1980), xi–xxx; N. Katherine Hayles, *How We Became Posthuman: Virtual Bodies in Cybernetics, Literature, and Informatics* (Chicago: University of Chicago Press, 1999), chap. 6; Eden Medina, *Cybernetic Revolutionaries: Technology and Politics in Allende's Chile* (Cambridge, MA: MIT Press, 2011); and Maturana, "Interview on Heinz von Foerster, Autopoiesis, the BCL, and Augusto Pinochet," in *An Unfinished Revolution? Heinz von Foerster and the Biological Computer Laboratory (BCL), 1958–1976*, ed. Albert Müller and Karl Müller (Vienna: Echoraum, 2007), 37–52.

91. Heinz von Foerster, "Cybernetics of Cybernetics," in *Communication and Control in Society*, ed. Klaus Krippendorff (New York: Gordon and Breach, 1979): 5–8, on 7, 8; Bernard Scott, "Second-Order Cybernetics: An Historical Introduction," *Kybernetes* 33 (2004): 1365–1378; and Albert Müller, "A Brief History of the BCL: Heinz von Foerster and the Biological Computer Laboratory," in *An Unfinished Revolution?*, ed. Müller and Müller, 253–276, on 289.

92. Francisco J. Varela, "Introduction: The Ages of Heinz von Foerster," in von Foerster, *Observing Systems* (Seaside, CA: Intersystems, 1981), xi–xvi, on xi.

93. Maturana and Varlea, *Autopoiesis and Cognition*, 85.

94. Gregory Bateson, "Form, Substance, and Difference" (1970), rpt. in Bateson, *Steps to an Ecology of Mind: Collected Essays in Anthropology, Psychiatry, Evolution, and Epistemology* (San Francisco: Chandler, 1972), 448–465, on 461; and Ryan, *Cybernetics of the Sacred*, 31, 65–66.

95. Turner, *From Counterculture to Cyberculture*, chaps. 2, 4.

96. See, e.g., Gregory Bateson, "Reality and Redundancy," *CoEvolution Quarterly* 6 (Summer 1975): 132–135; Bateson, "Invitational Paper," for Mind/Body Dualism Conference, *CoEvolution Quarterly* 11 (Fall 1976): 56–57; Francisco Varela, "On Observing Natural Systems," an interview with Donna Johnson, *Co-Evolution Quarterly*, no. 10, Summer 1976, 26–31; and Varela, "Not One, Not Two," *CoEvolution Quarterly* 11 (Fall 1976): 62–68.

97. Heinz von Foerster, "Two Cybernetics Frontiers," letter to the editor, *Co-Evolution Quarterly* 6 (Summer 1975): 143; and von Foerster, "Gaia's Cybernetics Badly Expressed," *CoEvolution Quarterly* 7 (Fall 1975): 51.

98. Christina Dunbar-Hester, "Listening to Cybernetics: Music, Machines, and Nervous Systems, 1950–1980," *Science, Technology, and Human Values* 35 (2010): 113–139; and Pickering, *Cybernetic Brain*, 78–89.

99. Stuart A. Umpleby, "Interview on Heinz von Foerster, the BCL, Second-Order Cybernetics and the American Society for Cybernetics," in *An Unfinished Revolution?*, ed. Müller and Müller, 77–87, on 87.

100. Conversation with C. Richard Johnson, Ithaca, NY, spring 2009. On

renaming the group, see William R. Ferrell and John N. Warfield, "Editorial," *IEEE Trans SMC* 1 (1971): 1.

101. Umpleby, "Interview on Heinz von Foerster," 84, 85.

102. Scott, "Second-Order Cybernetics"; Ranulph Glanville, "Cybernetics," in *Encyclopedia of Science, Technology, and Ethics*, ed. Carl Mitcham, 4 vols. (New York: Thomson, 2005), 1:455–459; Stuart A. Umpleby, "A History of the Cybernetics Movement in the United States," *Journal of the Washington Academy of Sciences* 91, no. 2 (Summer 2005): 54–66; and Karl H. Müller, "The BCL—an Unfinished Revolution of an Unfinished Revolution," in *An Unfinished Revolution?*, ed. Müller and Müller, 407–466.

103. This meaning of cybernetics was celebrated by the IEEE group in 1975; see unsigned, "Report on the Norbert Wiener Commemorative Symposium," *IEEE Trans SMC* 5 (1975): 359–375.

104. Hunter Crowther-Heyck, "Patrons of the Revolution: Ideals and Institutions in Postwar Behavioral Science," *Isis* 97, no. 3 (September 2006): 420–446.

105. Ibid., 436.

106. A prominent exception is computer scientist Michael Arbib at the University of Massachusetts—Amherst, a former student of McCulloch's at MIT, who used the term *cybernetics* as late as 1983 to "refer to a conjoined study of brain theory and AI." See Arbib, "Cybernetics: The View from Brain Theory," in *Study of Information*, ed. Machlup and Mansfield, 459–465, on 459. He adopted this usage from an early date; see Arbib, *Brains, Machines, and Mathematics* (New York: McGraw-Hill, 1964), chap. 4.

107. Frickel, "Building an Interdiscipline," 273.

108. While Hayles, *How We Became Posthuman*, 6, 13–18, stresses the intellectual morphing of first-order cybernetics into second-order cybernetics, I emphasize their coexistence as subfields.

### Chapter 8: Inventing an Information Age

1. James R. Beniger, *The Control Revolution: Technological and Economic Origins of the Information Society* (Cambridge, MA: Harvard University Press, 1986), introduction.

2. Based on a Google Ngram search of these terms conducted in July 2013.

3. On technological narratives, see David E. Nye, *America as Second Creation: Technology and Narratives of New Beginnings* (Cambridge, MA: MIT Press, 2003), chap. 1.

4. John Lamb, "IT 82—A Critical Year for Britain," *New Scientist* 93 (January 28, 1982): 221–224, on 221.

5. Based on an online search of the *NYT* conducted in September 2010, and the Ngram search described in note 2. The usage of all terms drops after 2001, the year of the second dot-com bust, but the usages of *information technology* and *information age* outpace the other terms through 2009.

6. Fred Turner, *From Counterculture to Cyberculture: Stewart Brand, the*

*Whole Earth Network, and the Rise of Digital Utopianism* (Chicago: University of Chicago Press, 2006). Ronald E. Day, *The Modern Invention of Information: Discourse, History, and Power* (Carbondale: Southern Illinois University Press, 2001), argues that the current information discourse comes from the merger of three historical strands: the European documentation movement; cybernetics and information theory; and today's talk about the "virtual." I expand the first two strands to include many more groups in science, engineering, the social sciences, business, policy circles, and the popular media.

7. See Hans Wellisch, "From Information Science to Informatics: A Terminological Investigation," *Journal of Librarianship* 4 (1972): 157–187; Alvin M. Schrader, "In Search of a Name: Information Science and Its Conceptual Antecedents," *Library and Information Science Research* 6 (1984): 227–271; Fritz Machlup, "Semantic Quirks in Studies of Information," in *The Study of Information: Interdisciplinary Messages*, ed. Machlup and Una Mansfield (New York: John Wiley, 1983), 641–671; and Rafael Capurro and Birger Hjorland, "The Concept of Information," *Annual Review of Information Science and Technology* 37 (2003): 343–411. On the concept of discourse communities, see Ruth Oldenziel, *Making Technology Masculine: Men, Women, and Modern Machines in America, 1870–1950* (Amsterdam: Amsterdam University Press, 1999), introduction.

8. N. Katherine Hayles, *How We Became Posthuman: Virtual Bodies in Cybernetics, Literature, and Informatics* (Chicago: University of Chicago Press, 1999), chap. 3.

9. Machlup, "Semantic Quirks in Studies of Information," 642–649; and Thomas Haigh, "Inventing Information Systems: The Systems Men and the Computer, 1950–1968," *Business History Review* 75 (2001): 15–61, on 46–47.

10. See, e.g., Edmund C. Berkeley, *Giant Brains: Or Machines That Think* (New York: John Wiley, 1949); and Irwin Pollack, "Information Theory," in *International Encyclopedia of the Social Sciences*, 2nd ed. (New York: Macmillan, 1968), 7:331–337.

11. See, e.g., *NWCyb*, 155; and John Diebold, *Automation: The Advent of the Automatic Factory* (New York: D. van Nostrand, 1952), 2, 90.

12. On the information crisis, see Mark D. Bowles, "Crisis in the Information Age: How the Information Explosion Threatened Science, Democracy, the Library, and the Human Body, 1945–1999" (Ph.D. diss., Case Western Reserve University, 1999).

13. On the analyses of keywords, see Raymond Williams, *Keywords: A Vocabulary of Culture and Society*, rev. ed. (New York: Oxford University Press, 1983).

14. Ronald Kline, "Technological Determinism," in *IESB*, 23:15495–15498.

15. For a fuller account of this genealogy, see Ronald Kline, "Cybernetics, Management Science, and Technology Policy: The Emergence of 'Information Technology' as a Keyword, 1948–1985," *Technology and Culture* 47 (July 2006): 513–535, upon which this section is based.

16. Eric Schatzberg, "*Technik* Comes to America: Changing Meanings of *Technology* before 1930," *Technology and Culture* 47 (2006): 486–512.

17. Harold J. Leavitt and Thomas L. Whisler, "Management in the 1980s," *Harvard Business Review* 36 (November–December 1958): 41–48, on 41, their emphasis.

18. Thomas L. Whisler and George P. Shultz, "Information Technology and Management Organization," in *Management Organization and the Computer*, ed. Shultz and Whisler (Glencoe, IL: Free Press, 1960), 3–36, on 3, 10–12.

19. Herbert A. Simon, *The New Science of Management Decision* (New York: Harper and Row, 1960), 35; and John F. Burlingame, "Information Technology and Decentralization," *Harvard Business Review* 39 (November–December 1961): 121–126.

20. Edward A. Tomeski, *The Computer Revolution: The Executive and the New Information Technology* (New York: Macmillan, 1970), 10, his emphasis.

21. William B. Levine, "Developments in Public Administration," *Public Administrative Review* 21 (1961): 45–54, on 54; and Ida Hoos, "Automation, Systems Engineering, and Public Administration," *Public Administrative Review* 26 (1966): 311–319, on 311. On Hoos, see Jennifer S. Light, *From Warfare to Welfare: Defense Intellectuals and Urban Problems in Cold War America* (Baltimore: Johns Hopkins University Press, 2003), 72–73, 88, 154–155.

22. William J. Gore, "Developments in Public Administration," *Public Administrative Review* 22 (1962): 165–169, on 167.

23. Gilbert Burck, "Management Will Never Be the Same Again," *Fortune* 70, no. 2 (August 1964): 125–126, 199, 200, 202, 204, on 125.

24. Harold L. Wilensky, "The Uneven Distribution of Leisure: The Impact of Economic Growth on 'Free Time,'" *Social Problems* 9, no. 1 (Summer 1961): 32–56, on 50.

25. Thomas L. Whisler, *Information Technology and Organizational Change* (Belmont, CA: Wadsworth, 1970), 11.

26. Diebold, *Automation*; and Diebold, *Man and the Computer: Technology as an Agent of Social Change* (New York: Frederick A. Praeger, 1969). For his praise of Wiener, see Diebold to Wiener, May 6, 1953, NW, box 12-172; and September 21, 1953, NW, box 12-178. For his criticism, see *Automation*, 151, 154, 155, 157.

27. See John Diebold, "Computers, Program Management, and Foreign Affairs," *Foreign Affairs* 45 (October 1966): 125–134.

28. *Datamation* 13 (May 1967): 9 (advertisement); Diebold, *Man and the Computer*, 6 (quotation), 149; and Diebold, "What's Ahead in Information Technology," *Harvard Business Review* 43 (September–October 1965): 76–82.

29. Peter F. Drucker, *The Age of Discontinuity: Guidelines to Our Changing Society* (New York: Harper and Row, 1969), 27.

30. Carl Heyel, *Computers, Office Machines and the New Information Technology* (New York: Macmillan, 1969), 10.

31. Haigh, "Inventing Information Systems," 35–36.

32. Robert V. Head, "Management Information Systems: A Critical Appraisal," *Datamation* 13 (May 1967): 22–27, on 27.

33. Alan F. Westin, ed., *Information Technology in a Democracy* (Cambridge, MA: Harvard University Press, 1971), 1–11, 149–151; and Harold Wilensky, "The Road from Information to Knowledge," in ibid., 277–286. On the Harvard program, see Matthew Wisnioski, *Engineers for Change: Competing Visions of Technology in 1960s America* (Cambridge, MA: MIT Press, 2012), 51–54.

34. Ulric Neisser, "Computers as Tools and as Metaphors," in *The Social Impact of Cybernetics*, ed. Charles Dechert (New York: Simon and Schuster, 1966), 71–93, on 92; and Robert L. Pranger, "The Clinical Approach to Organizational Theory," *Midwest Journal of Political Science* 9 (1965): 215–234, on 216–217.

35. Samuel A. Miles, "An Introduction to the Vocabulary of Information Technology," *Technical Communications* 14 (Fall 1967): 20–24.

36. Jerome M. Clubb and Howard Allen, "Computers and Historical Studies," *Journal of American History* 54 (1967): 599–607, on 604.

37. See Kline, "Cybernetics, Management Science, and Technology Policy," 527–529.

38. Michael Riordan and Lillian Hoddeson, *Crystal Fire: The Birth of the Information Age* (New York: Norton 1997); Janet Abbate, *Inventing the Internet* (Cambridge, MA: MIT Press, 1999); and Alfred D. Chandler, Jr., and James W. Cortada, eds., *A Nation Transformed by Information: How Information Has Shaped the United States from Colonial Times to the Present* (Oxford: Oxford University Press, 2000), chaps. 6–8.

39. John Diebold, "Application of Information Technology," *Annals of the American Academy of Political and Social Science* 340 (March 1962): 38–45, on 39, 45.

40. Lewis C. Bohn, *Information Technology in Development* (Croton-on-Hudson, NY: Hudson Institute, 1968), 5.

41. Edwin B. Parker, "Information and Society," *Annual Review of Information Science and Technology* 8 (1973): 345–373, on 347, 356–357; Parker and Donald A. Dunn, "Information Technology: Its Social Potential," *Science*, n.s., 176 (1972): 1392–1399; and Parker, "Implications of New Information Technology," *Public Opinion Quarterly* 37 (1973–1974): 590–600, on 593–594.

42. Nicholas L. Henry, "Copyright, Public Policy, and Information Technology," *Science* 183 (1974): 384–391, on 384.

43. Arthur R. Miller, "Personal Privacy in the Computer Age: The Challenge of a New Technology in an Information-Oriented Society," *Michigan Law Review* 67 (1969): 1089–1246.

44. Emanuel G. Mesthene, "Some General Implications of the Research of the Harvard University Program on Technology and Society," *Technology and Culture* 10 (1969): 489–513, on 491; and Anthony G. Oettinger and Nikki Zapol, "Will Information Technologies Help Learning?," *Annals of the American Academy of Political and Social Science* 412 (March 1974): 116–126.

45. Parker, "Implications of New Information Technology," 592.

46. Donald M. Lamberton, "Preface" to "The Information Revolution," a special issue of *Annals of the American Academy of Political and Social Science* 412 (March 1974): xi; and Lamberton, "National Information Policy," ibid.: 147–148.

47. Theodore J. Lowi, "The Third Revolution, Politics, and the Prospect for an Open Society," *IEEE Transactions on Communications* 23 (1975): 1019–1028, on 1019.

48. Anatol Rapoport, "War and Peace," *Annals of the American Academy of Political and Social Science* 412 (March 1974): 152–162, on 152, 153.

49. Based on a search in the Cornell University Library catalog and in the JSTOR collection of journals, conducted in September 2010, and on the online search described in note 5 above.

50. See, for example, Philip Sadler, "Welcome Back to the 'Automation' Debate," in *The Microelectronics Revolution: The Complete Guide to the New Technology and Its Impact on Society*, ed. Tom Forester (Cambridge, MA: MIT Press, 1981), 290–296, on 293; and Juan Rada, *The Impact of Micro-electronics: A Tentative Appraisal of Information Technology* (Geneva: UNESCO, 1980).

51. Alec M. Lee, "A Tale of Two Countries: Some Systems Perspectives on Japan and the United Kingdom in the Age of Information Technology," *Journal of the Operations Research Society* 34 (1983): 753–763.

52. Tom Forester, ed., *The Information Technology Revolution* (Cambridge, MA: MIT Press, 1985), xiii.

53. See Christopher Dede, "Educational and Social Implications," in *Information Technology Revolution*, ed. Forester, 242–257, on 243; David A. Buchanan, "Using the New Technology," in ibid., 454–465, on 457, 465; Juan Rada, "Information Technology and the Third World," in ibid., 571–589, on 571, 573; and Margaret A. Boden, "The Social Impact of Thinking Machines," in ibid., 95–103, on 99.

54. See Bowles, "Crisis in the Information Age." On the belief that technology is unstoppable, see Langdon Winner, *Autonomous Technology: Technics-Out-of-Control as a Theme in Political Thought* (Cambridge, MA: MIT Press, 1977).

55. Beniger, *The Control Revolution*, 4–5 (table), 21 (quotation), 26. In contrast, British journalist John Lamb traced the concept to Norbert Wiener's early writings; see Lamb, "IT 82," 221.

56. Michael Marien, "Some Questions for the Information Society" (1983), rpt. in *Information Technology Revolution*, ed. Forester, 648–660, on 649.

57. Based on online searches of *NYT*, JSTOR, and Google Books, conducted in July 2013. For early usages of the term in its modern sense in American English, see Eugene J. Kelley, "From the Editor," *Journal of Marketing* 34, no. 3 (July 1970): 1–2, on 1; and Jean Gottman, "When in Milwaukee, Do as Bostonians Do," *NYT*, June 2, 1975, 25.

58. Henry M. Boettinger, "The Place of Automation in History," *Data Processing* 2, no. 11 (December 1960): 22–25, on 22.

59. See, e.g., Edmund C. Berkeley, *The Computer Revolution* (New York:

Doubleday, 1962), 3; Reed C. Lawlor, "Information Technology and the Law," *Advances in Computers* 3 (1962): 299–352, on 304; Charles R. Dechert, "The Development of Cybernetics," in *Social Impact of Cybernetics*, ed. Dechert, 11–37, on 17; and Erich Fromm, "Humanizing a Technological Society" (1968), rpt. in *Information Technology in a Democracy*, ed. Westin, 198–206, on 199.

60. Donald N. Michael, *Cybernation: The Silent Conquest* (Santa Barbara, CA: Center for the Study of Democratic Institutions, 1962), 6n; and Amy Sue Bix, *Inventing Ourselves out of Jobs: America's Debate over Technological Unemployment, 1929–1981* (Baltimore: Johns Hopkins University Press, 2000), chap. 8.

61. H. Marshall McLuhan, *Understanding Media: The Extensions of Man* (New York: McGraw-Hill, 1964), 207, 248, 298, his emphasis.

62. H. Marshall McLuhan, "Cybernation and Culture," in *Social Impact of Cybernetics*, ed. Dechert, 95–108, on 102.

63. Donald Agger et al., "The Triple Revolution" (1964), rpt. in *The New Left: A Collection of Essays*, ed. Pricilla Long (Boston: Porter Sargent, 1969): 339–354, on 339, 348.

64. Howard Brick, "Optimism of the Mind: Imagining Postindustrial Society in the 1960s and 1970s," *American Quarterly* 44 (1992): 348–380, on 349, 350.

65. Daniel Bell, *The Coming of Post-industrial Society: A Venture in Social Forecasting* (New York: Basic Books, 1973), xi–xiii.

66. Daniel Bell, "The Post-industrial Society," in *Technology and Social Change*, ed. Eli Ginzburg (New York: Columbia University Press, 1964), 44–59; and Bell, "Notes on the Post-industrial Society," *Public Interest* 6 and 7 (Winter and Spring 1967): 24–35, 102–118.

67. Bell, *Coming of Post-industrial Society*, x, 366–367.

68. Ibid., 13 (quotations, his emphasis), 16.

69. Ibid., 14.

70. Ibid., 117.

71. Ibid., 116, my emphasis.

72. Ibid., 127, 128.

73. Ibid., 29, 347; and Philip Mirowski, *Machine Dreams: Economics Becomes a Cyborg Science* (Cambridge: Cambridge University Press, 2002), 11–18.

74. Daniel Bell, "The Year 2000: The Trajectory of an Idea," in *Toward the Year 2000: Work in Progress*, ed. Bell (Boston: Beacon Press, 1968), 1–13, on 9; Lawrence K. Frank, "The Need for a New Political Theory," in ibid., 177–184, on 182; and "Discussion" in ibid., 46–47.

75. Margaret Mead, "The Life Cycle and Its Variations: The Division of Roles," in *Toward the Year 2000*, ed. Bell, 239–243; John R. Pierce, "Communication," in ibid., 297–309; and George A. Miller, "Some Psychological Perspectives on the Year 2000," in ibid., 251–264.

76. Daniel Bell, "A Summary by the Chairman," in *Toward the Year 2000*, ed. Bell, 63–69, on 65.

77. Bell, *Coming of Post-industrial Society*, 263.

78. Ibid., 38. On their work, see Brzezinski, "The American Transition" (1967), rpt. in *Information Technology in a Democracy*, ed. Westin, 161–167; and Herman Kahn and Anthony J. Wiener, "The Next Thirty-Three Years: A Framework for Speculation," in *Toward the Year 2000*, ed. Bell, 73–100.

79. Bell, *Coming of Post-Industrial Society*, 176, 212, his emphasis.

80. Ibid., 37.

81. Peter F. Drucker, *The Age of Discontinuity: Guidelines to Our Changing Society* (New York: Harper and Row, 1969), 11n, 24–27, 261, 263, 264, 353.

82. Bell, "Foreword: 1976," in *Coming of Post-industrial Society* (1976; rpt., New York: Basic Books, 1999), lxxxvii–c, on lxxxvii.

83. Youichi Ito, "Cross Cultural Perspectives on the Concept of an Information Society," in *Information Societies: Comparing the Japanese and American Experiences*, ed. Alex S. Edelstein, John E. Bowes, and Sheldon M. Harsel (Seattle: School of Communications, 1978), 253–258; Ito, "Birth of Joho Shakai and Johoka Concepts in Japan and Their Diffusion outside Japan," *Keio Communication Review* 13 (1991): 3–12; Arisa Ema, "Systems Thinking and National Policy," unpub. mss., Science and Technology Studies Dept., Cornell University, May 2010; and personal communications with Ema, September 2010 and July 2012. The term *johoka shakai* is often translated as "informationized society" to draw an analogy to "industrialized society."

84. Japan Computer Usage Development Institute, Computerization Committee, *The Plan for an Information Society: A National Goal toward Year 2000* (Tokyo: Japan Computer Usage Development Institute, final report, May 1972), 8 (quotations), 10, 17–26. By 1980, the Japanese government, local governments, telecommunications firms, and private companies had made progress on about one-half of the projects in the plan; see Yoneji Masuda, *The Information Society as Post-industrial Society* (Washington, DC: World Future Society, 1981), 12–15.

85. Ito, "Birth of Joho Shakai and Johoka Concepts," 9; Parker, "Information and Society," 373; and Parker, "Implications of New Information Technology," 598, 600.

86. *Educational Researcher* 2, no. 2 (February 1973): 22.

87. *Science* 190 (1975): 656–657, 1192.

88. Edelstein, Bowes, and Harsel, eds., *Information Societies*, xi–xiii; and Youichi Ito and Koichi Ogawa, "Recent Trends in Johoka Shakai and Johoka Policy Studies," *Keio Communication Review* 5 (March 1984): 15–28, on 17.

89. John McHale, "The Future of Art and Mass Culture," *Leonardo* 12, no. 1 (Winter 1979): 59–64, on 63.

90. Kelley, "From the Editor," 1; Joseph Coates, "Why Think about the Future? Some Administrative-Political Perspectives," *Public Administration Review* 36 (1976): 580–585, on 580, 585; and Pauline Wilson, "[Review of] *The Information Society: Issues and Answers*, edited by E. J. Josey . . . 1978," *Library Quarterly* 49 (1979): 325–327, on 325.

91. Parker, "Information and Society," 351.

92. Ema, "Systems Thinking and National Policy," 7–9, fig. 1.

93. Ito, "Cross Cultural Perspectives on the Concept of an Information Society," 253–254. Ito noted that Professor Ithiel de Sola Pool at MIT had visited Japan and was adopting its method of studying information flow, as were the conference's organizers at the University of Washington.

94. Marc U. Porat, *Information Economy*, 9 vols. (Washington, DC: Department of Commerce, 1977), 1:viii, quotations on 4 (his emphasis), 8, and 204. On the international context of this report, especially the measurement work of the OECD (Organisation for Economic Co-operation and Development), see Benoît Godin, "The Information Economy: The History of a Concept through Its Measurement, 1945–2005," *History and Technology* 24 (2008): 255–287.

95. Porat, *Information Economy*, 1:iii.

96. Ibid.; and Marc U. Porat, "Global Implications of the Information Society," *Journal of Communication* 28 (1978): 70–80, on 72.

97. Robert H. Anderson, "Joseph Becker, In Memoriam," *Information Society* 11 (1995): 241; *Information Society* 1, no. 1 (1981): frontispiece; Edwin B. Parker, "Information Services and Economic Growth," ibid., 71–78; Wilson P. Dizard, Jr., "The Coming Information Age," ibid., 91–112; and Dizard, *The Coming Information Age: An Overview of Technology, Economics, and Politics* (New York: Longman, 1982), chap. 1, quotation on 2.

98. Quoted in Theodore Roszak, *The Cult of Information: The Folklore of Computers and the True Art of Thinking* (New York: Pantheon, 1986), 23–24.

99. Daniel Bell, "The Social Framework of the Information Society," in *The Computer Age: A Twenty-Year View*, ed. Michael L. Dertouzos and Joel Moses (Cambridge, MA: MIT Press, 1979), 163–211, on 163 (quotations), 177, 193. The paper was part of a larger report prepared for the Computer Science Laboratory at MIT; see Bell, "The Future World Disorder: The Structural Control of Crisis," *Foreign Policy* 27 (Summer 1977): 109–135, on 112.

100. Bell, "Social Framework of the Information Society," 166–167, 168, 171, 174, 209n11. While heading the Commission on the Year 2000, Bell was more appreciative of Shannon's theory as applied by Harvard psychologist George Miller to the information-processing abilities of human memory. See Bell, "Summary by the Chairman," 68.

101. He defended his position twenty years later; see Bell, "The Axial Age of Technology Foreword: 1999," in Bell, *Coming of Post-industrial Society*, 1999, ix–lxxxiv, on xviii.

102. Joseph Weizenbaum, "Once More the Computer Revolution," in *Computer Age*, ed. Dertouzos and Moses, 439–458, on 440, 443. British information theorist Colin Cherry also noted the "extravagant claims" made about how information technology was creating a second industrial revolution; see Cherry, *The Age of Access: Information Technology and Social Revolution: Posthumous Papers of Colin Cherry*, ed. William Edmonson (London: Croom Helm, 1985), 12.

103. Daniel Bell, "A Reply to Weizenbaum," in *Computer Age*, ed. Dertouzos and Moses, 459–462, on 461.

104. Langdon Winner, "Mythinformation" (1984), rpt. in Winner, *The Whale and the Reactor: A Search for Limits in an Age of High Technology* (Chicago: University of Chicago Press, 1986), 98–117, on 105, 108, 112–113.

105. Theodore Roszak, *The Making of a Counterculture: Reflections on the Technocratic Society and Its Youthful Opposition* (New York: Doubleday, 1969).

106. Roszak, *Cult of Information*, ix–x.

107. Ibid., 12–13.

108. Frank Webster, *Theories of the Information Society* (London: Routledge, 1995), chap. 2, quotation on 28.

109. Ibid., chaps. 4–6.

110. Ibid., chap. 3, quotation on 50, his emphasis. He went further in the third edition, saying that Bell's "equation of 'post-industrial' and 'information' society is untenable," because the foundations of his theory of postindustrial society are insecure. See Webster, *Theories of the Information Society*, 3rd ed. (London: Routledge, 2006), 53.

111. Webster, *Theories of the Information Society*, 3rd ed.; and Webster, "Information Society," in *IESB*, 11:7464–7468, on 7468.

112. Based on the online searches described in notes 2 and 5 above. On Gingrich and Toffler, see Roszak, *Cult of Information*, 23–27; and Turner, *From Counterculture to Cyberculture*, 215–216.

113. Michael A. Arbib, *Computers and the Cybernetic Society* (New York: Academic Press, 1977). On Lem, see Patricia S. Warrick, *The Cybernetic Imagination in Science Fiction* (Cambridge, MA: MIT Press, 1980), 191–198.

### Chapter 9: Two Cybernetic Frontiers

1. Stewart Brand, ed., "For God's Sake, Margaret: Conversation with Gregory Bateson and Margaret Mead," *CoEvolution Quarterly* 10 (Summer 1976): 32–44, on 32. On the meeting and Bateson's teaching and celebrity, see David Lipset, *Gregory Bateson: The Legacy of a Scientist* (Boston: Beacon Press, 1982), chap. 15.

2. Brand, "For God's Sake, Margaret," 36, 37, 43–44. In 1955, Mead had also proposed that cybernetics could be used to explain emotions. See Jamie Cohen-Cole, "The Creative American: Cold War Salons, Social Science, and the Cure for Modern Society," *Isis* 100 (2009): 219–262, on 248.

3. Brand, "For God's Sake, Margaret," 37.

4. Although Wiener did not formally include the observer in cybernetic systems, Bateson may have referred to Wiener's criticism of social scientists for not recognizing their role as observers in their own research; see *NWCyb*, 189–191. Ironically, Mead criticized Bateson during the interview for intruding himself as an observer when he filmed human behavior during their fieldwork in Bali by walking around with a handheld movie camera instead of setting the camera on

a tripod and letting it roll. See Brand, "For God's Sake, Margaret," 40. On Bateson's diagrams as an example of second-order cybernetics, see N. Katherine Hayles, *How We Became Posthuman: Virtual Bodies in Cybernetics, Literature, and Informatics* (Chicago: University of Chicago Press, 1999), 74–75.

5. William Aspray, *John von Neumann and the Origins of Modern Computing* (Cambridge, MA: MIT Press, 1990), chap. 3; and Atsuhi Akera, *Calculating a Natural World: Scientists, Engineers, and Computers during the Rise of U.S. Cold War Research* (Cambridge, MA: MIT Press, 2006), chap. 3.

6. Stewart Brand, ed., *II Cybernetic Frontiers* (New York: Random House, 1974), 7, 9, 38.

7. On this method, see Bateson, *Mind and Nature: A Necessary Unity* (New York: E. P. Dutton, 1979), 142–144; and Robert I. Levy and Roy Rappaport, "Gregory Bateson, 1904–1980," *American Anthropologist* 84 (1982): 379–394, on 384.

8. An example is a car sliding on an icy road, where the system is the driver, car, and road. See Gregory Bateson, "The Cybernetics of 'Self': A Theory of Alcoholism" (1971), rpt. in Bateson, *Steps to an Ecology of Mind: Collected Essays in Anthropology, Psychiatry, Evolution, and Epistemology* (1972; rpt., Northvale, NJ: Jason Aronson, 1987), 309–337, on 330. For a popular version of the general approach and a cybernetic theory of evolution, see Bateson, *Mind and Nature*.

9. Brand, *II Cybernetic Frontiers*, 28.

10. Ibid., 7.

11. Bateson, "Cybernetics of 'Self,'" 317.

12. Brand, *II Cybernetic Frontiers*, 39–79.

13. See, e.g., Janet Abbate, *Inventing the Internet* (Cambridge, MA: MIT Press, 1999); Paul Ceruzzi, *A History of Modern Computing*, 2nd ed. (Cambridge, MA: MIT Press, 2003), chaps. 7–8; and Thierry Bardini, *Bootstrapping: Douglas Engelbart, CoEvolution, and the Origins of Personal Computing* (Stanford, CA: Stanford University Press, 2000), chaps. 3–6. For the Dynabook sketch, see Brand, ed., *II Cybernetic Frontiers*, 69.

14. Brand, *II Cybernetic Frontiers*, 63.

15. Bardini, *Bootstrapping*, introduction, chap. 1, 140.

16. "The Computer Society," *Time*, February 20, 1978, 44–59; and Otto Friedrich, "The Computer Moves In," *Time*, January 3, 1983, 14–24.

17. "The Cybernated Generation," *Time*, April 2, 1965, 88ff.

18. "The Age of Miracle Chips," *Time*, February 20, 1978, 44–45, on 45.

19. Fred Turner, *From Counterculture to Cyberculture: Stewart Brand, the Whole Earth Network, and the Rise of Digital Utopianism* (Chicago: University of Chicago Press, 2006), 118–128.

20. Stewart Brand, ed., "Caring and Clarity: Conversation with Gregory Bateson and Edmund G. Brown, Governor of California," *CoEvolution Quarterly* 7 (Fall 1975): 32–47.

21. Gregory Bateson, "Invitational Paper," *CoEvolution Quarterly* 11 (Fall

1976): 56–57; Francisco J. Varela, "Not One, Not Two," *CoEvolution Quarterly* 11 (Fall 1976): 62–67; and Mary Catherine Bateson, *Our Own Metaphor: A Personal Account of a Conference on the Effects of Conscious Purpose on Human Adaptation* (New York: Knopf, 1972).

22. Gregory Bateson, "Number Is Different from Quantity," *CoEvolution Quarterly* 17 (Spring 1978): 44–46; Bateson, "The Pattern Which Connects," *CoEvolution Quarterly* 18 (Summer 1978): 5–15; and Stephen Nachmanovitch, "Gregory Bateson: Old Men Ought to Be Explorers," *CoEvolution Quarterly* 35 (Fall 1982): 34–44. For a listing of articles and interviews by Bateson in *Co-Evolution Quarterly*, see Levy and Rappaport, "Gregory Bateson," 392–394. On his illness and death, see Lipset, *Gregory Bateson*, 301–304.

23. Turner, *From Counterculture to Cyberculture*, 128–140, chap. 5.

24. Stewart Brand and Matt Heron, "'Keep Designing': How the Information Economy Is Being Created and Shaped by the Hacker Ethic," *Whole Earth Review* 46 (May 1985): 44–55, on 44.

25. Turner, *From Counterculture to Cyberculture*, 208–217. The articles were Brand, "Stream of Consciousness: Camille Paglia Speaks," *Wired* 1, no. 1 (March/April 1993); and Brand, "The Physicist [interview with Nathan Myhrvoid at Microsoft]," *Wired* 3, no. 9 (September 1995).

26. Stewart Brand, *The Media Lab: Inventing the Future at MIT* (New York: Viking, 1987), chap. 1. See also Turner, *From Counterculture to Cyberculture*, 176–180.

27. Brand, *Media Lab*, 134–137; and Nicholas Negroponte, "The Origins of the Media Lab," in *Jerry Wiesner: Scientist, Statesman, Humanist; Memories and Memoirs*, ed. Walter Rosenblith (Cambridge, MA: MIT Press, 2003): 149–156.

28. Brand, *Media Lab*, 195, 212, 242–243.

29. Ibid., 78–79, 246. Brand also likened one interviewee's way of talking to a Batesonian metalogue (45). On Bateson's definition of information, see, e.g., Bateson, "Double Bind" (1969), rpt. in *Steps to an Ecology of Mind*, 271–278, on 272.

30. Brand, *Media Lab*, 8, chap. 12.

31. Nicholas Negroponte, *Being Digital* (New York: Knopf, 1995): 108. On Negroponte's funding of *Wired*, see Turner, *From Counterculture to Cyberculture*, 211.

32. See, e.g., Abbate, *Inventing the Internet*; Arthur L. Norberg and Judy E. O'Neill, *Transforming Computer Technology: Information Processing for the Pentagon, 1962–1986* (Baltimore: Johns Hopkins University Press, 1996), chap. 4; and Katie Hafner, *Where Wizards Stay Up Late: The Origins of the Internet* (New York: Simon and Schuster, 1996). None of these authors mention the Cybernetics Technology Office.

33. Richard H. van Atta, et al., *DARPA Technical Accomplishments*, 2 vols. (Alexandria, VA: Institute for Defense Analysis, 1991), 2:16-2–16-8.

34. Judith Ayres Daly, "Social Science Research at the Defense Advanced Re-

search Projects Agency," *PS: Political Science and Politics* 13 (1980): 416–418; and "[Interview with] Paul J. Werbos," in *Talking Nets: An Oral History of Neural Networks*, ed. James A. Anderson and Edward Rosenfeld (Cambridge, MA: MIT Press, 1998), chap. 15, quotation on 350.

35. Judith Ayres Daly and Craig Fields, "Close-Coupled Man/Machine Systems Research (Biocybernetics)," Program Completion Report (Alexandria, VA: DARPA Cybernetics Technology Division, 1981), http://www.dod.gov/pubs/foi /Science_Technology/DARPA/877.pdf, accessed August 10, 2011; and "Mind-Reading Computer," *Time*, July 1, 1974, 73. On the dissolution of the CTO, see Alex Roland and Philip Shiman, *Strategic Computing: DARPA and the Quest for Machine Intelligence, 1983–1993* (Cambridge, MA: MIT Press, 2002), 215.

36. The Media Lab also proposed the map to solve the urban crisis of the time; see Aubrey Anable, "The Architecture Machine Group's *Aspen Movie Map*: Mediating the Urban Crisis of the 1970s," *Television and New Media* 13 (2012): 498–519.

37. Yoneji Masuda, *The Information Society as Post-industrial Society* (Washington, DC: World Future Society, 1981); Tom Forester, ed., *The Information Technology Revolution* (Cambridge, MA: MIT Press, 1985); and Manuel Castells, *The Information Age: Economy, Society and Culture*, 3 vols. (London: Blackwell, 1996–1998).

38. Michael L. Dertouzos and Joel Moses, eds., *The Computer Age: A Twenty-Year View* (Cambridge, MA: MIT Press, 1979), 46, 116, 165–166, 170–171, 401–402.

39. Robert Lucky, *Silicon Dreams: Information, Man, and Machine* (New York: St. Martin's Press, 1989), chap. 2.

40. John Seely Brown and Paul Duguid, *The Social Life of Information* (Boston: Harvard Business School Press, 2000), 138 (quotation), 225.

41. Kathleen Woodward, "Introduction," in *The Myths of Information: Technology and Postindustrial Culture*, ed. Woodward (Madison, WI: Coda Press, 1980), xiii–xxvi, on xxv.

42. Theodore Roszak, *The Cult of Information: The Folklore of Computers and the True Art of Thinking* (New York: Pantheon, 1986), 9–11; and Roszak, *The Making of a Counter Culture: Reflections on the Technocratic Society and Its Youthful Opposition* (New York: Doubleday, 1969), 294.

43. *Whole Earth Review* 44 (January 1985): cover; E. M. Forster, "The Machine Stops," rpt. in ibid., 40ff.; Langdon Winner, "Mythinformation," ibid., 22–28; Stewart Brand, "Biting the Hand That Feeds," ibid., 5 (quotation); Brand, introduction to "Whole Earth Software Catalog Version 1.1," ibid., 74 (quotation); and Brand, *Media Lab*, 227.

44. James R. Beniger, *The Control Revolution: Technological and Economic Origins of the Information Society* (Cambridge, MA: Harvard University Press, 1986), chaps. 1–2.

45. R. Felix Geyer and Johannes van der Zouwen, "Introduction," in *Socio-*

*cybernetics: An Actor-Oriented Social Systems Approach*, ed. Geyer and van der Zouwen, 2 vols. (Leiden: Martinus Nijhof, 1978), 1:1–14; James R. Beniger, "Control Theory and Social Change: Toward a Synthesis of the System and Action Approaches," ibid., 2:15–28; and Beniger and Clifford I. Nass, "Preprocessing and Societal Control: Neglected Component of Sociocybernetics," *Kybernetes* 38 (1984): 173–177. Sociocybernetics was recognized as a subfield of sociology by the International Sociological Association in 1980. See Geyer and van der Zouwen, "John Rose and the Early Years of Sociocybernetics," *Kybernetes* 38 (2009): 61–64, on 62.

46. F. Geyser, "Sociocybernetics," in *IESB*, 21:14,549–14,554, on 14,549.

47. M. E. Maron, "Cybernetics," in *International Encyclopedia of the Social Sciences*, 2nd ed. (New York: Macmillan, 1968) 4:3–62.

48. J. Law and I. Moser, "Cyborg," in *IESB*, 3202–3204; and entries in ibid. on "Cognitive Science, History," 2154–2158; "Deutsch, Karl," 3553–3555; "Family Therapy, Clinical Psychology of," 5401–5406; and "Jakobson, Roman," 7945–7949.

49. Beniger has called cybernetics "an important early contribution [to sociology] largely spent by the 1960s." See James R. Beniger, "Communication—Embrace the Subject, Not the Field," *Journal of Communication* 43 (1993): 18–25, on 20.

50. Andrew Pickering, *The Cybernetic Brain: Sketches of Another Future* (Chicago: University of Chicago Press, 2010), 60–64. On his robots, Cog and Kismet, see Evelyn Fox-Keller, "Booting Up Baby," in *Genesis Redux: Essays in the History of Artificial Life*, ed. Jessica Riskin (Chicago: University of Chicago Press, 2007), 334–345; and Sherry Turkle, *Alone Together: Why We Expect More from Technology and Less from Each Other* (New York: Basic Books, 2011), part 1.

51. Pickering, *Cybernetic Brain*, chaps. 1–2, quotation on 22.

52. An important exception is Turkle, *Alone Together*.

53. Steve Lohr, "Google Goes Hardware Shopping," *NYT*, August 21, 2011, Sunday Review, 9.

54. Greg Downey, "Virtual Webs, Physical Technologies, and Hidden Workers: The Spaces of Labor in Information Internetworks," *Technology and Culture* 42 (2001): 209–235.

technocracy movement, 90–91, 183
technological determinism, 204, 210, 216, 218, 225, 226
technological narratives, 5–8, 203. *See also* information, discourses of
Teleological Society, 23, 41–42, 43
*Terminator, The* (movie), 169, 177
Teuber, Hans Lukas: and ASC, 307n47; at Macy conferences on cybernetics, 44, 45, 49, 51, 53, 63; Morison on, 159
Thatcher, Margaret, 203, 212
Toffler, Alvin, 227
Tomeski, Edward, 206, 207
Tsien, Hsue Shen, 95
Tukey, John, 17, 306n36
Tuller, William, 114–15
Turing, Alan, 32, 40, 154–56
Turner, Fred, 184, 198, 203, 236, 276n77

Umpleby, Stuart, 198, 199
United Aircraft Corporation (UAC), 174, 175
Unity of Science movement in the US, 94, 98–100, 268n109
US Commerce Department, 220, 222
Uttley, Albert, 154–56, 160, 164–65, 180

Varela, Francisco, 197, 198, 236
Viterbi algorithm, 130
vocoder, 31–32, 71, 74
von Bertalanffy, Ludwig, 192, 194–95, 196, 218
von Foerster, Heinz: and ARTORGA, 180, 184; and ASC, 188, 189–90, 192, 194, 308n64; at BCL, 101, 165–67, 181, 200; and bionics, 87, 181, 165–68; editor of Macy conferences on cybernetics, 2–3, 44, 51; and general systems theory, 196; and second-order cybernetics, 5, 196–99, 232, 236
Vonnegut, Kurt, 88, 89, 211
von Neumann, John: and AI, 154, 156, 158; and Bigelow, 3, 162, 232; and game theory, 84, 136, 206; at Macy conferences on cybernetics, 3, 39,

41–43, 47–49, 65, 232; and NW, 23, 25, 86–87, 94; on origin of information theory, 10; work on computers, 36, 47, 77
Voris, Frank, 176

Wallman, Henry, 36
Walter, W. Grey: and AI, 10, 154–55; at Macy conferences on cybernetics, 2, 4, 61, 65, 67; tortoises automata, 61, 78–79, 84, 153, 169, 243; work on cybernetics, 107, 180, 243
Weaver, Warren: awarding grants at the Rockefeller Foundation, 27, 58, 97, 98, 109, 141–42; and CES, 30–31, 121–25; concept of organized complexity, 195, 217; funding the Dartmouth AI conference, 158, 160–61; *The Mathematical Theory of Communication* (with Shannon), 121–23, 125, 131, 141; and NW, 19–21, 23, 97, 126–27, 255n45; popularization of CES's information theory, 121–23, 129, 131–33, 257n70; and Ridenour, 122–23; work for the NDRC, 30, 39
Webster, Frank, 225–26
Wecter, Dixon, 96
Weizenbaum, Joseph, 224
WELL (Whole-Earth 'Lectronic Link), 237
Werbos, Paul, 239
Westin, Alan, 208–9
Weyl, Hermann, 29
Whisler, Thomas, 205–6, 207–8, 211
*Whole Earth Catalog.* See Brand, Stewart, works of
Widrow, Bernard, 163
Wiener, Anthony, 218, 219
Wiener, Barbara (daughter of NW), 66, 71, 87
Wiener, Margaret (wife of NW), 66
Wiener, Norbert (NW): and Aiken, 79, 85; and Bigelow, 19–22, 25–26, 31, 40, 43; and CES, 8, 10–11, 13, 19–20, 26, 35–36; against cold war military research, 79, 83–87, 185, 188, 275n71; cutting off relations with McCulloch, Lettvin, Pitts, and Wiesner, 65–67; on cybernetics and social sciences, 90,

Wiener, Norbert (NW) (*continued*)
135, 190, 291n2; cybernetics as a
universal discipline, 81, 94, 153;
death of, 152, 180, 188; and Deutsch,
82, 143–44; development of cybernetics, 11–12, 18, 21, 23, 25, 45 (*see also
under* cybernetic machines; prosthetics); development of information
theory, 10–11, 12, 13–14, 21–26,
118–19, 132; and Diebold, 88, 207,
210; and Fano, 24, 124; and FBI, 87,
188; and GB, 96, 148–50, 320n4; on
interdisciplinarity, 64, 103; and Yuk-
Wing Lee, 18, 21, 101, 181, 251n1;
and Lettvin, 9, 66, 275n71; at Macy
conferences on cybernetics, 43–44, 45,
48–49, 56–57, 64; and McCulloch, 3,
65–67, 85, 87; in Mexico City, 3, 80,
86; and Pitts, 9–10, 12, 23–26, 42,
119, 159; publicizing cybernetics,
69–73, 74, 77–78, 80–83, 101, 126;
and Reuther, 73; and Rosenblueth, 11,
22–23, 49, 80, 158–59, 161 (*see also*
"Behavior, Purpose and Teleology");
and Stibitz, 79–90; taking credit for
information theory, 35–36, 103, 124;
and the Teleological Society, 40–42;
trip to the Soviet Union, 185; and
von Neumann, 23, 25, 86–87, 94;
and Weaver, 19–21, 23, 97, 126–27,
255n45; and Barbara Wiener, 66, 71,
87; and Margaret Wiener, 66; and
Peggy Wiener, 66; and Wiesner, 74,
124; work at RLE, 65, 74, 76, 85, 95,

115, 275n71; work on antiaircraft
problem, 11, 18–23, 25–26, 39–40,
84; and Zemurray, 80. *See also under*
second industrial revolution
Wiener, Norbert, works of: *Cybernetics*,
10, 11–14, 161; *God and Golem*,
151–53, 177; *The Human Use of
Human Beings*, 70, 80–83, 86, 88,
138, 214; reception of *Cybernetics*,
68–69, 70–73, 96–100, 127
Wiener, Peggy (daughter of NW), 66
Wiesner, Jerome (Jerry): and Cherry,
111; and Cybernetics Panel of PSAC,
185–87, 237; and MacKay, 59; at
Macy conferences on cybernetics, 53;
and Mead, 113–14; and NW, 65, 74,
76, 124; at RLE, 53, 111, 121, 238
Wilensky, Harold, 207, 209
Wilson, E. O., 129
Winner, Langdon, 225, 242
*Wired* (magazine), 237, 239
Wolfe, Bernard, 88
Woodward, Kathleen, 241

Xerox Palo Alto Research Center
(Xerox PARC), 233–34, 236, 241

Young, John F. (cybernetician), 183
Young, John Z. (anatomist), 54
Young, Robert, 239

Zemurray, Margot (secretary to NW),
80
Zipf, George, 136